KT-231-780

Exploring Numerical Methods: An Introduction to Scientific Computing using MATLAB

Peter Linz
University of California at Davis

Richard Wang
University of California at Davis

JONES AND BARTLETT PUBLISHERS
Sudbury, Massachusetts
BOSTON TORONTO LONDON SINGAPORE

World Headquarters
Jones and Bartlett Publishers
40 Tall Pine Drive
Sudbury, MA 01776
978-443-5000
info@jbpub.com
www.jbpub.com

Jones and Bartlett Publishers
Canada
2406 Nikanna Road
Mississauga, ON L5C 2W6
CANADA

Jones and Bartlett Publishers
International
Barb House, Barb Mews
London W6 7PA
UK

Library of Congress Cataloging-in-Publication Data

Linz, Peter.
Exploring numerical methods: an introduction to scientific computing using MATLAB / Peter Linz, Richard Wang.
p. cm.
Includes index.
ISBN 0-7637-1499-2
1. Numerical analysis—Data processing. 2. MATLAB. I. Wang, Richard. II. Title.

QA297 . L54 2002
519.4—dc21

2002073036

Editor-in-Chief, College: J. Michael Stranz
Production Manager: Amy Rose
Editorial Assistant: Theresa DiDonato
Associate Production Editor: Karen C. Ferreira
Production Assistant: Jenny L. McIsaac
Senior Marketing Manager: Nathan J. Schultz
V.P., Manufacturing and Inventory Control: Therese Braüer
Cover Design: Night and Day Design
Composition: Northeast Compositors, Inc.
Printing and Binding: Malloy, Inc.
Cover Printing: Malloy, Inc.

Printed in the United States of America
06 05 04 03 02 10 9 8 7 6 5 4 3 2 1

Preface

This book is intended for an introductory, one-year course in numerical analysis for students in mathematics, engineering, and the physical sciences. Historically, such a course has been offered by mathematics departments mostly to junior and senior level students. In recent years a more elementary course has achieved some popularity. This course is often taught in engineering or computer science departments and has an audience of freshmen and sophomores. These two groups have different skill levels and different needs, but there is a great deal that they have in common. It is our view that their separate needs can be addressed effectively in combination, and that the subject matter can indeed profit from such a combination.

To serve the needs of these two different audiences, each topic is carefully developed in a graduated manner. The first few sections of each chapter present motivation and simple algorithms in an intuitive fashion. Later sections show the underlying theory and introduce the more complicated, less common methods. These later sections are starred to indicate their more advanced nature. The unstarred sections are independent of the starred ones, so the elementary course can be taught relying only on the simple material. For a more challenging course, much of the material will come from the starred sections, with the elementary part providing insight and motivation.

The material selected for the book is, for the most part, standard and traditional. Only in the last three chapters were some choices and compromises made. What to do about partial differential equations is not clear-cut for any author of an introductory text, since any reasonable treatment presumes more theoretical knowledge than the prospective audience can be expected to have. Additionally, there are many practical complications that make it impossible to do justice to the topic at the undergraduate level. But partial differential equations are so important in practice that one cannot see what numerical analysis is all about without some exposure to the main issues. Our way of dealing with this dilemma is to present some simple prototype partial differential equations, with a quick and intuitive overview of the difficulties of implementation and the more theoretical question of stability. While this does not prepare students to solve real-life partial differential equations, it does present them with the flavor of the subject matter in preparation for more advanced courses.

One quite nontraditional topic is in the last chapter: The solution of inverse and ill-posed problems. This inclusion not only reflects the author's special interests, but also gives an introduction to an increasingly more important topic. The matter is an advanced subject, so only the more intuitive aspects are presented. It shows that applying numerical methods arbitrarily does not always work.

This book discusses the most important numerical algorithms, but it does not attempt to be a reference work. Instead, it concentrates on the limited subject matter that is most commonly presented to undergraduate students and stresses pedagogical issues rather than completeness. This book emphasizes:

- providing insight and motivation for the construction of numerical methods
- understanding the strengths and limitations of such methods
- evaluating the effectiveness of available numerical software
- modifying existing software for specific purposes
- gaining experience in choosing between alternative approaches
- experimenting with numerical software in settings that mirror real-world situations, and
- building a strong experiential base for continuing study and more specialized courses.

These aims are greatly aided by the close connection between the discussion of methods and algorithms and their implementation in MATLAB. While the MATLAB library is much less extensive than many industrial libraries (such as IMSL), it does give the student experience working with ready-made software whose internal structure may not always be entirely clear. MATLAB's numerical analysis functions are well constructed but are quite automatic and are used essentially as black boxes. Since most libraries have similar characteristics, students learn how to use numerical methods libraries effectively.

To complement the standard MATLAB functions, we provide an extension: NASOFT. This set of functions does two things. First, it extends the MATLAB functionality to problems such as two-point boundary value problems and some prototype partial differential equations, allowing the student to experiment with fairly complex algorithms. Second, NASOFT functions are coded in a straightforward manner and the MATLAB source is available. This gives the student the opportunity to critique the implementation and modify to improve or adapt it to different purposes. The source code for the NASOFT functions and example scripts are available online at: http://math.jbpub.com/numericalmethods.

Many traditional numerical analysis courses are primarily lecture courses, with perhaps a lab of secondary importance. This book envisions a different emphasis in which lab work is at least as important as the lectures. For this purpose, we have added a chapter called "Explorations." The problems in this exploration section deal with the very practical issues of software evaluation, selection, modification and the solution of not-entirely-specified, open-ended problems. Students should conduct investigations using their own methodology, and should be expected to write informative reports on their observations and conclusions. This is the part of the course that most closely models real life situations and should therefore be considered the heart of the course.

This book is designed for an undergraduate numerical analysis course that stresses insight and hands-on experience over detailed knowledge of a host of numerical methods and their mathematical justification.

Peter Linz
Richard L.C. Wang

Contents

Chapter 1

Numerical Analysis and Scientific Computing

Numerical analysis is a branch of applied mathematics that deals with methods for solving problems by purely numerical computations. To understand what this means and why such an approach is needed, let us look at a simple example familiar from calculus.

Example 1.1 A three-dimensional solid of revolution can be created by revolving the curve

$$y = r(x)$$

about the x-axis (Figure 1.1).

As we know from calculus, the volume of such a solid is given by

$$V = \pi \int_0^1 r^2(x)dx. \tag{1.1}$$

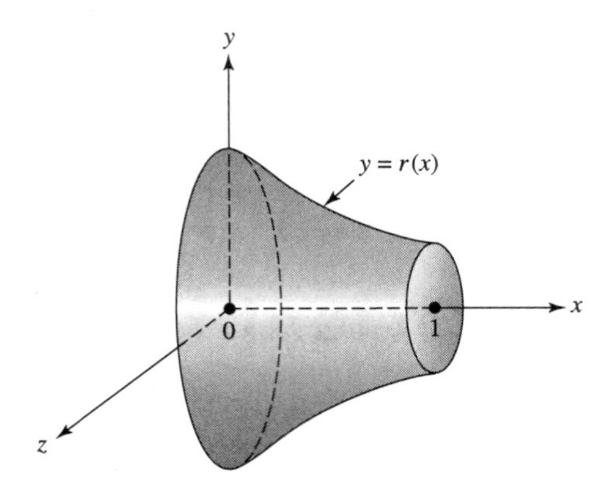

Figure 1.1
A solid created by revolving a curve around the x-axis.

If $r(x)$ is simple, the definite integral can be computed by finding the antiderivative of the integrand and then substituting the limits. Therefore, if $r(x) = e^x$, then

$$\begin{aligned} V &= \pi \int_0^1 e^{2x} dx \\ &= \frac{\pi}{2} e^{2x} \Big|_0^1 \\ &= \frac{\pi}{2}(e^2 - 1). \end{aligned}$$

We call the explicit formula for the integral in terms of elementary functions a *closed form* solution. When a closed form solution can be found, the value of the integral is easily obtained by putting in appropriate numerical values; in this case, $V = 10.0359$.

Unfortunately, closed form solutions are often impossible to get. If we complicate the problem just slightly by taking $r(x) = e^{x^2}$, the integral for the volume

$$V = \pi \int_0^1 e^{2x^2} dx \tag{1.2}$$

can no longer be found by elementary methods because no simple closed form solution exists. Here is where numerical analysis comes in.

The answer we are looking for is just a number, so we can try using numerical techniques to find it. To do so, we must reduce the problem to a sequence of steps that can be performed numerically, either on a pocket calculator for simple cases, or on a computer for more typical problems. Suppose we subdivide the interval $[0,\ 1]$ into n equal parts, using the partition points

$$x_i = \frac{i}{n}, \quad i = 0,\ 1,\ \ldots,\ n-1$$

Table 1.1
Approximating the integral (1.2) by the sum (1.3).

n	V_n
10	6.5016
100	7.3286
1000	7.4181
10000	7.4271

and form the sum

$$V_n = \frac{\pi}{n} \sum_{i=0}^{n-1} e^{2x_i^2}. \tag{1.3}$$

We know from calculus that a definite integral is the limit of such sums as $n \to \infty$, so we can claim that V_n approximates V. We write this as

$$\pi \int_0^1 e^{2x^2} dx \cong \frac{\pi}{n} \sum_{i=0}^{n-1} e^{2x_i^2}.$$

Table 1.1 shows how this works. We see that each n gives us a different value and that the sequence of values slowly tends to approximately 7.43. We might guess from the results that for $n = 10000$ we have at least three correct digits.[1]

■

This specific example is easily generalized. For a given function $f(x)$, the integral

$$I = \int_a^b f(x)dx \tag{1.4}$$

can be approximated by

$$I_n = \frac{b-a}{n} \sum_{i=0}^{n-1} f(x_i), \tag{1.5}$$

with

$$x_i = a + i\frac{(b-a)}{n}.$$

[1]All computations in this book were done on a computer that carries about 16 decimal digit accuracy. Normally we do not display all computed digits, but only those relevant to the specific discussion.

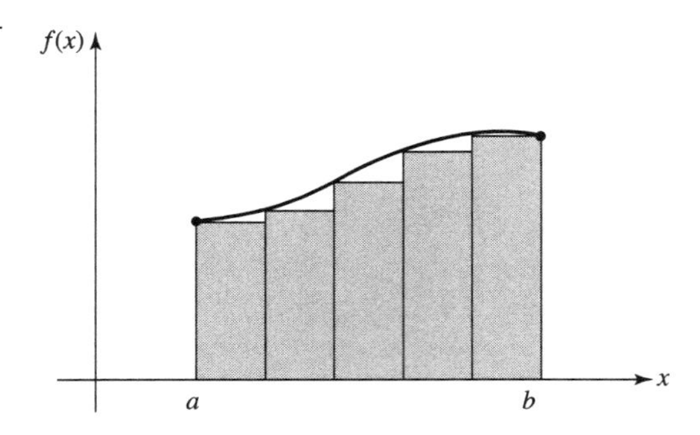

Figure 1.2 Approximating an integral by the rectangular rule. The area under the curve represents the true value of the integral. The shaded area is the approximation by the rectangular rule.

This method of approximating integrals is called the *rectangular rule*, since we approximate the area under a curve by a set of rectangles (Figure 1.2). The rectangular rule is one of the simplest numerical algorithms.

The problems we encounter in practice are usually much more complicated than computing definite integrals. To study physical phenomena, scientists and engineers construct *mathematical models* that embody the physical laws in order to describe the real situation. These models frequently involve differential equations, which are the mainstay of applied mathematics and engineering computations. The solutions of the modeling equations allow us to understand real-life systems and predict their behavior. Different areas of mathematics, from calculus to the theory of partial differential equations, were developed for studying and solving such mathematical models. Arriving at a successful solution often requires a great deal of insight and inventiveness.

Classical applied mathematics of the nineteenth and early twentieth centuries emphasized analytical methods for finding closed form solutions. Many methods, such as the separation of variables, were devised for this purpose and are still studied in courses on differential equations. Analytical techniques are often useful for getting insight into the general nature of a problem, and can give closed form solutions for simple cases. In most instances, though, they break down somewhere and have to be supplemented with numerical methods.

Example 1.2

The motion of a pendulum is a problem from elementary physics. An object of mass m is attached to a pivot by a string of length L, as shown in Figure 1.3. When the object is released from rest at an angle θ_0 the pendulum, following the laws of motion, will oscillate about the equilibrium position $\theta = 0$. A mathematical model allows us to predict the motion.

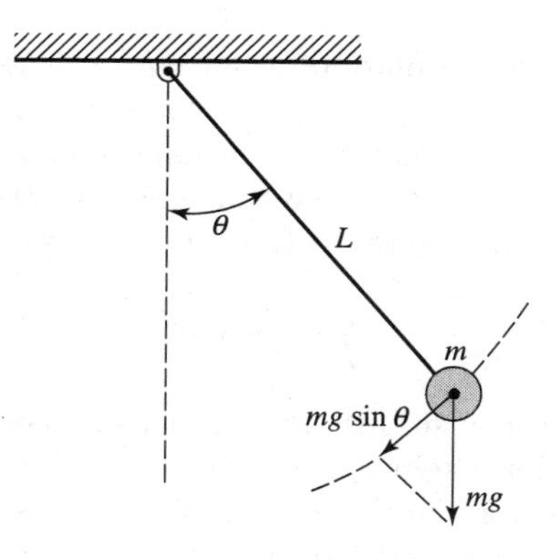

Figure 1.3 A simple pendulum.

Using Newton's second law and the force diagram in Figure 1.3, we are led to the modeling equation

$$mL\frac{d^2\theta}{dt^2} = -mg\sin\,\theta,$$

or

$$\frac{d^2\theta}{dt^2} = -\frac{g}{L}\sin\,\theta. \tag{1.6}$$

If at time $t = 0$ the pendulum is at rest in position θ_0, the conditions

$$\theta(0) = \theta_0 \tag{1.7}$$

and

$$\theta'(0) = 0 \tag{1.8}$$

must also be satisfied. The solution of (1.6), subject to (1.7) and (1.8), will give us the dependence of the angle θ in terms of the time t, the length of the pendulum L, and the constant of gravity g.

If θ is very small, then sin θ is very close to θ and we can replace (1.6) by the small-angle approximation

$$\frac{d^2\theta}{dt^2} = -\frac{g}{L}\,\theta. \tag{1.9}$$

Equation (1.9), with conditions (1.7) and (1.8), has the known closed form solution

$$\theta\,(t) = \theta_0\cos\left(\sqrt{\frac{g}{L}}\,t\right).$$

Therefore the pendulum executes a periodic motion with period

$$T = 2\pi\sqrt{\frac{L}{g}}.$$

The small-angle approximation is inaccurate when θ_0 is more than a few degrees, so if high accuracy is required we may have to work with (1.6). In real-life situations there may be additional complications, such as the drag created by air resistance. If we can represent the effect of drag by a term proportional to some power of the velocity, an equation of the form

$$\frac{d^2\theta}{dt^2} = -\frac{g}{L}\sin\theta + c\left(\frac{d\theta}{dt}\right)^\alpha$$

may represent the physical situation much better than (1.9). This equation has no known closed form solution so will have to be solved numerically.

Intuitive methods for the numerical solution of such differential equations are not hard to invent, but the subject is complicated and we defer any discussion to later chapters. For the moment, we just want to understand the need for numerical methods that can solve various kinds of differential equations.

■

Let us return briefly to the numerical integration in Example 1.1. Even though numerical integration is a very simple problem, it has all the major characteristics of more complicated numerical methods.

- A continuous and infinitesimal operation, in this case integration, is replaced by a finite sequence of arithmetic operations. This process is called *discretization*. In a sense, discretization reverses the limit process that is the basis of calculus.

- The result of a numerical calculation is not exact. The error that it has comes from discretization, and is called the *discretization error*. The statement that I_n approximates I is intuitively clear, but we can make it a little more precise by saying that I_n becomes closer and closer to I as n increases or, as we normally state it, that I_n *converges* to I. The approximation will never give us an exact answer, but we can get increasingly better accuracy by simply taking more terms in the sum in (1.5).[2]

- If the result is to be useful, we need to have some idea of how large the discretization error is. From Table 1.1 we guessed that the final result had about a three digit accuracy, but we did not give a rigorous justification for this claim.

- Numerical methods may involve thousands, perhaps millions, of simple individual operations. This raises the question of *efficiency* and the relative merits of different algorithms. Example 1.1 shows that the

[2]Shortly we will see that there is a practical limit to the accuracy we can get, but this is not important at the moment.

rectangular rule requires a great deal of work to get even moderate accuracy, so we should look for better ways. As we will see, there are methods that give the same accuracy with just a few computations.

These issues need to be addressed in every numerical computation, and they represent the focus of our discussion in this book. To solve a problem numerically, we first have to be able to discretize it so that it can be put on a computer. Conceptually, discretization itself may not be a major difficulty. Even for partial differential equations there are fairly obvious and intuitive discretization methods. The more demanding task comes when we try to put the concepts into practice. The production of a computer program for a complicated mathematical model is a lengthy and tedious matter. Another challenge is to construct methods that are effective and efficient, and whose performance can be predicted using convincing arguments. If possible, we also want to have ways for assessing the errors so that we can have some faith in the numbers that are produced. The analysis of numerical methods is not always easy and often requires a blend of mathematical sophistication, insight, and experience.

Although some simple numerical methods date back to the eighteenth and nineteenth centuries, the development of numerical analysis is closely tied to the history of digital computers. The first digital computers of the 1940s and early 1950s were created specifically to aid in the complex calculations connected with the design of nuclear weapons, and their use was limited to a few universities and government institutions. By current standards, these early machines were primitive and their use required much tedious work by experts writing programs in machine language. Nevertheless, some fairly significant problems were attacked and solved by these early computers. The next phase, starting around 1950, saw a huge increase in the power of computers and a proliferation of these machines into the industrial and academic world. The advent of the so-called higher-level programming languages, Fortran, Algol, and Cobol, coupled with the increased capacity of the new computers, made numerical computation a powerful tool for scientists and engineers. Numerical analysis became a standard topic of applied mathematics and computer science, and the ensuing activity led to the invention of many new numerical algorithms. Libraries of commonly used numerical methods were created that saved users much effort. However, doing numerical work, while certainly less tedious, continued to call for detailed programming and much routine work. Since the most important numerical techniques were developed during this period, numerical analysis is often still viewed in this light.

In the last decade or so another transformation has taken place. The large, centrally-located computer systems have given way to networks of powerful desktop systems. At the same time, the old style procedure-oriented languages are being replaced by high level object-oriented languages. Instead of submitting decks of punched cards, we now type our commands directly into the computer, interacting with computer programs to guide the solution process and explore options. Digital scanners and direct links to measuring devices are used to get data into the system, and powerful graphical communication interfaces have replaced the reams of computer printout. Several existing systems, including MATLAB, now give engineers the ability to solve complex problems in a few minutes. As users of modern numerical methods we are faced by very different challenges from our counterparts forty years ago. We are no longer entirely pre-occupied with the minutiae of algorithm design or the incidentals of programming language syntax. Instead, we are concerned with how to use available resources to solve problems quickly and reliably. We have to learn how to select the appropriate programs from a large body of available software. Sometimes we find existing software that solves our problem, but most of the time we need to integrate existing routines into a larger program. This means that we have to be able to evaluate software and learn how to select that which is appropriate for our purpose. We also have to be concerned with the best way of getting voluminous data into the computer, and how to display the results most effectively. The term *scientific computing* is used to denote the use of numerical methods in a complex hardware and software environment to solve models from various scientific disciplines. The tools of scientific computing give engineers and scientists the ability to deal effectively with very complicated and realistic mathematical models of real-world situations.

EXERCISES

1. Based on the results of Table 1.1, estimate what value of n would be required to evaluate the integral (1.2) to an accuracy of eight decimal digits.

2. Write a computer program that evaluates integrals by the rectangular rule (1.5). Use this program to find an approximate value of the volume for the solid of revolution given by (1.1) in Example 1.1 with $r(x) = \dfrac{1}{1+\sqrt{x^5}}$.

3. Use the program from Exercise 2 to correctly compute

$$\int_0^{\pi/2} \frac{x \cos^2(x)}{\sqrt{1+x}}\, dx$$

to three decimal digits. Provide arguments that lead you to believe that your answer meets the accuracy requirement.

4. Use the rectangular method to approximate the area of the ellipse with boundary

$$x^2 + 2y^2 = 1$$

to four correct decimal digits.

5. Use the rectangular method to approximate the circumference of the ellipse in Exercise 4 to three correct decimal digits.

6. The taper of a beam with the rectangular cross-section shown below is given by the function

$$y = \frac{e^{-\sqrt{x}}}{1+x}.$$

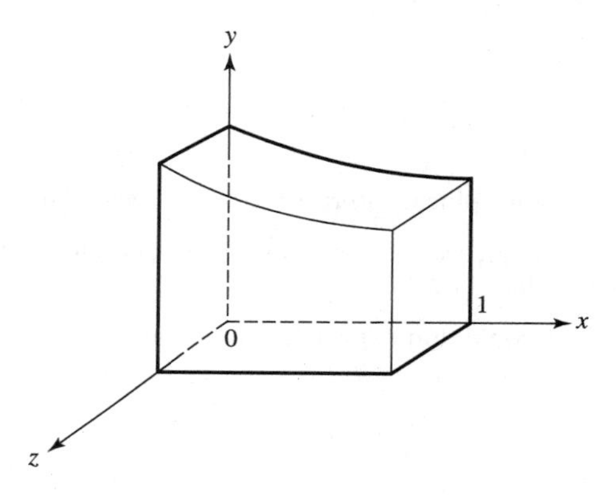

Find the x-coordinate of the center of gravity of the beam to three correct decimal digits.

7. The differential equation

$$\frac{dy}{dt} = \frac{t\sqrt{t+1}}{ye^t},$$

with initial condition $y(0) = 0$, has a particular solution

$$y(x) = \sqrt{2\int_0^x \frac{t\sqrt{t+1}}{e^t}\,dt}$$

for $x > 0$. Use the rectangular method to approximate $y(1)$ to four decimal digits.

8. Intelligence Quotient (IQ) scores are distributed normally with mean 100 and standard deviation 15 by the probability density function

$$f(x) = \frac{1}{15\sqrt{2\pi}} e^{-(x-100)^2/450}.$$

Use the program from Exercise 2 to compute the percentage of the probability that has an IQ score between 80 and 120 by taking the integral

$$p(80 \le x \le 100) = \int_{80}^{120} f(x)\,dx.$$

Justify that your answer is correct to three decimal digits.

9. Write a computer program to approximate

$$S = \sum_{i=0}^{\infty} \frac{1}{1+i^4}$$

to six decimal digit accuracy. Defend your claim that the desired accuracy is achieved.

10. In another version of the rectangular rule, the integral I in (1.4) is approximated by

$$\overline{I}_n = \frac{b-a}{n} \sum_{i=1}^{n} f(x_i).$$

(a) Give a graphical interpretation for this formula.

(b) In general, would you expect this modification to give better results than (1.5)?

(c) What could you expect if you averaged these two approximations and used

$$\widetilde{I}_n = \frac{1}{2}(I_n + \overline{I}_n)$$

as an approximation to the integral? Explore this suggestion numerically.

Chapter 2

Computing with Numbers

No matter how complicated a numerical algorithm is, we always end up having to carry out a sequence of simple arithmetic operations. Since virtually all numerical calculations are performed on some kind of digital computer, we begin our exploration of numerical methods by looking at the details of computer arithmetic. First, we review how numbers are represented in a computer.

2.1 Number Representation

The decimal notation we use every day is so familiar, and we work with it so easily, that we rarely think about the fact that it is only one of many ways to represent numbers. The decimal system is a *positional* system, using the *base* 10. This means that the value of a number not only depends on the individual digits, but also on their position relative to a fixed point, the *decimal point.* The value of each digit is multiplied by a power of 10, with increasing powers to the left of the decimal point and decreasing powers to the right. Thus, the digit string 23.56, interpreted as a decimal number, stands for

$$(23.56)_{10} = 2 \times 10^1 + 3 \times 10^0 + 5 \times 10^{-1} + 6 \times 10^{-2}.$$

Here we use a subscript that shows the base 10 explicitly.

Because of the binary, on–off nature of digital computers, base 10 is not efficient and is replaced by a base-2 system. A base-2 system is similar to decimal notation, except that the base is 2, the allowable binary digits (bits) are 0 and 1, and each bit is multiplied by a power of 2, depending on its position relative to the *binary point.* Conversion from binary to decimal and vice versa is easy:

$$\begin{aligned}(101.011)_2 &= 1 \times 2^2 + 0 \times 2^1 + 1 \times 2^0 + 0 \times 2^{-1} + 1 \times 2^{-2} + 1 \times 2^{-3} \\ &= (5.375)_{10} \,.\end{aligned}$$

One difficulty with this notation, often called *fixed-point*, is its inefficiency for very large or very small numbers. The string 0.00000056 has nine digits altogether, but only two of them carry actual information. The rest are zeros that are needed to scale the number. This difficulty can be remedied with *scientific notation* in which every number has two parts. The first is a decimal number with the decimal point in a fixed position, and the second part is a *scale factor* that allows us to represent a wide range of numbers. The digits in the first part, the *significant digits*, have a fixed length. In this representation, the decimal number 0.00000056 can be written as 5.6×10^{-7}, indicating that we have two significant digits. Scientific notation can handle a wide range of numbers, while at the same time clearly showing the number of significant digits. A binary version of this notation is widely used in computers and will be discussed in detail in the next section.

Another point to consider is how many digits to retain during computation. If we multiply two numbers with three significant digits, we may get a result with six digits. If we now multiply that product by another three digit number, we may get nine digits, and so on. This quickly becomes cumbersome, especially since most of the digits are not significant.[1] We need to throw away information that is of no relevance, retaining only a finite, fixed number of digits. This is normally done by eliminating some of the least significant digits. Using such a *finite-precision* system limits the set of numbers that can be represented, and finite–precision arithmetic may create a small error in each step.

Example 2.1

Take as representable numbers the set of all decimal numbers with exactly three digits. Thus 1.23 and 15.6 are representable numbers. If we add them, we get the sum

$$1.23 + 15.6 = 16.83$$

[1]If we have a number with three significant digits we must assume that the fourth digit is in doubt. We cannot increase this information content beyond three digits no matter what operations we perform.

which is not representable in our system. To make it fit into our convention, we have to eliminate one digit. Sensibly, we discard the least significant one, giving the approximate result

$$1.23 + 15.6 \cong 16.8.$$

In adjusting the result, we are forced to make a small error. ■

EXERCISES

1. Convert the following decimal numbers to binary: 105, 12.625, 32.5, 0.003.
2. Convert the following binary numbers to decimal: 1.001, 1001, 0.0011, 11.0011.
3. Can every binary number be represented exactly by a decimal number? Conversely, can every decimal number be converted exactly to a finite length binary number? Explain.
4. What is the largest positive integer that can be represented in binary with 16 bits?
5. Bases other than 2 and 10 are also possible. Explain how this works, and express $(135)_{10}$, using a base 3 representation.
6. Convert $(4.55)_6$ to decimal.
7. Convert $(10.2)_{10}$ to base 4. Can this be done without error?
8. Under what condition can one convert from base b_1 to base b_2 without error?

2.2 Floating-Point Representation

Since the early days of computers, a standard way of expressing finite-precision numbers has been a binary version of scientific notation called *floating-point*. The floating-point representation of a number has two parts: a *mantissa* f, often between 0 and 1, that represents the significant digits, and an *exponent* e that gives the scale factor. Both f and e are encoded as a fixed number bits, and the scale factor is interpreted as some power of two. Within this general framework, there are many possible choices.

Before 1980 there was no accepted convention for floating-point numbers and computer designers chose whatever form they considered to be most advantageous. As a result there were many floating-point systems and a great deal of confusion. In the 1980s the Institute for Electric and Electronics Engineering (IEEE) convened a panel of experts to bring some order into this chaos. This panel produced a set of specifications and requirements for floating-point arithmetic that are known as the *IEEE Standard*.

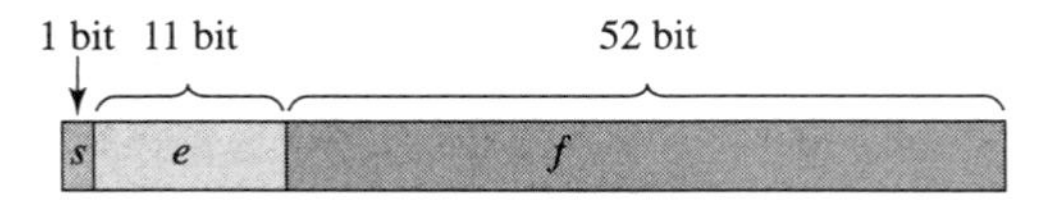

Figure 2.1 The IEEE double format.

While there is no compulsion for computer manufacturers to adhere to this standard, the IEEE recommendations were readily accepted and followed by most hardware designers. It is very likely that the computer you are working with uses these standards.

The most commonly used scheme for floating-point is the IEEE *double format*, shown schematically in Figure 2.1. The IEEE double format utilizes a 64-bit field (or 8 bytes); one bit is used for the sign of the number, and 52 bits are devoted to the mantissa. The remaining 11 bits represent the exponent e. The actual value denoted by this arrangement is

$$x = (-1)^s \times 2^E \times 1.f.$$

The exponent field e is a biased representation of its actual value E, with e and E related by

$$E = e - 1023.$$

The mantissa f is considered a fraction with the binary point immediately to its left, so that

$$f = 0101000000000000000000000000000000000000000000000000$$

has the decimal value $f = 2^{-2} + 2^{-4} = 0.3125$. e and s are considered integers.

Example 2.2 To represent 13.5 in IEEE double format, we start with

$$\begin{aligned} 13.5 &= 1.6875 \times 2^3 \\ &= (1 + 0.5 + 0.125 + 0.0625) \times 2^3 \\ &= \left(1 + 2^{-1} + 2^{-3} + 2^{-4}\right) \times 2^3 \end{aligned}$$

Because $2^{-1} + 2^{-3} + 2^{-4}$ has the binary representation 0.1011, f is the binary string 1011.The value of E is 3, so that

$$\begin{aligned} e &= 1026 \\ &= 2^{10} + 2. \end{aligned}$$

The binary expression for $e = 1026$ is 10000000010. Putting all this together, we see that the decimal number 13.5 is represented by the bit string

$$0100000000101011000000000000000000000000000000000000000000000000.$$

Normally, such long bit strings are presented in a condensed hexadecimal form. In this form each group of 4 bits is represented by a single symbol from the list 0, 1, ... , 9, A, B, C, D, E, F. Zero represents 0000, 1 stands for 0001, A stands for 1010, and F is 1111. In hexadecimal, the internal representation of 13.5 then is

402B000000000000.

■

Precise reasoning about floating-point numbers must be done taking the binary representation into account. However, in numerical work, order of magnitude arguments suffice most of the time and are conveniently made in decimal notation. Since $2^{-52} \cong 2.2 \times 10^{-16}$, the mantissa represents about 16 significant decimal digits. The relative value of the unit in the last place, commonly referred to as the *machine epsilon* or *ulp*, is therefore roughly 2×10^{-16}. The 11 bit exponent allows for scale factors of two to the power 2^{11}, so representable numbers range from about $2^{1023} \cong 10^{308}$ to $2^{-1024} \cong 10^{-308}$. While the relative accuracy of this is high and the range of numbers is large, the set of all floating-point numbers is finite.

The IEEE recommendations also call for reserving special bit patterns for exceptional values: *positive infinity* to represent very large positive values, and *negative infinity* to represent large negative numbers. Computations that are indeterminate or invalid are to yield another special pattern, the *not-a-number*. These special patterns can be used by the hardware to signal exceptional conditions, such as an overflow, without invalidating the entire computation. Software designers can then use these conditions as is suitable for the computation. For details on the IEEE standard, see [3].

While the IEEE standard is widely used, there is a possibility that you will encounter a computer with a different floating-point format. It is hard to find out the details of the representation without a user manual; however, the important question of the size on an ulp is relatively easy to answer using the following MATLAB program.[2]

Example 2.3 To find the approximate value of an ulp, we add a sequence of decreasing numbers to the value 1 until the result shows no change, indicating that the limit of significance has been reached. This will give the largest number that is negligible relative to a unit value. That largest number is roughly one-half ulp. Here is a simple program that will do the job.

[2]Scripts for specific programs in this book are given in the MATLAB language. The programs are always written so that they are easily converted to other languages, such as C or Fortran.

```
eps=1;
while (1+eps)>1
  eps=eps/2;
end
ulp=2*eps
```

When this program was run on our computer, the value printed was ulp= $2.2204e-16$, exactly what one expects from the IEEE convention. ■

In most computations, the details of floating-point representation, and the fact that computers use binary representation while we think in decimal, are of little consequence. What is more important is what can happen in a long sequence of simple arithmetic operations.

EXERCISES

1. What are the hexadecimal representations of the decimal numbers 1.25, −0.06, and 1000 in IEEE double format?
2. In IEEE double format, what real numbers are represented by the hexadecimal strings 402C000000000000, 402BB00000000000, and 702B000000000000?
3. IEEE single format specifies four bytes (or 32 bits) for floating-point representation. One bit is reserved for the number sign, eight bits for the exponent, and 23 bits for the mantissa. Discuss what this means in terms of the significant digits and the range of representable numbers. What is the value of an ulp in this format?
4. Some computers use a *hexadecimal floating-point*. In this form the scaling is done in terms of powers of 16; that is, the value of a number is something like

$$x = (-1)^s \times 16^e \times f.$$

What do you see as advantages and disadvantages of this representation?

5. Some early computers provided *binary-coded decimal* representation. In this form each decimal digit is encoded with four bits, so that the decimal integer 163 appears as 000101100011. Discuss possible advantages or disadvantages of this representation.
6. Another convention for floating-point representation is to express numbers in the form

$$s_1 s_2 \, e \, f$$

where s_1 is the sign of the number, s_2 is the sign of the exponent, e is the magnitude of the exponent, and $\frac{1}{2} \leq f < 1$ is the magnitude of the mantissa. Compare this with IEEE conventions to determine its advantages and disadvantages.

7. Many decimal numbers, such as 0.1, cannot be converted exactly to binary, because the binary expansion involves an infinite number of bits. What is the smallest possible error when $(12.4)_{10}$ is converted to IEEE double format?

8. The IEEE double format has about 16 significant decimal digits, considerably more than are needed in most engineering applications. How many correct (and hence significant) digits could you expect in the following circumstances?

   (a) Measuring the length of a desk with a ruler.

   (b) Measuring a time interval of about an hour with a wristwatch.

   (c) Timing a 100m sprint with a stopwatch.

   (d) Weighing an object of about 10 pounds on a bathroom scale.

9. Predict what the results from the following arithmetic operations will be: $1/0$, $-1/0$, $0/0$, $1/0 - 1/0$. Compare your predictions with what happens on your computer.

10. In IEEE double format, which of the following statements could result in a nonzero answer ?

    (a) $(100 + 1/3) - 1/3 - 100$.

    (b) $(1/3 + 100) - 100 - 1/3$.

    (c) $(1/3 + 100) - (100 + 1/3)$.

    Check your predictions using the computer.

2.3 Rounding Errors: A Practical View

When we develop numerical algorithms, such as the rectangular method in Chapter 1, we always base our reasoning on the real-number system. However, when we run the methods on a computer we work with finite-precision numbers. The finite-precision numbers that can be represented in floating-point format are only a finite subset of the real numbers (Figure 2.2); as a consequence some of the rules that apply to the real numbers system do not hold for computer-representable numbers. For example, the sum of two computer-representable numbers is not necessarily a computer-representable number. Furthermore, many of the laws that hold for real numbers, such as the associative and distributive laws, do not carry over to finite-precision arithmetic. The discrepancy that arises from the theoretical development of methods and their practical implementation as computer algorithms sometimes creates difficulties.

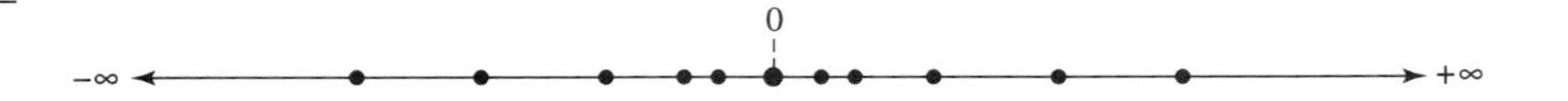

Figure 2.2
Real numbers and finite-precision numbers. The real numbers are represented by the solid line; the computer-representable numbers are shown as •.

When an arithmetic operation is performed with two floating-point numbers, the result is a real number which is not necessarily representable in floating-point. To work with the result, some adjustment has to be made. Since there is always a floating-point number close to the exact value, we make only a small error if we replace the result with a close-by floating-point number. This process is called *rounding*, and the small error committed by it is the *rounding error*. There are several possible rounding strategies.

The most plausible way of rounding is to pick the floating-point number that is closest to the true result. We call this *rounding to the nearest* and will use $round(a)$ to denote the floating-point number associated with any real number a by rounding to the nearest. The error created by rounding to the nearest is easily expressed in terms of the specifics of the floating-point system. For example, in IEEE double format and in most other conventions

$$round(a) = a(1 + \eta),$$

with

$$|\eta| \leq \frac{1}{2}\epsilon, \tag{2.1}$$

where ϵ has the value of an ulp. This is the normal way of rounding, so that when we use the term without a qualifier we will mean rounding to the nearest.

Another kind of rounding is *rounding toward zero*, denoted by $round_0$. Here we pick the closest floating-point number on the side toward zero from the true result. In essence, this means just chopping off the bits that do not fit into the limited field width for f, so this is also called *chopping* or *truncating*. Truncating is easy to implement in the hardware, but gives larger errors than rounding and is not often used. For reasons that will become apparent in Section 2.4, IEEE also recommends two other kinds of rounding: *Rounding toward infinity* or *rounding up* gives the smallest floating-point number greater than a, while *rounding toward negative infinity* or *rounding down* gives the largest floating-point number smaller than a. We denote these two rounded values by $round_\uparrow(a)$ and $round_\downarrow(a)$, respectively. The different rounding strategies are illustrated in Figure 2.3.

Rounding is necessary for most floating-point operations, and the error so created carries through any sequence of arithmetic operations. We use $fl(x)$ to denote the result we get when we evaluate an expression x with floating-point operations. If floating-point arithmetic is implemented carefully and properly, each individual operation yields maximum possible accuracy; that is, the result of an arithmetic operation is the rounded true

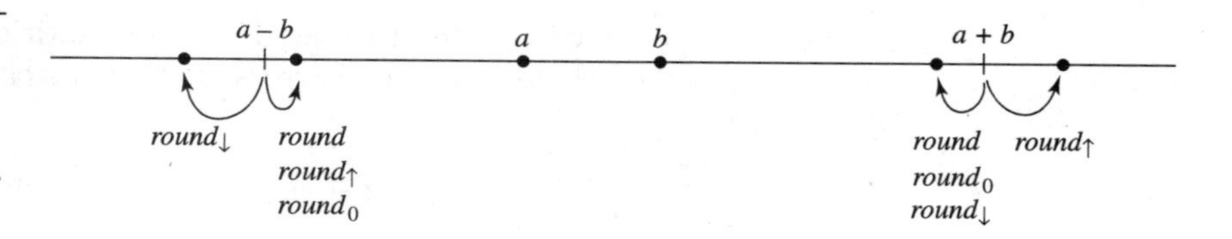

Figure 2.3 Floating-point numbers and rounding strategies. The straight line represents the real numbers; the dots are the floating-point numbers.

result, so that

$$\begin{aligned} fl(a+b) &= round(a+b) \\ &= (a+b)(1+\eta), \end{aligned} \tag{2.2}$$

and

$$\begin{aligned} fl(ab) &= round(ab) \\ &= ab(1+\eta), \end{aligned} \tag{2.3}$$

where η satisfies (2.1).

These rules assume that the result of the operation after rounding is computer-representable; that is, we do not create a number so large that it exceeds the range of representable numbers. If a result is too large we have an *overflow* that leads to an exceptional situation. Some computers simply terminate computations when this happens, but in IEEE format the provision for positive infinity, negative infinity, and not-a-number make it possible for the software to deal with exceptional cases in a more controlled fashion.

While it is easy to describe the effect of rounding in one operation, it is much more difficult to see what happens in a lengthy computation. Every step creates a small error, and these individual errors add up. The important question is to what extent the accumulated rounding error affects the accuracy of the final result.

At first it may seem that we need not worry about rounding at all. The number of significant digits used by computers is so much larger than what is needed and what the data accuracy warrants (see Exercise 8 in Section 2.1), that even if error accumulates over millions of operations it will never become important enough to affect the answers. There is some validity to this point of view, but there are some computations where small errors can grow very rapidly and as a result such an optimistic attitude is not justified.

One of these situations comes from *catastrophic cancellation.* Suppose we compute the difference

$$z = a - b$$

where a and b are very close in value. If these quantities arise from some computation they could have small errors, so that instead of z we actually get $\hat{z}$, with

$$|z - \hat{z}| \leq \eta,$$

where η is small. This makes $\hat{z}$ accurate in an absolute sense, but its relative error

$$\left| \frac{z - \hat{z}}{z} \right| \leq \frac{\eta}{|a - b|}$$

can be quite large. Another way of saying this is that when we subtract two close quantities, many of the correct digits cancel, leaving only those seriously contaminated by error. When such quantities are used in further computations, one small error can cause errors many orders of magnitude larger in subsequent computations.

Example 2.4 Consider the quadratic equation

$$ax^2 + bx + c = 0,$$

with $a \neq 0$. As we know, the two solutions to this equation can be found in analytic form as

$$x_1 = \frac{-b + \sqrt{b^2 - 4ac}}{2a},$$

and

$$x_2 = \frac{-b - \sqrt{b^2 - 4ac}}{2a}.$$

While these formulas are commonly used, they are not always the best to use. Take the specific values $a = 1$, $b = 10.00001, \text{and}\, c = 0.0001$, corresponding to the roots $x_1 = -10^{-5}$ and $x_2 = -10$. When these coefficients were used in actual computations,[3] the second root x_2 was obtained to 15 significant digits, but the other root was found as

$$x_1 = -1.000000000050960 \times 10^{-5}$$

showing a loss of five significant digits. The cause for it is cancellation between the very close values of b and $\sqrt{b^2 - 4ac}$.

[3]Unless otherwise stated, all computations in this book are done with IEEE double format floating-point, using MATLAB on an IBM PC under Windows 95. Other systems may give results differing from these by a few ulps, but the general conclusions should be unaffected.

In this case, we can get around the difficulty by using the mathematically equivalent expression

$$x_1 = \frac{c}{ax_2}.$$

This gives the root x_1 correct to 15 digits.

This example shows that rounding errors can grow rapidly in just a few steps. We say such a computation is *unstable*.[4] In this case we can overcome the instability by changing the algorithm, but this option is not always available. Certain kinds of recursive computations can be unstable, but it is not easy to see how to reformulate them in a more stable manner.

Example 2.5 Let the sequence $f(n)$ be defined by

$$f(n+1) = 2.9f(n) - 1.8f(n-1), \quad n = 2,\ 3,\ \ldots, \tag{2.4}$$

with $f(1) = 1$, $f(2) = 0.9$. It can be verified by a simple substitution that the solution to this recurrence relation is

$$f(n) = 0.9^{n-1}.$$

When results were computed using (2.4), the results in Table 2.1 were obtained. As we can see, the error accumulates rapidly, completely overshadowing the true value by the time we reach $n = 45$.

These examples show that rounding errors can become significant even in short computations. To guard against errors that may affect our conclusions, we would like to have a mathematically rigorous theory by which we can predict rounding effects. Unfortunately, we will see in the next section that there is no completely satisfactory theory that we can use. As a result, we normally have to deal with rounding errors in a pragmatic way. Here are some rules of thumb that work in most circumstances.

The first step is to look at an algorithm carefully for possible places where cancellations could cause instability. Whenever possible, we should look for a reformulation that will increase stability. If all parts of the computation look stable, the chances are good that the entire process is stable, but this is not an absolute guarantee. If there is any doubt, rearranging the details of the method may give an indication of the loss of significant digits.

[4]In numerical analysis, the term *stability* has various definitions, several of which we encounter later. Intuitively, a computation is stable if small errors have small effects on the result and unstable if small errors can be significantly magnified. The computation in Example 2.4 could be classified as mildly unstable.

Table 2.1
True and computed results for Example 2.5.

n	$fl(f(n))$	0.9^{n-1}
3	0.8100	0.8100
4	0.7290	0.7290
5	0.6561	0.6561
6	0.5905	0.5905
⋮	⋮	⋮
43	0.0111	0.0120
44	0.0091	0.0108
45	0.0064	0.0097

Example 2.6 The sum

$$S = \sum_{i=1}^{N} (-1)^{i+1} \frac{1}{1+i^2} \tag{2.5}$$

can be done by adding the terms in forward order

$$S = \frac{1}{2} - \frac{1}{5} + \frac{1}{10} - \ \dots ,$$

or in reverse order

$$S = (-1)^{N+1} \frac{1}{1+N^2} + (-1)^N \frac{1}{1+(N-1)^2} + \ \dots$$

While these two ways are mathematically equivalent, the rounding is different so they may not give exactly the same results. In Table 2.2 we show the values computed in these two ways for several N. Since in absence of rounding we would get identical results, the difference must be attributed to rounding. In this case, the accumulated rounding error appears to be small.

■

Sometimes it is not clear how to rearrange the computations suitably. In that case, we may want to rerun the problem, perturbing the data and other parameters by a small amount. If the new results show a significant change, we can conclude that there is a stability problem. This is useful even in stable calculations. When the input data is experimental and has

Table 2.2
The sum (2.5) computed in two different ways. Digits in doubt are underlined.

N	Forward summation	Backward summation
1000	$0.33984973009432\underline{0}$	$0.3384973009431\underline{9}$
5000	$0.3398545251293\underline{35}$	$0.3398545251293\underline{42}$
10000	$0.3398546750943\underline{25}$	$0.3398546750943\underline{34}$

an inherent measurement error, perturbing the initial data by an amount that reflects this data error, the computations will show how the data error affects the final results.

Example 2.7 To examine the stability of the algorithm in Example 2.5 we ran the computations, using an initial condition

$$f(1) = 1 + 1 \times 10^{-15}.$$

The results that arise from this small perturbation are denoted by $\hat{f}(n)$ and are shown in Table 2.3 alongside the results of the unperturbed computation.

The difference between the two calculations is a clear indication that instability has affected the computations to the point where the results are no longer of any value. ■

These examples give us hints as to what to do in practice about rounding errors. First, we should carefully examine the algorithm for places where there could be significant loss of significance, and for repetitive computations in which there could be error growth in each step. If possible, the implementation should be changed to produce more stable results. If, in spite of our best efforts, excessive rounding error still seems possible, recomputations with perturbed data is advisable. If small changes in the data cause large changes in the results, the computations are not trustworthy. At that point, a complete, in-depth re-examination of the problem is advisable.

Table 2.3
Differences between perturbed and unperturbed computations.

n	$fl(f(n))$	$fl(\hat{f}(n))$
43	0.0111	0.0073
44	0.0091	0.0013
45	0.0064	−0.0092

EXERCISES

1. Why is the error in truncating normally larger than in rounding?
2. Show that for all a in IEEE format $|round_{\uparrow}(a) - round_{\downarrow}(a)| \leq \varepsilon\,|a|$, where ε denotes an ulp.
3. Give the hexadecimal representations for two IEEE double-format numbers whose addition will produce an overflow.
4. If the result of an arithmetic operation is smaller in magnitude than the smallest representable number, we have an *underflow* situation. Give the hexadecimal representations of two IEEE double-format numbers whose multiplication will produce an underflow. Is it a reasonable strategy to replace an underflow result by zero?
5. Compare the stability of computing the expression

$$y = 1 - \sqrt{1 - x^2}$$

with the mathematically equivalent form

$$y = \frac{x^2}{1 + \sqrt{1 - x^2}}$$

for small values of x.

6. A mathematical equivalent form for complex division

$$w = \frac{u + iv}{x + iy} = \frac{(xu + yv) + i(xv - yu)}{x^2 + y^2}$$

is

$$w = \begin{cases} \dfrac{\left(u + \dfrac{vy}{x}\right) + i\left(v - \dfrac{uy}{x}\right)}{x\left(1 + \dfrac{y^2}{x^2}\right)} & \text{if } |x| \geq |y|\,, \\[2ex] \dfrac{\left(v + \dfrac{ux}{y}\right) + i\left(-u + \dfrac{vx}{y}\right)}{y\left(1 + \dfrac{x^2}{y^2}\right)} & \text{if } |x| < |y|\,. \end{cases}$$

Show by an example that the two methods actually result in different numerical answers. Which method do you think is more stable?

7. To keep the rounding error small in computing

$$S_n = \sum_{i=1}^{n} \frac{1}{i^2},$$

is it better to add the larger terms first? Or is there some advantage in doing this in reverse order by

$$S_n = \frac{1}{n^2} + \frac{1}{(n-1)^2} + \ldots ?$$

8. What happens in computing the roots by the algorithm in Example 2.4 if a is very small compared to b and c?

9. The instability in linear recurrences can be predicted using the following well-known formula. The solution of the recurrence

$$f(n+k) = a_0 f(n) + a_1 f(n+1) + \ldots + a_{k-1} f(n+k-1)$$

is

$$f(n) = c_1 \rho_1^n + c_2 \rho_2^n + \ldots + c_{k-1} \rho_{k-1}^n + c_k \rho^n,$$

where the ρ_i are the solutions of the characteristic equation

$$a_0 + a_1 \rho + \ldots + a_{k-1} \rho^{k-1} - \rho^k = 0,$$

and the c_i are constants determined by the given values $f(0),\ f(1),\ \ldots,\ f(k-1)$. Note that if any of the solutions $|\rho_i| > 1$, $f(n)$ will approach ∞, so the linear recurrence will be unstable. Use this to explain the observed instability in Example 2.5.

10. Suggest a stable method for computing

$$c = \frac{\sin(x)}{x}$$

for small x.

11. Investigate the stability of the following recurrences.

    (a) $f(n) = \frac{3}{2} f(n-1) - \frac{1}{2} f(n-2),\ \ f(0) = 2,\ \ f(1) = 1.5$

    (b) $f(n) = 3f(n-1) - 2f(n-2),\ \ f(0) = 0.9,\ \ f(1) = 0.9$

    (c) $f(n) = f(n-1) + f(n-2),\ \ f(0) = 1,\ \ f(1) = 1$

12. When $f(50)$ was computed by the recursive expression

$$f(n) = 0.6 f(n-1) + 0.5 f(n-2), \quad n = 3,\ 4,\ \ldots\ 50,$$

with $f(1) = 1,\ f(2) = 2$, the value obtained was $f(50) = 40.5680$. How many digits of this answer do you think are correct?

2.4 Theories of Rounding Error*

While the heuristic and ad hoc attacks on rounding that we just discussed are very useful in practice, a more rigorous treatment would obviously help to complement that point of view. Many attempts have been made to develop a rigorous approach based on bounding the errors in each step and using inequalities to carry the bounds forward. On the whole, this has not been particularly successful. While it is not hard to write the basic rules of rounding, the analysis tends to become intractable for even moderately complicated problems.

In a computation involving several operations, the rounding error incurred at one step is carried along to subsequent steps. To analyze the entire effect of rounding we have to see how all of the small errors propagate and accumulate. While we can easily find a bound on the rounding error in each step, when these bounds are propagated things get messy quite quickly. Suppose we add three floating-point numbers a, b, c. Even in this simple case, we have to state the order of the operations, because we cannot assume that the associative law holds, and the conclusions may be affected by the order in which the steps are carried out. For simplicity, let us assume that all three numbers are positive, and that we add a and b first. Then

$$\begin{aligned} fl((a+b)+c) &= fl(fl(a+b)+c) \\ &= fl((a+b)(1+\eta_1)+c) \\ &= ((a+b)(1+\eta_1)+c)(1+\eta_2), \end{aligned}$$

where η_1 and η_2 are no larger than one-half ulp in magnitude. From this we can produce a bound on the error involved in the two consecutive floating-point additions by

$$|a+b+c-fl((a+b)+c)| \;\leq\; |(a+b+c)\eta \;+\; (a+b)(\eta+\eta^2)|,$$

where $\eta = \max(|\eta_1|, |\eta_2|)$. If we ignore the very small term η^2 and assume that $a, b,$ and c are all roughly the same size, we see that, approximately, the relative error in the sum of three numbers is one ulp—not an unexpected result.

With a little work and a few judicious assumptions one can carry this sort of argument another step forward.

Example 2.8 Estimate the effect of rounding on the sum

$$S_n = \sum_{i=1}^{n} x_i, \tag{2.6}$$

where $0 \leq x_i \leq 1$.

Again, we have to be specific about the order in which the additions are performed. We do this by defining the partial sums recursively as

$$S_i = S_{i-1} + x_i,$$

with $S_0 = 0$. In a floating-point computation, this becomes

$$\begin{aligned} fl(S_i) &= fl(fl(S_{i-1})+x_i) \\ &= (fl(S_{i-1})+x_i)(1+\eta_i) \\ &= fl(S_{i-1})+x_i+(fl(S_{i-1})+x_i)\eta_i, \end{aligned}$$

where $|\eta_i| \leq \varepsilon/2$. The errors

$$e_i = S_i - fl(S_i)$$

then satisfy the recurrence

$$e_i = e_{i-1} + (fl(S_{i-1}) + x_i)\eta_i,$$

with $e_1 = 0$. If the accumulated effect of rounding is not too severe, we can expect that

$$fl(S_{i-1}) + x_i \leq n,$$

so that

$$|e_i| \leq |e_{i-1}| + \frac{1}{2}\varepsilon n.$$

From this it follows that

$$|e_n| \leq \frac{1}{2}\varepsilon n^2,$$

and we might expect that after a few thousand operations five or six significant digits could be lost. ■

The result of this example suggests, then, that rounding errors may accumulate quickly enough to grow significantly even in fairly short computations. But this is a worst-case situation, and in reality the problem is much less serious. First, because S_n can be expected to be of order n, the relative error[5]

$$\left| \frac{S_n - fl(S_n)}{S_n} \right| = O(n)$$

grows only linearly and is more manageable. Furthermore, since rounding errors behave like random variables with mean zero, some individual roundings will increase the propagated error while others will decrease it. This tends to make the accumulated effect much smaller than the worst-case prediction. In reality, in this kind of simple computation the rounding error rarely grows quickly enough to become significant.

Example 2.9 The MATLAB program below generates 10,000 random numbers between 0 and 1 and computes the sum (2.6), first summing forward, then backward. We expect that the difference is a good measure of the accumulated rounding errors.

[5] Here we use the traditional O-notation to denote orders of magnitude.

```
for i=1:10000
  x(i)=rand(1,1);
end
sum1=0;
for i=1:10000
  sum1=sum1+x(i);
end
sum2=0;
for i=10000:-1:1
  sum2=sum2+x(i);
end
(sum1-sum2)/sum1
```

Repeated runs with the program indicated that the average loss of accuracy in summing these 10,000 numbers was only one significant digit. In no cases were more than two significant digits lost.

■

The difficulty with such a rigorous error analysis[6] is clear: not only is it cumbersome, but it also tends to give overly pessimistic results.

Virtually from the first use of floating-point arithmetic, numerical analysts have sought alternatives that would give a handle on the rounding problem. In particular, mathematicians have looked at number representations that can contain in the notation an assessment of the accuracy. Among the various alternatives that have been proposed, *interval arithmetic* has been the most productive. In interval arithmetic a real number a is represented by two floating-point numbers, a_1 and a_2, such that

$$a_1 \leq a \leq a_2.$$

We write this as $a \sim [a_1,\ a_2]$, indicating that the real number a is represented by the two floating-point numbers that enclose it. The width of the interval $[a_1,\ a_2]$ gives the uncertainty about the true value of a. The rules of interval arithmetic are chosen so that these uncertainties, including those caused by rounding, are carried correctly through a computation of many steps, and so that the true result is always guaranteed to be included in the interval. The rules are fairly straightforward. For example, if all arithmetic is precise, interval addition uses the rule

$$[a_1,\ a_2] + [b_1,\ b_2] = [a_1 + b_1,\ a_2 + b_2]. \qquad (2.7)$$

---

[6] What we have described here is a somewhat brute-force method. More sophisticated approaches have been suggested; for example the backward error analysis of Wilkinson ([28]) has had some success in linear algebra. However, the success has been limited and is of little help in complicated situations.

Since the sum of any number in the interval $[a_1,\ a_2]$ and any number in the interval $[b_1,\ b_2]$ will lie in the interval $[a_1 + b_1,\ a_2 + b_2]$, the inclusion rule for interval arithmetic is satisfied. Interval multiplication is a little less obvious, but it is still not hard to see that the correct rule is

$$\begin{aligned}[a_1,\ a_2] \times [b_1,\ b_2] = [&\min(a_1b_1,\ a_1b_2,\ a_2b_1,\ a_2b_2), \\ &\max(a_1b_1,\ a_1b_2,\ a_2b_1,\ a_2b_2)].\end{aligned} \tag{2.8}$$

Example 2.10 Take $a \sim [-0.1,\ 0.2]$ and $b \sim [1,\ 1.4]$. Then

$$a + b \sim [0.9,\ 1.6],$$

and

$$ab \sim [-0.14,\ 0.28].$$

■

These rules work only when all operations are carried out without error. In finite precision arithmetic where rounding is necessary, the rounding has to be done so that the inclusion requirement is not violated. For example, if $a \sim [a_1,\ a_2]$ and $b \sim [b_1,\ b_2]$, then

$$a + b \sim [round_{\downarrow}(a_1 + b_1),\ round_{\uparrow}(a_2 + b_2)], \tag{2.9}$$

and the computed interval will enclose the true value of $a + b$. Similar rules can be established for subtraction, multiplication, and division, leading to a complete set of rules for interval arithmetic in an environment of computer arithmetic. Notice that directed rounding is essential to assure that the effects of rounding are properly accounted for. With interval arithmetic we have complete control of the rounding error. If it grows rapidly, the intervals become very large; if the intervals stay small we can be sure that the rounding error is not significant.

Although interval arithmetic was proposed nearly forty years ago, only recently has it become accepted as a serious tool for numerical computations. There are several reasons for this. It is certainly slower than ordinary floating-point arithmetic, because every interval computation requires several floating-point operations. But with rapidly increasing computer speeds, this objection has lost much of its force. A second difficulty is that interval arithmetic requires directed rounding. For most of the earlier years this was not available on many machines, and as a result it was quite difficult to implement interval arithmetic. The IEEE recommendations have gone a long way toward remedying this situation. The most serious objection to interval arithmetic is that the intervals tend to grow much more rapidly than necessary, and that we often end up with intervals that are so large that we cannot get any useful information from them. This is still a valid point, but more sophisticated analysis has been able to overcome it in many cases. There are now many important applications where interval arithmetic is able to solve a problem that defies treatment by standard floating-point computation. As a result of this progress, it is possible to find compilers for several programming languages, such as C and Fortran, that provide an interval data type and the supporting functions.

EXERCISES

1. In Example 2.6 we could compute S by

$$S_p = \text{sum of all positive terms,}$$
$$S_n = \text{sum of all negative terms,}$$

and

$$S = S_p + S_n.$$

Do you expect this alternative to show a smaller or larger accumulated rounding error than the results in Table 2.2?

2. Give arguments why the interval in (2.8) encloses the correct result.

3. Give a set of rules for interval subtraction and division in the presence of rounding. Justify all of your answers.

4. If $a \sim [-0.1,\ 0.2]$ and $b \sim [-1,\ 1]$, what are $a + b$ and ab? What can you say about a/b?

5. Suppose that you have a program that computes $\sin(x)$ correctly within an absolute error of 5×10^{-16} for all $0 \leq x \leq a$. Show how you could use such a program to define an interval sin program for this range of x.

6. Which of the following implications are true for IEEE floating-point arithmetic?

   (a) $fl(ab) = 0$ implies $fl(ba) = 0$

   (b) $fl(a - b) = 0$ implies $a = b$

   (c) $fl(ab) = 0$ implies $a = 0$ or $b = 0$

   (d) $fl(a + b) = a$ implies $b = 0$

   (e) $fl(x) > 0$ implies $fl(-x) < 0$

   Wherever the implication is false, give specific examples that show this.

7. In a well-designed floating-point system, which of the associative, commutative, and distributive laws could reasonably be satisfied?

Chapter 3

The Solution of Systems of Linear Equations

The need to solve sets of simultaneous linear equations arises in many numerical computations. Linear systems can come directly from physical situations, or may be incidental to the solution of more complex problems, such as partial differential equations. Here is an example that leads to a small linear system.

Example 3.1 A simple electrical circuit that consists of a source of voltage and some resistors in a network will have some current flowing through the branches of the network. The amount of current in each branch is determined by the following laws: (a) conservation of current, so that at any junction the amount of current flowing in is equal to the amount of current flowing out, and (b) *Ohm's Law*, which states that the voltage drop across a resistor is equal to the product of the resistance and the current flowing through the resistor. If we apply this to the circuit in Figure 3.1, we get a system of equations that relate the currents, the resistances, and the applied voltage.

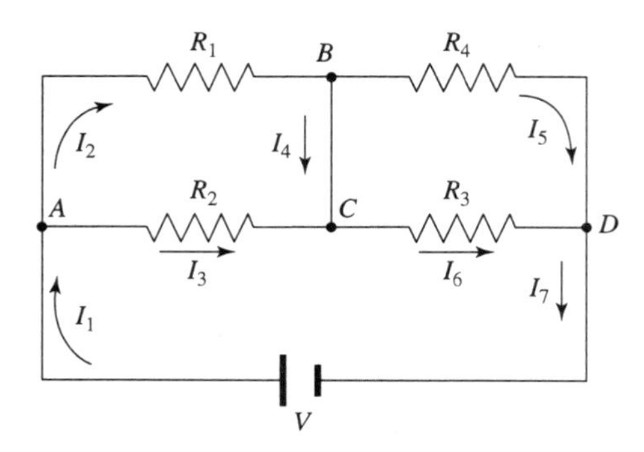

Figure 3.1
A simple electrical circuit.

Balancing currents at the junctions A, B, C, and D, respectively, we get

$$\begin{aligned} I_1 &= I_2 + I_3, \\ I_2 &= I_4 + I_5, \\ I_6 &= I_3 + I_4, \\ I_7 &= I_5 + I_6. \end{aligned}$$

Then, from Ohm's Law,

$$\begin{aligned} R_1 I_2 &= R_2 I_3, \\ R_3 I_6 &= R_4 I_5, \\ R_1 I_2 + R_4 I_5 &= V. \end{aligned}$$

If V and the four resistances have known values, we have a system of seven equations for the seven unknown currents.

Small systems such as the one in this example can be solved by hand, but the work gets excessive quickly. For typical systems with 50 to 100 unknowns a computer is essential. For large systems involving thousands of unknowns (not a rare occurrence), we need not only very fast computers, but also very efficient algorithms.

3.1 The Gauss Elimination Method

All of the familiar methods for solving linear systems combine equations to eliminate variables, and reduce the system to a form in which the solution becomes easy to see. The next example illustrates the most common form of this approach.

Example 3.2 Consider the system

$$
\begin{aligned}
3x_1 + 4x_2 + 5x_3 &= 4,\\
6x_1 + 2x_2 + 3x_3 &= 7,\\
x_1 + 3x_2 + 3x_3 &= 1.
\end{aligned}
$$

First, we subtract twice the first row from the second, and $\frac{1}{3}$ of the first row from the third. This gives

$$
\begin{aligned}
3x_1 + 4x_2 + 5x_3 &= 4,\\
-6x_2 - 7x_3 &= -1,\\
\frac{5}{3}x_2 + \frac{4}{3}x_3 &= -\frac{1}{3}.
\end{aligned}
$$

Next, we multiply the second row by $\frac{5}{18}$ and add it to the third row. We now have

$$
\begin{aligned}
3x_1 + 4x_2 + 5x_3 &= 4,\\
-6x_2 - 7x_3 &= -1,\\
-\frac{11}{18}x_3 &= -\frac{11}{18}.
\end{aligned}
$$

At this point we have reduced the system to a form that can be solved easily. The matrix associated with the reduced system is called an *upper-triangular* matrix, indicating that all nonzero elements are located on or above the main diagonal. Triangular systems can be solved by a simple substitution. From the third equation, we first get that $x_3 = 1$. From the second equation, substituting the value just obtained for x_3, we get $x_2 = -1$. Finally, using the first equation and substituting the values for x_2 and x_3, we find $x_1 = 1$. It should be quite obvious that this pattern can be used for any number of equations.

■

The method illustrated in this example is easily extended to the $n \times n$ general system

$$
\begin{aligned}
a_{11}x_1 + a_{12}x_2 + \cdots + a_{1n}x_n &= b_1,\\
a_{21}x_1 + a_{22}x_2 + \cdots + a_{2n}x_n &= b_2,\\
\vdots \qquad\quad \vdots &\\
a_{n1}x_1 + a_{n2}x_2 + \cdots + a_{nn}x_n &= b_n.
\end{aligned}
\tag{3.1}
$$

The coefficients a_{ij} and the right sides b_i are given and the system is to be solved for the unknowns $(x_1, x_2, \ldots, x_n)$. We will use standard matrix notation to write (3.1) as

$$
\mathbf{Ax} = \mathbf{b}, \tag{3.2}
$$

where $\mathbf{A}$ is a square $n \times n$ matrix, and $\mathbf{x}$ and $\mathbf{b}$ are vectors.[1] Small boldface letters are used to denote vectors and large boldface letters for matrices. A special subscript will be used for individual elements, so $[\mathbf{x}]_i$ denotes the ith component of the vector $\mathbf{x}$ and $[\mathbf{A}]_{ij}$ stands for the element in the ith row and jth column of the matrix $\mathbf{A}$. The conventional definitions of matrix addition, multiplication, and equality are used.

To visualize the solution process for the system (3.2), we first construct the *augmented matrix* by appending $\mathbf{b}$ as a column to the right of $\mathbf{A}$,

$$[\mathbf{A}|\mathbf{b}] = \left[\begin{array}{cccc|c} a_{11} & a_{12} & \cdots & a_{1n} & b_1 \\ a_{21} & a_{22} & \ldots & a_{2n} & b_2 \\ \vdots & \vdots & & \vdots & \vdots \\ a_{n1} & a_{n2} & \cdots & a_{nn} & b_n \end{array}\right]. \tag{3.3}$$

If a_{11} is not zero, we multiply the first row of this augmented matrix by suitable multiples and subtract this from the other rows so as to reduce all elements of the first column below the main diagonal to zero. This gives

$$\left[\begin{array}{cccc|c} a_{11} & a_{12} & \cdots & a_{1n} & b_1 \\ 0 & \bar{a}_{22} & \ldots & \bar{a}_{2n} & \bar{b}_2 \\ \vdots & \vdots & & \vdots & \vdots \\ 0 & \bar{a}_{n2} & \cdots & \bar{a}_{nn} & \bar{b}_n \end{array}\right]$$

where

$$\bar{a}_{22} = a_{22} - a_{12}\frac{a_{21}}{a_{11}}, \tag{3.4}$$

$$\bar{b}_2 = b_2 - b_1\frac{a_{21}}{a_{11}}, \tag{3.5}$$

with similar expressions for the other elements. The system represented by this new augmented matrix clearly has the same solution as (3.2).

In subsequent stages, a similar process that assumes the main diagonal will always be nonzero, is used to reduce other elements below the main

[1] We use the term vector, or n-vector, for any n-tuple of numbers.

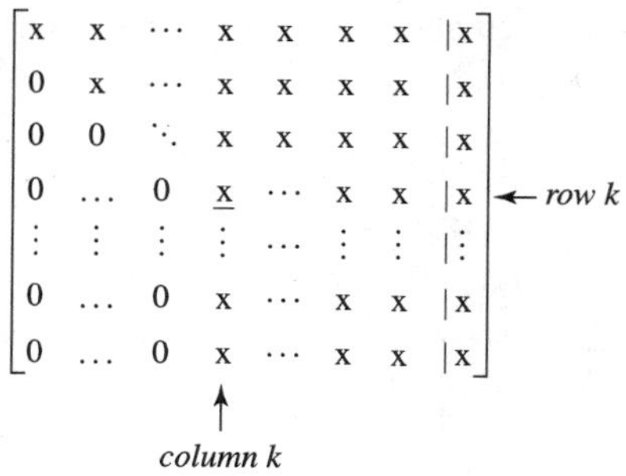

$$\left[\begin{array}{ccccccc|c} x & x & \cdots & x & x & x & x & x \\ 0 & x & \cdots & x & x & x & x & x \\ 0 & 0 & \ddots & x & x & x & x & x \\ 0 & \cdots & 0 & \underline{x} & \cdots & x & x & x \\ \vdots & \vdots & \vdots & \vdots & \cdots & \vdots & \vdots & \vdots \\ 0 & \cdots & 0 & x & \cdots & x & x & x \\ 0 & \cdots & 0 & x & \cdots & x & x & x \end{array}\right] \leftarrow \textit{row } k$$

↑ *column k*

Figure 3.2 Pattern of elements in the reduction of a matrix.

diagonal to zero. Figure 3.2 shows the process schematically at the kth stage, the symbol x representing elements not necessarily zero.

At the kth stage, shown in Figure 3.2, columns 1 to $k-1$ have been reduced so that all elements below the diagonal in these columns are zero. The underlined element on the diagonal is next used to reduce the kth column. Eventually, after $n-1$ such steps, we end up with an upper triangular system, at which point we solve for the unknowns, starting with x_n.

A formal description of the process is straightforward. We think of it as a sequence of transformation, producing augmented matrices

$$[\mathbf{A}^{(1)}|\mathbf{b}^{(1)}] \rightarrow [\mathbf{A}^{(2)}|\mathbf{b}^{(2)}] \rightarrow \ldots \rightarrow [\mathbf{A}^{(n)}|\mathbf{b}^{(n)}] \tag{3.6}$$

so that at the kth stage, the matrix $\mathbf{A}^{(k)}$ has zeros below the diagonal in columns 1 to $k-1$, and so that the solution of

$$\mathbf{A}^{(k)}\mathbf{x} = \mathbf{b}^{(k)}$$

is identical to the solution of the system (3.2). Starting with $\mathbf{A}^{(1)} = \mathbf{A}$ and $\mathbf{b}^{(1)} = \mathbf{b}$ we end with the upper triangular matrix $\mathbf{A}^{(n)}$ in the last stage. Specific formulas that achieve this *triangularization* are, for $k = 2, 3, \ldots, n$,

$$\begin{aligned}[\mathbf{A}^{(k)}]_{i,j} &= [\mathbf{A}^{(k-1)}]_{i,j} - m_{i,k}[\mathbf{A}^{(k-1)}]_{k-1,j} \\ &\qquad\qquad i = k, k+1, \ldots, n, \; j = k-1, \; k, \; \ldots, \; n \\ &= [\mathbf{A}^{(k-1)}]_{i,j}, \quad \text{otherwise,}\end{aligned} \tag{3.7}$$

$$\begin{aligned}[\mathbf{b}^{(k)}]_i &= [\mathbf{b}^{(k-1)}]_i\,, \quad i = 1, 2, \ldots, k-1, \\ &= [\mathbf{b}^{(k-1)}]_i - m_{i,k}[\mathbf{b}^{(k-1)}]_{k-1}\,, \quad i = k, \; k+1, \; \ldots, \; n,\end{aligned} \tag{3.8}$$

with

$$m_{i,k} = \frac{[\mathbf{A}^{(k-1)}]_{i,k-1}}{[\mathbf{A}^{(k-1)}]_{k-1,k-1}}. \tag{3.9}$$

We leave it as an exercise to show that this gives the desired triangularization. In the second part, the *backsubstitution*, the solution $\mathbf{x}$ is produced by

$$[\mathbf{x}]_n = \frac{[\mathbf{b}^{(n)}]_n}{[\mathbf{A}^{(n)}]_{n,\,n}}, \tag{3.10}$$

$$\begin{aligned}[\mathbf{x}]_{n-i} = \frac{1}{\left[\mathbf{A}^{(n)}\right]_{n-i,n-i}} & \left\{ \left[\mathbf{b}^{(n)}\right]_{n-i} \right. \\ & \left. - \sum_{j=0}^{i-1} \left[\mathbf{A}^{(n)}\right]_{n-i,n-j} [\mathbf{x}]_{n-j} \right\}, i = 1, 2, \ldots, n-1\end{aligned} \tag{3.11}$$

Equations (3.7) to (3.11) define the classical *Gauss Elimination Method* (GEM) for solving linear systems of equations. With these formulas, implementation of the GEM is very straightforward and writing a computer program for it is an elementary exercise.[2] But there are a few practical wrinkles.

The GEM terminates after a finite number of steps. If there were no rounding errors, it would give the exact solution of (3.2). But rounding affects the accuracy of the results, so we need to keep the method as stable as possible. The underlined element in Figure 3.2 that is used to reduce the kth column is called the *pivot*. If by chance this element is zero it cannot be used, and we must rearrange the computations to get a suitable pivot. The simplest way is to interchange rows to get a nonzero element in the pivot position. Since exchanging rows just permutes the order in which we write the equations, it does not change the solution. While in theory any nonzero element can be used as a pivot, stability considerations are important. A very small element may be the result of cancellation of two larger numbers and therefore relatively inaccurate. If it is used as the pivot we could get large magnification of rounding errors. We do better by using the largest (in magnitude) element as the pivot. This is the *row pivoting* strategy. At the kth stage, we search column k, rows k to n, for the element of largest magnitude. We then interchange row k with the row in which this largest element was found; after that the reduction proceeds as described. We do not want to use small pivots, because cancellation may have affected their relative accuracy. But because any row can be multiplied by an arbitrary constant, it is possible that a small pivot is quite accurate, while a pivot of unit magnitude may be the result of cancellation of two large quantities. To avoid this, we must scale the system so that coefficients do not range over many orders of magnitude. The easiest way is to multiply each row so that its maximal element has order of magnitude one. We refer

[2]It is unlikely, though, that you will have to implement it as virtually every system now has some sort of library routine for it.

to this as *equilibration*.[3] While failure to choose a proper pivot can result in large errors, it is known that equilibration with row pivoting produces a very stable algorithm. The GEM is a good example of how one can avoid instability by a proper arrangement of the computation.

An important question is the amount of work that is involved in applying the GEM. At the ith stage there are $(n-i)$ rows to reduce. Each of the reductions requires $(n-i+1)$ additions and multiplications. To go from stage 1 to stage n we need to do $n-1$ such reductions. Therefore, the number of additions and multiplications[4] to be performed is roughly

$$N_1 = \sum_{i=1}^{n-1} (n-i+1)(n-i) = \frac{1}{3}n^3 + O(n^2). \tag{3.12}$$

There are also some divisions and interchanges if row pivoting is used, but the number of these is $O(n^2)$. Furthermore, it is not difficult to show that the number of additions and multiplications for the backsubstitution is about

$$N_2 = \sum_{i=1}^{n-1} i = \frac{1}{2}n^2 + O(n). \tag{3.13}$$

For large n then, the solution of n equations in n unknowns by the GEM takes roughly $n^3/3$ additions and multiplications. Present computer speeds can handle $n = 100$ or $n = 200$ without any difficulty, but $n = 10{,}000$ will give even the fastest computers trouble.

EXERCISES

1. For the system

$$\begin{aligned} x_1 + x_2 + 2x_3 &= 3, \\ 3x_1 - x_2 + x_3 &= 2.5, \\ x_1 + 2x_2 - 4x_3 &= -1, \end{aligned}$$

compute $[\mathbf{A}^{(2)}|\mathbf{b}^{(2)}]$ and $[\mathbf{A}^{(3)}|\mathbf{b}^{(3)}]$. Then backsubstitute to find the solution of the system.

---

[3] Equilibration is effective only if all matrix elements have similar relative accuracy. In some situations, for example when the elements come from experimental measurements, this may not be so and the smaller elements could have a much larger relative error. In this case, equilibration may do more harm than good. For this reason, many implementations of the GEM do not equilibrate automatically, but leave it to the user to do so when appropriate.

[4] We use the term *operation* to loosely refer to a single addition or multiplication, or even a multiplication followed by an addition. Since we are only interested in orders of magnitude, this imprecision is of no consequence.

2. Prove that

$$\sum_{i=1}^{n} i = \frac{n(n+1)}{2},$$

and

$$\sum_{i=1}^{n} i^2 = \frac{n(n+1)(2n+1)}{6}.$$

Use these to show that (3.12) holds.

3. Show that the number of operations for backsubstitution in the GEM is $O(n^2)$, as claimed in (3.13).

4. In *complete pivoting*, both rows and columns are searched for the largest element (in magnitude), and rows and columns are interchanged to get it into the pivot position. This somewhat increases the stability of the method, but is normally not used. Can you think of a reason why row pivoting is usually preferred?

5. The *Gauss-Jordan* method is a variant of the GEM in which the matrix $\mathbf{A}$ is changed to diagonal form by reducing the elements above the diagonal, as well as those below, to zero. This makes the reduction more expensive, but eliminates backsubstitution. Give explicit formulas, analogous to (3.7) to (3.11), for this method.

6. Make an operations count for the Gauss-Jordan method to determine if this variant is competitive with the GEM.

7. Show that the determinant of a square $n \times n$ matrix A can be calculated from the result of GEM as $\det(\mathbf{A}) = [\mathbf{A}^{(1)}]_{1,1} \times [\mathbf{A}^{(2)}]_{2,2} \ \ldots \ \times [\mathbf{A}^{(n)}]_{n,n}$; in other words, the determinant is the product of the pivots.

8. The age distribution in a population of female beetles can be modeled as follows: Every female beetle has a survival rate of 1/3 in the first year after its birth, and a survival rate of 1/2 from the second to third year, before giving birth to six new females and dying at the end of the third year.

   Let $\mathbf{x}_k$ denote the distribution of ages in year k, with $[\mathbf{x}_k]_1$ standing for the number of beetles of age one in year k, and so on. Then the relation between populations in successive years can be expressed as

$$\mathbf{x}_{k+1} = \mathbf{A}\mathbf{x}_k,$$

   where

$$\mathbf{A} = \begin{bmatrix} 0 & 0 & 6 \\ 1/3 & 0 & 0 \\ 0 & 1/2 & 0 \end{bmatrix}.$$

   If, after five years, the beetle population has the distribution (300, 60, 30), what was the original distribution?

9. An economic system consists of three sectors: Energy, Materials, and Service, that are labeled by E, M, and S, respectively. Each sector produces output to supply the demand of customers, as well as the demands of the other sectors, including itself. Suppose that for any sector, with output x the demand of customers is constant, and the demands of a sector are fractions of its output x. Moreover, when the economic system is "closed," we require that the outputs and demands for the system are always balanced. The following table lists the data for the closed economic system under consideration.

		Demands			
		E	M	S	Consumers
Outputs	E	0.2	0.3	0.1	2
	M	0.2	0.7	0.2	1
	S	0.4	0.2	0.5	3

If we denote the outputs for the sectors E, M, and S by x_1, x_2, and x_3 respectively, then, from the above table, we arrive at the following system of linear equations:

$$\begin{aligned} x_1 &= 0.2x_1 + 0.3x_2 + 0.1x_3 + 2, \\ x_2 &= 0.2x_1 + 0.7x_2 + 0.2x_3 + 1, \\ x_3 &= 0.4x_1 + 0.2x_2 + 0.5x_3 + 3. \end{aligned}$$

What are the outputs of this system in its closed state?

10. Find the point of intersection of the following three planes in an x, y, z coordinate system:

$$\begin{aligned} x + y + z &= 4, \\ 2x + 2y + 5z &= 11, \\ 3x + 3y + 4z &= 12. \end{aligned}$$

11. In Example 3.1, find the currents if $V = 12$, $R_1 = R_2 = 5$, $R_3 = 15$, and $R_4 = 20$.

3.2 Computing the Inverse of a Matrix

There are some occasions where (3.2) has to be solved for several right sides. While the matrix **A** remains the same, the right side **b** can have several possible values. The GEM is easily adapted for this purpose.

The reduction of **A** to upper triangular form does not depend on the right side **b**, so when we have to solve a set of equations several times with different right sides, we can save some work. Suppose we want to solve (3.2) twice, first with right side $\mathbf{b}$, then with right side $\mathbf{c}$. We form the augmented matrix

$$[\mathbf{A}|\mathbf{b}|\mathbf{c}]$$

and carry out the reduction of $\mathbf{A}$ to upper triangular form, as before, making the appropriate changes in the columns of $\mathbf{b}$ and $\mathbf{c}$. Backsubstitution is then done for $\mathbf{b}$ and $\mathbf{c}$ separately, giving the two solutions.

Example 3.3 If in Example 3.2 we want to solve the system with a second right side, say, $\mathbf{c} = (3, 2, 2)$, the first step in the reduction of the augmented matrix gives

$$\begin{bmatrix} 3 & 4 & 5 & | & 4 & | & 3 \\ 0 & -6 & -7 & | & -1 & | & -4 \\ 0 & 5/3 & 4/3 & | & -1/3 & | & 1 \end{bmatrix}.$$

Note that the part corresponding to the matrix $\mathbf{A}$ is unaffected by the second right side.

Modifying the analysis leading to equations (3.12) and (3.13), we see that for k right sides, the operation counts are

$$N_3 = \sum_{i=1}^{n-1} (n - i + k)(n - i) = \frac{1}{3}n^3 + \frac{kn^2}{2} + O(n^2) \tag{3.14}$$

for triangularization, and

$$N_4 = k \sum_{i=1}^{n-1} i = \frac{kn^2}{2} + O(n) \tag{3.15}$$

for backsubstitution. If k is much smaller than n, the main part of the work lies in the triangularization of $\mathbf{A}$, and the process is quite efficient.

Gauss elimination can also be used to compute the inverse of a matrix. As we know, a square matrix $\mathbf{A}$ has an inverse if there exists another square matrix $\mathbf{A}^{-1}$, such that

$$\mathbf{A}^{-1}\mathbf{A} = \mathbf{A}\mathbf{A}^{-1} = \mathbf{I},$$

where $\mathbf{I}$ is the identity matrix, namely a matrix in which all diagonal elements have value 1 and all off-diagonal elements have value 0. From this we notice that the first column of $\mathbf{A}^{-1}$ can be derived from (3.2) with the first column of $\mathbf{I}$ as $\mathbf{b}$, the second column of $\mathbf{A}^{-1}$ with the second column of $\mathbf{I}$, and so on. In other words, the inverse is the solution of the augmented system $[\mathbf{A}|\mathbf{I}]$. Using (3.14) and (3.15) with $k = n$, we see that matrix inversion takes about

$$N_5 = \frac{4}{3}n^3 + O(n^2) \tag{3.16}$$

operations. With a little care in the implementation and a more closely reasoned analysis, this can be reduced to approximately n^3. Even so, matrix inversion is much more expensive than solving a single linear system. For this reason, single systems are usually solved by the scheme in Section 3.1, rather than by the mathematically equivalent

$$\mathbf{x} = \mathbf{A}^{-1}\mathbf{b}.$$

Nevertheless, as we will see shortly, there are some instances where knowledge of $\mathbf{A}^{-1}$ is useful.

EXERCISES

1. Make an operations count for the Gauss-Jordan method to determine if this variant is competitive with the GEM for matrix inversion.
2. Give the justification for the estimates (3.14) and (3.16).
3. Suppose you need to solve the equation

   $$\mathbf{A}^3\mathbf{x} = \mathbf{b},$$

   where $\mathbf{A}$ is a large invertible matrix. Which of the following options is likely to take the least computer time?

   (a) Compute $\mathbf{A}^3$ first, then solve the system $\mathbf{A}^3\mathbf{x} = \mathbf{b}$.

   (b) Compute $\mathbf{A}^{-1}$ first, followed by $\mathbf{x} = \mathbf{A}^{-1}(\mathbf{A}^{-1}(\mathbf{A}^{-1}\mathbf{b}))$.

   (c) Solve three consecutive systems: Solve $\mathbf{Au} = \mathbf{b}$ first, then solve $\mathbf{Ay} = \mathbf{u}$, followed by solving $\mathbf{Ax} = \mathbf{y}$.

   Substantiate you conclusion by an operations count.
4. Examine the pattern of zeros in matrix inversion to demonstrate that in inversion some operations that are necessary when (3.1) is solved with n arbitrary right sides can be omitted. Use this to show that the number of operations in matrix inversion is approximately n^3.
5. Find the inverse of the matrix in Example 3.2 by applying the GEM to the augmented system $[\mathbf{A}|\mathbf{I}]$.

3.3 Least Squares Solutions

In some cases, we are led to linear systems that either have no solution or have several possible answers. In such cases, we may want to get a solution that fits the given information as well as possible.

Example 3.4

A race car accelerates at a constant, but unknown, rate along a straight track. We can estimate the car's rate of acceleration by observing its distance from a given point as a function of time. If at time $t = 0$ we take the distance traveled as $s = 0$, the relation between s and t is obtained by the well-known formula

$$s = v_0 t + \tfrac{1}{2} a t^2. \tag{3.17}$$

We can compute the initial velocity v_0 and the acceleration rate a by observing the distance s at two points in time. Table 3.1 represents a set of observations one might get in this situation.

Table 3.1
Observations of the distance for a race car.

t	s
5	209
10	510
15	909
20	1409
25	2001
30	2709

In (3.17) we can use any two time points t_1 and t_2 to satisfy the equation to get the unknowns v_0 and a. If, in Table 3.1, we use the data with $t_1 = 5$ and $t_2 = 25$, we find values

$$v_0 = 32.24, \quad a = 3.82.$$

If, on the other hand, we use data with $t_1 = 10$ and $t_2 = 30$, we get slightly different values

$$v_0 = 31.35, \quad a = 3.94.$$

It is not clear which of these should be preferred. Such discrepancies often arise in experimental situations. They can come from measurement errors, from disagreement between the model and the actual situation (here the car's acceleration may not be completely constant), or a variety of other factors. When this happens we may get better results by using all the data, fitting as best we can. Instead of using two arbitrary points, we can minimize the discrepancy at all points; that is, choose v_0 and a so as to minimize

$$\rho = \sum_{k=1}^{6} (v_0 t_k + \tfrac{1}{2} a t_k^2 - s_k)^2,$$

where the time values in Table 3.1 are now labeled $t_1, t_2, \ldots, t_6$. With a little algebra, we see that the two unknowns can be gotten from the system

$$\begin{pmatrix} a_{11} & a_{12} \\ a_{21} & a_{22} \end{pmatrix} \begin{pmatrix} v_0 \\ \frac{1}{2}a \end{pmatrix} = \begin{pmatrix} b_1 \\ b_2 \end{pmatrix}, \tag{3.18}$$

where

$$a_{ij} = \sum_{k=1}^{6} t_k^{i+j},$$

and

$$b_i = \sum_{k=1}^{6} s_k t_k^i.$$

Using all the measured data gives

$$v_0 = 31.10, \quad a = 3.94,$$

which in the absence of other information can be considered more reasonable than the previous results.

■

The generalization of this example is straightforward. Let us consider the nonsquare system

$$\begin{aligned} a_{11}x_1 + a_{12}x_2 + \cdots + a_{1n}x_n &= b_1, \\ a_{21}x_1 + a_{22}x_2 + \cdots + a_{2n}x_n &= b_2, \\ \vdots \quad\quad \vdots \quad\quad\quad\quad \vdots \quad &\quad \vdots \\ a_{m1}x_1 + a_{m2}x_2 + \cdots + a_{mn}x_n &= b_m, \end{aligned} \tag{3.19}$$

where $m > n$. Since we have more equations than unknowns, the system is overdetermined and normally has no solution, so the best we can do is to satisfy all of the equations approximately. One way to do this is to find the vector $\mathbf{x}$ that minimizes the sum of squares of residuals

$$\rho(\mathbf{x}) = \sum_{i=1}^{m} \left(\sum_{j=1}^{n} a_{ij} x_j - b_i \right)^2 . \tag{3.20}$$

To get the best solution we differentiate $\rho(\mathbf{x})$ with respect to x_k and set the result to zero. This, after a few simple manipulations, gives

$$\sum_{j=1}^{n} \left(\sum_{i=1}^{m} a_{ik} a_{ij} \right) x_j = \sum_{i=1}^{m} a_{ik} b_i, \quad k = 1,\ 2,\ \ldots,\ n. \tag{3.21}$$

In matrix form, this can be written as

$$\mathbf{A}^T \mathbf{A} \mathbf{x} = \mathbf{A}^T \mathbf{b}, \tag{3.22}$$

where $\mathbf{A}$ is the $m \times n$ matrix with entries a_{ij}. A solution of this $n \times n$ system is called a *least squares* solution to (3.19). The system (3.22) is usually referred to as the *normal equations*.

Equation (3.22) is an easy way to solve the least squares problem. For small, well-conditioned systems it is quite acceptable, but in other cases it is not always the best approach. We will see later why some less obvious methods are sometimes preferred.

EXERCISES

1. Show in detail how to derive (3.21) by minimizing the residual in (3.20).
2. If in (3.19) we have $m < n$, the system is called *underdetermined* and may have an infinite number of solutions. What happens to the normal equations for underdetermined systems?
3. Find the least squares solution of the system

$$\begin{aligned} x_1 + x_2 &= 1, \\ x_1 - x_2 &= 3, \\ 4x_1 + x_2 &= 2. \end{aligned}$$

4. Find a simple 2×2 system that does not have a unique least squares solution.
5. Show that if $\mathbf{A}$ is a square, nonsingular matrix, then (3.22) has a unique solution that satisfies (3.2).

6. A parachutist jumps from a plane and the distance of his drop is measured. Suppose that the distance of descent s as a function of time t can be modeled by

$$s = at + bt^2 e^{-0.1t}.$$

Find values of a and b that are reasonable in view of the data in the table below.

Time t	Distance s
5	30
10	83
15	126
20	157
25	169
30	190

7. In many applications, observed data is to be approximated by a straight line (a linear regression):

$$y = ax + b$$

with specific constants a and b.

(a) Use (3.22) to form the normal equations for the linear regression with a given set of measurements $(x_1, y_1), \ldots, (x_m, y_m)$.

(b) Assuming the normal equations have a unique solution, show that

$$a = \frac{m \sum_{i=1}^{m} x_i y_i - \sum_{i=1}^{m} x_i \sum_{i=1}^{m} y_i}{m \sum_{i=1}^{m} x_i^2 - \left(\sum_{i=1}^{m} x_i \right)^2}$$

and

$$b = \frac{\sum_{i=1}^{m} x_i^2 \sum_{i=1}^{m} y_i - \sum_{i=1}^{m} x_i y_i \sum_{i=1}^{m} x_i}{m \sum_{i=1}^{m} x_i^2 - \left(\sum_{i=1}^{m} x_i \right)^2}.$$

3.4 Testing the Accuracy of a Solution

Solving linear systems by Gaussian elimination involves only a finite number of arithmetic operations. If these could be done without rounding errors, the GEM would give the exact solution after a finite number of steps. But rounding does affect the solution and, because there are normally many operations involved, we must be concerned with the growth of the accumulated rounding errors and assure ourselves that rounding has not affected the accuracy beyond a tolerable level.

As we saw in Chapter 2, trying to bound the accumulated rounding errors in a long sequence of operations is generally not practical, so let us try a different approach. We first compute an approximate solution, then substitute it back in the original equation to see how well that equation is satisfied. If we had an exact solution, the equation would be satisfied exactly, so we can hope that if our approximate solution satisfies the equation very nearly, then it is a good approximation to the true solution.

To measure errors, we need to introduce the concept of a *norm*. For the moment, we use the definition

$$||\mathbf{x}|| = \max_i |[\mathbf{x}]_i| \tag{3.23}$$

as the norm of the vector $\mathbf{x}$, and the corresponding

$$||\mathbf{A}|| = \max_i \sum_j |[\mathbf{A}]_{ij}| \tag{3.24}$$

as the norm of the matrix $\mathbf{A}$. It is a simple matter to show that these definitions imply that

$$||\mathbf{A}\mathbf{x}|| \leq ||\mathbf{A}|| \; ||\mathbf{x}|| \tag{3.25}$$

for all matrices $\mathbf{A}$ and all vectors $\mathbf{x}$.

Suppose now that we have computed an approximate solution $\hat{\mathbf{x}}$ to (3.2). When we substitute this into the equation we get the residual

$$\mathbf{r}(\hat{\mathbf{x}}) = \mathbf{A}\hat{\mathbf{x}} - \mathbf{b}. \tag{3.26}$$

This residual will be zero only if $\hat{\mathbf{x}}$ is a solution of (3.2); otherwise it may be small, but nonzero. If the residual is very small, can we conclude that $\hat{\mathbf{x}}$ is a close approximation to the true solution? The answer is unfortunately not always yes. From (3.26) and (3.2) we see that

$$\mathbf{r}(\hat{\mathbf{x}}) = \mathbf{A}\hat{\mathbf{x}} - \mathbf{A}\mathbf{x},$$

$$\hat{\mathbf{x}} - \mathbf{x} = \mathbf{A}^{-1}\mathbf{r}(\hat{\mathbf{x}}),$$

and

$$||\hat{\mathbf{x}} - \mathbf{x}|| \leq ||\mathbf{A}^{-1}||\;||\mathbf{r}(\hat{\mathbf{x}})||. \tag{3.27}$$

If $||\mathbf{A}^{-1}||$ is not too large, then certainly a small residual will guarantee a good approximation. However, if $||\mathbf{A}^{-1}||$ is large, then it is quite possible that the residual is small even if the approximation is not very good. Equations with large $||\mathbf{A}^{-1}||$ are said to be *ill-conditioned.*

Example 3.5 The system

$$\begin{bmatrix} 1 & \frac{1}{2} & \frac{1}{3} & \frac{1}{4} & \frac{1}{5} \\ \frac{1}{2} & \frac{1}{3} & \frac{1}{4} & \frac{1}{5} & \frac{1}{6} \\ \frac{1}{3} & \frac{1}{4} & \frac{1}{5} & \frac{1}{6} & \frac{1}{7} \\ \frac{1}{4} & \frac{1}{5} & \frac{1}{6} & \frac{1}{7} & \frac{1}{8} \\ \frac{1}{5} & \frac{1}{6} & \frac{1}{7} & \frac{1}{8} & \frac{1}{9} \end{bmatrix} \begin{bmatrix} x_1 \\ x_2 \\ x_3 \\ x_4 \\ x_5 \end{bmatrix} = \begin{bmatrix} \frac{137}{60} \\ \frac{29}{20} \\ \frac{153}{140} \\ \frac{743}{840} \\ \frac{1879}{2520} \end{bmatrix}$$

has an exact solution that is $x_1 = x_2 = \ldots = x_5 = 1$. When we computed the solution by the GEM, the maximum absolute difference between the exact and computed solutions was found to be about 8×10^{-12}, indicating a loss of about four significant digits. The computed value of $||\mathbf{A}^{-1}||$ was about 4×10^5. The computed residual was zero to machine accuracy; that is, about 10^{-16} or smaller. These numbers are consistent with (3.27).

The matrix in this example is a 5×5 *Hilbert matrix.* A Hilbert matrix has elements

$$[\mathbf{A}]_{ij} = \frac{1}{i + j - 1}.$$

Hilbert matrices become rapidly ill-conditioned as their size increases. For a 10×10 Hilbert matrix, $||\mathbf{A}^{-1}||$ is about 10^{13}.

■

Checking the accuracy of a computed solution by (3.27) is straightforward, but somewhat expensive as the computation of $\mathbf{A}^{-1}$ takes about three times the work needed for the GEM itself. For this reason one sometimes uses the more cheaply obtained determinant of a matrix (See Exercise 7 in Section 3.1). If the original matrix is scaled so that its norm is of the order unity, then a very small determinant indicates ill-conditioning. As a rough measure of the accuracy of the computed solution of (3.2) we can take

$$||\mathbf{x} - \hat{\mathbf{x}}|| \approx \frac{||\mathbf{r}(\hat{\mathbf{x}})||}{|\det(\mathbf{A})|}$$

where $\det(\mathbf{A})$ denotes the determinant of matrix $\mathbf{A}$. This may considerably overestimate the error, but is useful as a rough guideline.

EXERCISES

1. Suppose that in (3.2) the right side $\mathbf{b}$ is perturbed by a small amount η. This could happen, for example, if $\mathbf{b}$ is a measured quantity so that η represents the measurement error. Show that this changes the solution by an amount not larger than $||\mathbf{A}^{-1}|| \, ||\eta||$.
2. Suppose that in Exercise 6, Section 3.3, the distance measurements are accurate to about 5 percent. What is the relative error in a and b that this entails?
3. For $n = 6,\ 7,\ 8$ find the value of $||\mathbf{A}^{-1}||$, where $\mathbf{A}$ is the $n \times n$ Hilbert matrix.
4. Using (3.27), estimate how close the computed solution in Example 3.4 is to the true least squares solution.

3.5 Error Analysis for Matrix Problems*

The simple error analysis in the previous section is readily generalized, starting with a precise definition of a norm. This allows us to deal with concepts such as size and distance, and the corresponding ideas of limits and convergence.

Definition 3.1

Suppose that with every vector $\mathbf{x}$ we associate a real number $||\mathbf{x}||$, such that

(i) $||\mathbf{x}|| \geq 0$ for all $\mathbf{x}$,

(ii) $||\mathbf{x}|| = 0$ if and only if $\mathbf{x}$ is the zero vector,

(iii) $||a\mathbf{x}|| = |a| \, ||\mathbf{x}||$ for all vectors $\mathbf{x}$ and real numbers a,

(iv) the *triangle inequality*

$$||\mathbf{x} + \mathbf{y}|| \ \leq \ ||\mathbf{x}|| \ + \ ||\mathbf{y}||$$

is satisfied for all vectors $\mathbf{x}$ and $\mathbf{y}$. We then say that $||\mathbf{x}||$ is the *norm* of $\mathbf{x}$. The *distance* between two vectors $\mathbf{x}$ and $\mathbf{y}$ is measured by $||\mathbf{x} - \mathbf{y}||$.

If $\mathbf{x}$ is a vector with components $x_1, x_2, \ldots, x_n$, we can define a norm by

$$||\mathbf{x}||_2 = \sqrt{\sum_{i=1}^{n} x_i^2}. \tag{3.28}$$

This is the *Euclidean* norm, also called the 2-*norm.* The Euclidean norm is the usual geometric length of the vector.

Another possibility has already been introduced in the previous section, where we use the magnitude of the largest component as a measure of size. The quantity

$$||\mathbf{x}||_\infty = \max_{1 \le i \le n} |x_i| \tag{3.29}$$

is called the *maximum* norm.[5] It can be shown that both (3.28) and (3.29) satisfy all the conditions of Definition 3.1 and are therefore proper norms. There are other possibilities for defining a vector norm, but these two are the main ones that we will use here.

We can also associate a norm with a matrix. How we define the matrix norm depends on what norm we have chosen for vectors. The matrix norm is said to be *induced* by the corresponding vector norm.

Definition 3.2

Let $\mathbf{A}$ be an $m \times n$ matrix. Then a nonnegative number $||\mathbf{A}||$ is the *norm* of the matrix $\mathbf{A}$ if

(i) $||\mathbf{Ax}|| \le ||\mathbf{A}||\ ||\mathbf{x}||$ for all n-vectors $\mathbf{x}$, and

(ii) there exists some $\mathbf{x} \neq 0$ such that $||\mathbf{Ax}|| = ||\mathbf{A}||\ ||\mathbf{x}||$.

Example 3.6 Let $[A]_{ij} = a_{ij}$ and $[\mathbf{x}]_i = x_i$. Then, for the vector norm

$$||\mathbf{x}||_\infty = \max_i |x_i| \tag{3.30}$$

the induced matrix norm is

$$||\mathbf{A}||_\infty = \max_i \sum_{j=1}^{n} |a_{ij}|. \tag{3.31}$$

[5]In some places this is referred to as the *uniform* or *infinity* norm.

To see this, consider

$$\begin{aligned}
||\mathbf{A}\mathbf{x}||_\infty &= \max_i \left| \sum_{j=1}^{n} a_{ij} x_j \right| . \\
&\leq \max_i \sum_{j=1}^{n} |a_{ij}|\,|x_j| \\
&\leq ||\mathbf{x}||_\infty \max_i \sum_{j=1}^{n} |a_{ij}| .
\end{aligned}$$

To complete the argument, we need to show that there exists a nonzero $\mathbf{x}$ such that equality is attained. Suppose that $\sum_{j=1}^{n} |a_{ij}|$ achieves its maximum value for $i = k$. Take $\mathbf{x}$ to be the vector such that

$$x_j = \begin{cases} 1 & \text{if } a_{kj} \geq 0 \\ -1 & \text{if } a_{kj} < 0 \end{cases} .$$

Then

$$\sum_{j=1}^{n} a_{kj} x_j = \sum_{j=1}^{n} |a_{kj}|$$

and

$$\begin{aligned}
||\mathbf{A}\mathbf{x}\,||_\infty &= \sum_{j=1}^{n} |a_{kj}| \\
&= ||\mathbf{x}\,||_\infty \sum_{j=1}^{n} |a_{kj}|.
\end{aligned}$$

■

Example 3.7 If we use the Euclidean norm as the vector norm, then the induced matrix norm is

$$||\mathbf{A}\,||_2 = \sqrt{\lambda_{\max}} \tag{3.32}$$

where $\lambda_{\max}$ is the largest eigenvalue[6] of $\mathbf{A}^T\mathbf{A}$. A proof of this is not quite so elementary and will be given later.

■

[6]Eigenvalues and their numerical computation will be discussed in Chapter 9.

Many different norms can be defined. In some instances, the conclusions we reach depend on which norm we use. In that case, the norm will be explicitly shown by the subscript. In other instances, definitions and results hold for all chosen norms; if that is the case, the subscript will be omitted.

Suppose now that we have an infinite sequence of vectors $\{\mathbf{x}_1, \mathbf{x}_2, \ldots\}$ and a single vector $\mathbf{x}$ such that

$$\lim_{i\to\infty} ||\mathbf{x} - \mathbf{x}_i\,|| = 0.$$

We then say that the sequence $\{\mathbf{x}_1, \mathbf{x}_2, \ldots\}$ *converges* to $\mathbf{x}$ or that $\mathbf{x}$ is the limit of this sequence. Sometimes we write this in abbreviated form as

$$\mathbf{x}_i \to \mathbf{x}.$$

A similar definition is made for matrices.

As we have seen in Section 3.4, the quantity $||\,\mathbf{A}^{-1}\,||$ is crucial in determining the accuracy with which the system (3.2) can be solved, but this quantity is scale-dependent. If we multiply $\mathbf{A}$ by a constant c, $||\,\mathbf{A}^{-1}\,||$ will be multiplied by $1/c$, but such a simple scaling does not affect the solution of (3.2) in any way. For this reason, it is more appropriate to use the scale-independent quantity

$$cond(\mathbf{A}) = ||\,\mathbf{A}\,||\;||\,\mathbf{A}^{-1}\,|| \tag{3.33}$$

as a measure of accuracy. This is the *condition number* of $\mathbf{A}$.

Another way of justifying $cond(\mathbf{A})$ as condition number is to look at what happens to $\mathbf{x}$ when we change the right side of (3.2) slightly. Suppose that $\hat{\mathbf{x}}$ is the solution of the perturbed equation

$$\mathbf{A}\hat{\mathbf{x}} = \mathbf{b} + \Delta\mathbf{b}. \tag{3.34}$$

Subtracting (3.2) from (3.34) and assuming that $\mathbf{A}$ is invertible and $\mathbf{b} \neq \mathbf{0}$, we have

$$\hat{\mathbf{x}} - \mathbf{x} = \mathbf{A}^{-1}\Delta\mathbf{b}.$$

Then

$$\begin{aligned} ||\,\hat{\mathbf{x}} - \mathbf{x}\,|| &\le ||\mathbf{A}^{-1}\,||\;||\Delta\mathbf{b}\,|| \\ &\le ||\mathbf{A}^{-1}\,||\;||\Delta\mathbf{b}\,||\;\frac{||\mathbf{A}\mathbf{x}\,||}{||\mathbf{b}\,||} \\ &\le ||\mathbf{A}^{-1}\,||\;||\Delta\mathbf{b}\,||\;\frac{||\mathbf{A}\,||\;||\mathbf{x}\,||}{||\mathbf{b}\,||}\;. \end{aligned}$$

From this it follows that

$$\frac{||\hat{\mathbf{x}} - \mathbf{x}||}{||\mathbf{x}||} \leq ||\mathbf{A}^{-1}|| \, ||\mathbf{A}|| \, \frac{||\Delta \mathbf{b}||}{||\mathbf{b}||}. \tag{3.35}$$

Therefore, the condition number determines the effect of a relative error in the right side on the relative accuracy of the result.

One thing to keep in mind is that, practically speaking, we cannot distinguish between matrices that are highly ill-conditioned and matrices that are exactly singular. In theory, there is a sharp dividing line between singular and nonsingular matrices, but numerical methods blur this distinction. A matrix whose condition number is higher than the reciprocal of the machine accuracy will look singular, while rounding may make a truly singular matrix appear just ill-conditioned.

Another point is that the condition number is the crucial piece of information needed to estimate the accuracy of the solution. The residual of an approximation is easily computed, and if it is large we very likely have a poor solution. On the other hand, if the residual is small we can tell very little without the condition number. The GEM tends to deliver solutions with small residuals, regardless of the conditioning of the problem. It is only when we have $||\mathbf{A}^{-1}||$ or $cond(\mathbf{A})$ that we can tell what a small residual means for the accuracy of the solution. If we are willing to expend the additional work to find the inverse of $\mathbf{A}$, then it seems straightforward to use (3.27) to bound the error. There is one hitch though—we cannot get $\mathbf{A}^{-1}$ but only an approximation to it. From a practical point of view this looks like a trivial objection. Even if our approximation differs from $\mathbf{A}^{-1}$ by a significant amount, it makes little difference to our error estimate. Still, there is a chance for error that may occasionally lead to an incorrect conclusion. If we want to remove this slight chance and produce something that is mathematically rigorous, we need to work a little harder. The following theorems are often helpful in putting a bound on $||\mathbf{A}^{-1}||$ or on the accuracy with which the inverse is computed.

Theorem 3.1 Suppose that $||\mathbf{A}|| < 1$. Then the matrix $(\mathbf{I} - \mathbf{A})$ is invertible and

$$||(\mathbf{I} - \mathbf{A})^{-1}|| \leq \frac{1}{1 - ||\mathbf{A}||}. \tag{3.36}$$

Proof: Consider the infinite sum

$$\mathbf{S} = \mathbf{I} + \mathbf{A} + \mathbf{A}^2 + \ldots \tag{3.37}$$

Such an expression makes sense only if it is interpreted as the limit of the finite sums

$$\mathbf{S}_n = \mathbf{I} + \mathbf{A} + \mathbf{A}^2 + \ \ldots \ + \mathbf{A}^n,$$

but the condition $||\mathbf{A}|| < 1$ permits this. We can then form

$$\begin{aligned} \mathbf{AS} &= \mathbf{SA} \\ &= \mathbf{A} + \mathbf{A}^2 + \ \ldots \end{aligned} \tag{3.38}$$

and, subtracting (3.38) from (3.37), get

$$\begin{aligned} (\mathbf{I} - \mathbf{A})\mathbf{S} &= \mathbf{S}(\mathbf{I} - \mathbf{A}) \\ &= \mathbf{I}. \end{aligned}$$

Therefore, $\mathbf{S}$ is the inverse of $\mathbf{I} - \mathbf{A}$, so that from (3.37),

$$\begin{aligned} ||(\mathbf{I} - \mathbf{A})^{-1}|| &\leq 1 + ||\mathbf{A}|| + ||\mathbf{A}||^2 + \ \ldots \\ &= \frac{1}{1 - ||\mathbf{A}||}, \end{aligned}$$

concluding the argument. ■

This theorem is useful in instances where matrices have off-diagonal elements much smaller than the diagonal. Such matrices are called *diagonally dominant*. As the following example shows, diagonally dominant matrices are always invertible.

Example 3.8 Suppose $\mathbf{A}$ can be written as

$$\mathbf{A} = \mathbf{D} + \mathbf{B}$$

where $\mathbf{D}$ is an invertible diagonal matrix and $\mathbf{B}$ has all zeros on the diagonal. Then

$$\mathbf{A} = \mathbf{D}(\mathbf{I} + \mathbf{D}^{-1}\mathbf{B}),$$

from which

$$\mathbf{A}^{-1} = (\mathbf{I} + \mathbf{D}^{-1}\mathbf{B})^{-1}\mathbf{D}^{-1},$$

provided $\mathbf{I} + \mathbf{D}^{-1}\mathbf{B}$ is invertible. By Theorem 3.1 the condition

$$||\mathbf{D}^{-1}\mathbf{B}\,|| < 1$$

is sufficient for this and can be taken as a definition of diagonal dominance. ■

We can also use Theorem 3.1 to establish a bound for $||\mathbf{A}^{-1}||$ from an approximation to it.

Theorem 3.2 Let $\mathbf{C}$ be an approximation to $\mathbf{A}^{-1}$ such that

$$\mathbf{I} - \mathbf{AC} = \mathbf{R} \tag{3.39}$$

with $||\mathbf{R}|| < 1$. Then

$$||\mathbf{A}^{-1}|| \le \frac{||\mathbf{C}||}{1 - ||\mathbf{R}||}. \tag{3.40}$$

Proof: Since from Theorem 3.1 we know that $\mathbf{I} - \mathbf{R}$ is invertible, we have

$$\mathbf{A}^{-1} = \mathbf{C}(\mathbf{I} - \mathbf{R})^{-1}.$$

Then

$$\begin{aligned} ||\mathbf{A}^{-1}|| &\le ||\mathbf{C}||\; ||(\mathbf{I} - \mathbf{R})^{-1}|| \\ &\le \frac{||\mathbf{C}||}{1 - ||\mathbf{R}||}, \end{aligned}$$

where the last step follows from Theorem 3.1. ■

Example 3.9 In Theorem 3.2, replace $\mathbf{A}$ by $\mathbf{A} + \Delta\mathbf{A}$ and C by$\mathbf{A}^{-1}$. Then

$$\begin{aligned} \mathbf{R} &= \mathbf{I} - (\mathbf{A} + \Delta\mathbf{A})\mathbf{A}^{-1} \\ &= -\Delta\mathbf{A}\mathbf{A}^{-1}. \end{aligned}$$

Then, if $||\Delta\mathbf{A}\,||\;||\mathbf{A}^{-1}\,|| < 1$,

$$||\,(\mathbf{A} + \Delta\mathbf{A})^{-1}\,|| \le \frac{||\,\mathbf{A}^{-1}\,||}{1 - ||\Delta\mathbf{A}\,||\,||\,\mathbf{A}^{-1}\,||}.$$

If $\mathbf{A}$ is well-conditioned, a small change in $\mathbf{A}$ has only a small effect on its inverse.

■

This theorem can also be used to get a bound on $||\mathbf{A}^{-1}\,||$ from a computed inverse. For this, we use the computed inverse as $\mathbf{C}$ and compute $\mathbf{R}$ in (3.39), then substitute into (3.40). For determining how close $\mathbf{C}$ is to $\mathbf{A}^{-1}$, refer to Exercise 7 at the end of this section.

EXERCISES

1. Prove that

$$(c\mathbf{A})^{-1} = \frac{1}{c}\mathbf{A}^{-1}.$$

2. Show that for any two matrices **A** and **B** for which addition and multiplication is defined

   (a) $||\mathbf{A}+\mathbf{B}|| \leq ||\mathbf{A}|| + ||\mathbf{B}||$,

   (b) $||\mathbf{AB}|| \leq ||\mathbf{A}|| \; ||\mathbf{B}||$.

3. Show that $cond(\mathbf{AB}) \leq cond(\mathbf{A})cond(\mathbf{B})$.

4. Let $\mathbf{x}$ satisfy (3.2) and let $\hat{\mathbf{x}}$ be the solution of the perturbed system

$$(\mathbf{A}+\Delta\mathbf{A})\hat{\mathbf{x}} = \mathbf{b}.$$

   Show that

$$||\mathbf{x}-\hat{\mathbf{x}}|| \leq ||\mathbf{A}^{-1}|| \; ||\Delta\mathbf{A}|| \; ||\hat{\mathbf{x}}||.$$

5. Let **A** and **B** be any two matrices, with $\mathbf{B} \to 0$. Show that then $\mathbf{A}+\mathbf{B} \to \mathbf{A}$.

6. Prove that for any real number a, with $0 < |a| < 1$,

$$\frac{1}{1-a} = \sum_{i=0}^{\infty} a^i.$$

7. Show that if **C** satisfies the conditions of Theorem 3.2 then

$$||\mathbf{A}^{-1}-\mathbf{C}|| \leq \frac{||\mathbf{C}|| \; ||\mathbf{R}||}{1-||\mathbf{R}||}.$$

8. The 1-norm of an n-vector is defined as

$$||\mathbf{x}||_1 = \sum_{i=1}^{n} |[\mathbf{x}]_i|.$$

   (a) Show that this definition satisfies the conditions of a norm.

   (b) Show that the induced matrix norm is

$$||\mathbf{A}||_1 = \max_j \sum_{i=1}^{n} |a_{ij}|.$$

9. Show that, for sufficiently small $||\Delta\mathbf{A}||$,

$$cond(\mathbf{A}+\Delta\mathbf{A}) \leq \frac{cond(\mathbf{A})}{1-||\Delta\mathbf{A}|| \; ||\mathbf{A}^{-1}||}\left(1+\frac{||\Delta\mathbf{A}||}{||\mathbf{A}||}\right).$$

10. For any n-vector $\mathbf{x}$, show that

(a) $\frac{1}{n}||\mathbf{x}||_1 \leq ||\mathbf{x}||_\infty \leq ||\mathbf{x}||_1$,

(b) $\frac{1}{\sqrt{n}}||\mathbf{x}||_2 \leq ||\mathbf{x}||_\infty \leq ||\mathbf{x}||_2$.

11. Does the inequality

$$cond(\mathbf{A} + \mathbf{B}) \leq cond(\mathbf{A}) + cond(\mathbf{B})$$

hold for all matrices of the same size?

Chapter 4

Polynomials and Polynomial Approximations

In constructing numerical methods we often encounter the need for approximating functions of one or more variables. For example, approximations may be needed to compute the value of an elementary function, to represent the solution to a differential equation, or to solve a system of nonlinear equations. There are many ways of approximating functions, and the challenge is to find a way that is suitable to a particular application. This usually means that the approximation has to be easy to work with, that it gives good accuracy with a small amount of work, and that there exists a body of knowledge about it that helps us analyze the results. Polynomials score high on all these points and some sort of polynomial approximations are used in most numerical computation.

4.1 Some Important Properties of Polynomials

A function of the form

$$p_n(x) = a_0 + a_1 x + a_2 x^2 + \ \dots \ + a_{n-1} x^{n-1} + a_n x^n \tag{4.1}$$

is a *polynomial of degree n*. The individual terms $a_i x^i$ are called *monomials.*

For a given x, the value of $p_n(x)$ can easily be computed from (4.1) by first forming x^i, then multiplying by a_i and adding all the monomials. This can obviously be done with only additions and multiplications, but is not often used in numerical work. A better way is *Horner's scheme*, which is based on the nested form

$$p_n(x) = (\ldots((a_n x + a_{n-1})x + a_{n-2})x + \ \ldots \ + a_1)x + a_0. \tag{4.2}$$

The polynomial in this form is conveniently evaluated recursively by

$$\begin{aligned} p_0 &= a_n, \\ p_i &= x\,p_{i-1} + a_{n-i}, \quad i = 1,\ 2,\ \ldots,\ n. \end{aligned} \tag{4.3}$$

The recursion (4.3) takes fewer operations than first forming the monomials, then adding them. It also tends to be less subject to accumulation of rounding errors, so it is generally preferred over (4.1). In any case, either form shows that polynomials are easily evaluated with just a few arithmetic operations. This makes them ideal for computer work. Polynomials have a number of other desirable properties. The sum, difference, and product of two polynomials yield other polynomials. Polynomials are easily differentiated and integrated, with polynomials as the result. There are few functions that have these properties.

Polynomials have been extensively studied and much is known about them. Since our main interest here is the use of polynomials to approximate nonpolynomial functions, we are most concerned about how this can be done effectively. This raises the question of what kinds of functions can be approximated accurately by polynomials. The important *Weierstrass Theorem* guarantees that, at least in principle, polynomials are suitable for a wide variety of approximations.

Theorem 4.1 Suppose that f is a continuous function on a finite interval $[a,\ b]$. Then, for any $\varepsilon > 0$, there exists a polynomial p_n such that

$$\max_{a \le x \le b} |f(x) - p_n(x)| \le \varepsilon.$$

Proof: All proofs of this theorem require nontrivial analysis that is of no concern here. The arguments can be found in books on approximation theory (e.g. Davis [6]) and some advanced texts in numerical analysis, such as Isaacson and Keller [12] and Hammerlin and Hoffmann [11]. ■

The Weierstrass theorem tells us that for any continuous function, there exists some polynomial that approximates that function arbitrarily closely. As stated, it does not tell us how to find the polynomial, but at least it assures us of its existence. To find an actual approximating polynomial one sometimes can use another classical and equally famous result, *Taylor's Theorem.*

Theorem 4.2 Assume that the function f is $n+1$ times continuously differentiable in some finite interval $[a,\ b]$ and let $s \in [a,\ b]$. Then for all $x \in [a,\ b]$

$$\begin{aligned} f(x) =& f(s) + (x-s)f'(s) + \frac{(x-s)^2}{2}f''(s) + \ \ldots \\ &+ \frac{(x-s)^n}{n!}f^{(n)}(s) + r_n(x,\ s), \end{aligned} \tag{4.4}$$

where

$$r_n(x,\ s) = \frac{1}{n!}\int_s^x (x-\tau)^n f^{(n+1)}(\tau)d\tau. \tag{4.5}$$

Equation (4.4) is the *Taylor expansion* of f about the point $x = s$. The function $r_n(x,\ s)$ is the *remainder* of the truncated expansion.

Proof: Start with the elementary identity

$$f(x) = f(s) + \int_s^x f'(t)dt.$$

If we now apply the same identity to $f'(t)$, we get

$$\begin{aligned} f(x) &= f(s) + \int_s^x \left\{ f'(s) + \int_s^t f''(\tau)d\tau \right\} dt \\ &= f(s) + (x-s)f'(s) + \int_s^x \int_s^t f''(\tau)d\tau dt. \end{aligned}$$

Interchanging the order of integration in the double integral

$$\begin{aligned} \int_s^x \int_s^t f''(\tau)\, d\tau dt &= \int_s^x \int_\tau^x f''(\tau)\, dt d\tau \\ &= \int_s^x (x-\tau)\, f''(\tau)\, d\tau \end{aligned} \tag{4.6}$$

verifies (4.5) for $n = 1$. Continuing in this manner, we can establish the Taylor expansion for any n. ■

To approximate a function by the Taylor expansion (4.4) we truncate the series at some point and ignore the remainder $r_n(x, s)$. The error committed in this is bounded by

$$|r_n(x,\ s)| \ \le\ \frac{1}{(n+1)!}|x - s|^{n+1} \max_{\xi \in [s,\, x]} |f^{(n+1)}(\xi)| \qquad (4.7)$$

which gives us an upper limit on the error and guarantees a minimum accuracy.

Example 4.1 Taking $f(x) = \sin(x)$ and applying Taylor's theorem with $s = 0$, we get the well-known power series

$$\sin\ x = x - \frac{x^3}{3!} + \frac{x^5}{5!} - \frac{x^7}{7!} + \ \cdots\ .$$

If we truncate this at the last term shown, we get the approximation

$$\sin\ x \cong x - \frac{x^3}{6} + \frac{x^5}{120} - \frac{x^7}{5040}.$$

The error in this can be immediately estimated from (4.7), which gives

$$\left|\sin\ x - \left\{x - \frac{x^3}{6} + \frac{x^5}{120} - \frac{x^7}{5040}\right\}\right| \le \frac{x^8}{40320}.$$

For $x = 0.1$ this is less than 2.5×10^{-13}, but for $x = 0.5$ the bound grows to 9.7×10^{-8}. This is typical of truncated Taylor expansions; the error grows quickly as we move away from the point of the expansion. ■

While Taylor's theorem plays an important role in the design of numerical methods, and Taylor series expansions are occasionally useful, the rule is not suitable as a general method for polynomial approximations. One problem is that it requires high derivatives of a function and this may be quite cumbersome. Another reason is that it does not necessarily give us improved accuracy as n increases, as the derivative of a function can easily increase at a rate faster than $n!$. Finally, as we noted, while a truncated Taylor polynomial can be quite accurate near the expansion point, its accuracy quickly deteriorates as we move away from this point.

EXERCISES

1. Show that (4.1) and (4.2) are equivalent ways of evaluating a polynomial.
2. Show why the recursive method (4.3) evaluates polynomials with fewer arithmetic operations than the method of forming and adding monomials.

3. The point z is said to be a *root* of the function $f(z)$ if $f(z) = 0$. Prove that every polynomial of odd degree has at least one real root.

4. Prove that if a function f is sufficiently differentiable and vanishes at k points on the real axis, then f' has at least $k-1$ roots on the real axis, f'' has at least $k-2$ roots, and so on. Use this to show that a polynomial of degree n cannot have more than n real roots.

5. Justify the interchange of order of integration in (4.6).

6. Show how (4.7) follows from (4.5).

7. In Example 4.1, estimate the accuracy of the approximation with the first three terms at $x = 0.1$.

8. Find the first three terms of the Taylor expansion of $\log_e(x)$ about $x = 1$.

9. Find the first three terms of the Taylor expansion of $\dfrac{1}{1+x^2}$ about $x = 0$. Estimate the accuracy of this approximation in $[0,\ 0.5]$.

10. Find a polynomial approximation to the function

$$g(x) = \int_0^x \frac{e^t}{1+\sin t}\, dt$$

that is accurate to within an error of 0.001 in the interval $[0,\ 0.2]$.

11. Why is it impossible to find a Taylor expansion of $\sqrt{x}$ about the point $x = 0$?

12. Extend the Weierstrass theorem to show that if a function $f(x)$is continuously differentiable in some finite interval $[a,\ b]$, then for any $\varepsilon > 0$ there exists a polynomial $p_n(x)$ such that

$$\max_{a \le x \le b} |f(x) - p_n(x)| \le \varepsilon$$

and

$$\max_{a \le x \le b} |f'(x) - p_n'(x)| \le \varepsilon.$$

13. Show that if a is a root of a polynomial p_n of degree n, then p_n must be of the form

$$p_n(x) = (x-a)p_{n-1}(x),$$

where p_{n-1} is a polynomial of degree $n-1$.

4.2 Polynomial Approximation by Interpolation

Taylor expansion gives us one way of approximating general functions by polynomials, but it is of limited usefulness. A more practical way of constructing polynomial approximations is by *interpolation*. In its simplest form the interpolation problem can be stated as follows. Suppose that $[a,\ b]$ is a finite interval and $x_0,\ x_1,\ \ldots,\ x_n$ are distinct points in this interval.

We are given the value of a function f at these points and want to find a polynomial $p_n(x)$ of degree n that agrees with f at all points; that is

$$p_n(x_i) = f(x_i), \quad i = 0,\ 1,\ \ldots,\ n. \tag{4.8}$$

We will call this type of interpolation *simple* interpolation.

There are many ways in which interpolating polynomials can be found. In the *method of undetermined coefficients* we consider the coefficients in (4.1) as unknowns that can be determined by imposing conditions on the polynomial. In simple interpolation we have $n+1$ conditions specified by (4.8) to be satisfied, and $n+1$ coefficients in (4.1) to select. When we write this down explicitly, we get the linear system

$$\begin{bmatrix} 1 & x_0 & x_0^2 & \cdots & x_0^n \\ 1 & x_1 & x_1^2 & \cdots & x_1^n \\ \vdots & \vdots & \vdots & & \vdots \\ 1 & x_n & x_n^2 & \cdots & x_n^n \end{bmatrix} \begin{bmatrix} a_0 \\ a_1 \\ \vdots \\ a_n \end{bmatrix} = \begin{bmatrix} f(x_0) \\ f(x_1) \\ \vdots \\ f(x_n) \end{bmatrix}. \tag{4.9}$$

Example 4.2 To approximate e^x in $[0,\ 1]$ by a second degree polynomial, we can use the interpolation points 0, $\frac{1}{2}$, and 1. The system to be solved then is

$$\begin{bmatrix} 1 & 0 & 0 \\ 1 & \frac{1}{2} & \frac{1}{4} \\ 1 & 1 & 1 \end{bmatrix} \begin{bmatrix} a_0 \\ a_1 \\ a_2 \end{bmatrix} = \begin{bmatrix} 1 \\ \sqrt{e} \\ e \end{bmatrix}.$$

This gives the approximation

$$p_2(x) = 1 + 0.8766x + 0.8417x^2.$$

The error in this approximation is plotted in Figure 4.1. For comparison we also show the error in the second degree Taylor expansion about $x = 0$,

$$e^x \cong 1 + x + \frac{x^2}{2}.$$

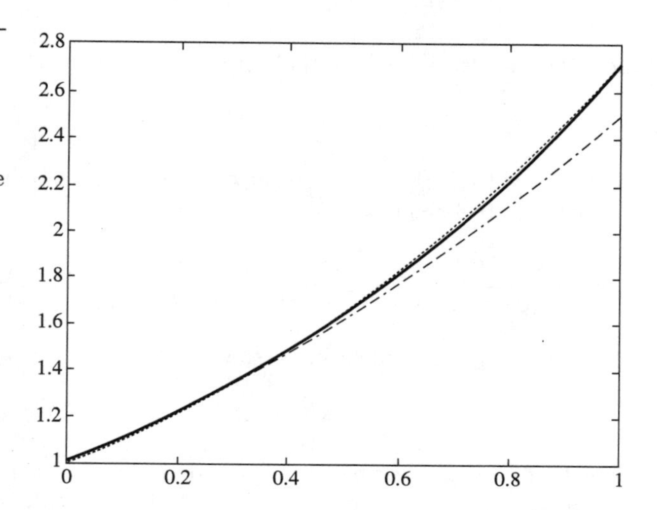

Figure 4.1 Approximation of e^x by a second degree polynomial. The solid line is the true function, the dashed line is the interpolation, and the line marked with -- · -- is the Taylor expansion.

As expected, the error in the Taylor expansion grows rapidly as we go away from $x = 0$. The interpolation approximation performs better throughout the interval.

■

The method of undetermined coefficients is simple and intuitive, and gives results with minimum effort. But there are reasons why it is not always suitable. One of the reasons is that the system (4.9) tends to become ill-conditioned quickly as n increases (see Exercise 1) and is not suitable for large n. Another reason is that it does not give a very explicit form of the polynomial, and so is hard to use for analysis.

A second way of getting the interpolating polynomial, one that overcomes some of these limitations of the undetermined coefficients method, is the *Lagrange* form

$$p_n(x) = \sum_{i=0}^{n} l_i(x) f(x_i), \tag{4.10}$$

where

$$l_i(x) = \frac{(x - x_0)(x - x_1)\ldots(x - x_{i-1})(x - x_{i+1})\ldots(x - x_n)}{(x_i - x_0)(x_i - x_1)\ldots(x_i - x_{i-1})(x_i - x_{i+1})\ldots(x_i - x_n)}. \tag{4.11}$$

The l_i are the *fundamental* or *Lagrange polynomials.* It is an easy matter to check that (4.10) and (4.11) define a polynomial of degree n that satisfies all interpolating conditions.

Example 4.3 Find a second degree polynomial approximation to

$$f(x) = e^x$$

in the interval [0, 1], using the set of interpolating points {0, 0.5, 1}. For this case, (4.10) becomes

$$p_2(x) = \frac{(x-0.5)(x-1)}{(0-0.5)(0-1)}e^0 + \frac{(x-0)(x-1)}{(0.5-0)(0.5-1)}e^{0.5} + \frac{(x-0)(x-0.5)}{(1-0)(1-0.5)}e^1.$$

After evaluating and collecting terms, we find

$$p_2(x) = 1 + 0.8766x + 0.8417x^2.$$

In spite of the low degree of the approximating polynomial the result is quite good, showing a maximum error of about 0.014.

Note that the answer we get here is identical to what we got in Example 4.2. This is as it should be; the different interpolation methods are theoretically equivalent, and if everything could be done precisely they would all give the same answer. In practice, though, this is not necessarily the case as rounding has varying effects.

■

A third way, that is very convenient for interpolation in tables,[1] is the *divided difference* formula. In this form, the interpolating polynomial is expressed as

$$\begin{aligned} p_n(x) =& f(x_0) + f[x_0, x_1](x - x_0) + f[x_0, x_1, x_2](x - x_0)(x - x_1) + \ldots \\ &+ f[x_0, x_1, \ \ldots, x_n](x - x_0)(x - x_1)\ldots(x - x_{n-1}). \end{aligned} \tag{4.12}$$

The quantities $f[x_0, x_1, \ \ldots, \ x_k]$, called the divided differences, can be computed recursively by

$$f[x_i, x_{i+1}, \ \ldots, \ x_k] = \frac{f[x_{i+1}, \ \ldots, \ x_k] - f[x_i, \ \ldots, \ x_{k-1}]}{x_k - x_i}, \tag{4.13}$$

with $f[x_i] = f(x_i)$.

Proving that the polynomial in (4.12) satisfies the stated interpolation conditions is not an entirely trivial matter; a full analysis can be found in more advanced texts. For small n we can verify the conditions by direct computations. The form, though, is very convenient for computation. The

[1]Before digital computers, interpolation in tables was a major topic in numerical analysis. Nowadays, it has lost much of its importance.

Table 4.1
Divided differences for Example 4.4.

x_i	$f(x_i)$	$f[x_i, x_{i+1}]$	$f[x_i, \ldots, x_{i+2}]$	$f[x_i, \ldots, x_{i+3}]$
0	1.0000			
		−0.1651		
1/3	0.9450		−0.4681	
		−0.4772		0.0788
2/3	0.7859		−0.3893	
		−0.7368		
1	0.5403			

divided differences are easy to evaluate and if we need to increase the degree of the polynomial, we only need to add a new term in (4.12), saving much work. For computation with divided differences we usually arrange the results in tabular form.

Example 4.4 Find a third degree polynomial approximation to $\cos(x)$ using interpolating points $\{0, \frac{1}{3}, \frac{2}{3}, 1\}$.

Table 4.1 shows the divided differences from which we get

$$p_3(x) = 1.0000 - 0.1651x - 0.4681x(x - \tfrac{1}{3}) + 0.0788x(x - \tfrac{1}{3})(x - \tfrac{2}{3}).$$

The value $p_3(0.5)$ =0.8773 differs from $\cos(0.5)$ by three units in the fourth decimal place. ■

Various other forms of constructing interpolating polynomials are known. These, with emphasis on interpolation in tables with equidistant abscissas, can be found in older numerical analysis literature. Here we use, for the most part, either the method of undetermined coefficients or the Lagrange form.

One question is, what happens when we increase the number of interpolation points and the degree of the interpolating polynomial? Does the polynomial converge to the function? A related question concerns the error of the approximation; in particular, how large the degree of the approximating polynomial needs to be to achieve a desired accuracy. The next theorem gives a partial answer to both these questions.

Theorem 4.3 Assume that f is $(n+1)$ times continuously differentiable. Let x, x_0, x_1, ..., x_n be distinct points in $[a, b]$ and let p_n denote the interpolating polynomial on these points. Then

$$f(x) - p_n(x) = \frac{(x - x_0)(x - x_1) \ \dots \ (x - x_n)}{(n+1)!} f^{(n+1)}(t), \tag{4.14}$$

where $a \leq t \leq b$.

Proof: The proof of this is most easily done by a repeated application of Rolle's theorem. For details, see Isaacson and Keller [12], p. 190. ■

From (4.14) we can draw some useful conclusions. If f is $(n+1)$ times continuously differentiable, then it follows easily that

$$\max_{a \leq x \leq b} |f(x) - p_n(x)| \leq \frac{|b-a|^{n+1}}{(n+1)!} \max_{a \leq t \leq b} |f^{(n+1)}(t)|. \tag{4.15}$$

If we can find a bound for the $(n+1)^{st}$ derivative of f, then (4.15) gives us a way of guaranteeing the accuracy of the approximation. Such inequalities are called *error bounds*, because they allow us to put a mathematically rigorous limit on the error.[2] Keep in mind, though, that (4.15) is only a bound that guarantees that the error is smaller than the right side; in an actual computation the true error may be in fact much smaller. If that is the case, we say that the error bound is *pessimistic* or *unrealistic*. If a bound is close to the actual error we say that it is *realistic*. For many numerical methods finding realistic error bounds is quite complicated.

Example 4.5 For the case of Example 4.2 we find easily that

$$\max_{a \leq x \leq b} |e^x - p_2(x)| \leq \frac{e}{6} \cong 0.45.$$

Since the observed error is about 30 times smaller, this bound is quite pessimistic. Exercise 6 at the end of this section explores how to improve this bound somewhat.

■

Note that the bound (4.15) can be used only if f is $(n+1)$ times differentiable. If the function is less smooth (in the sense that it has fewer derivatives), then (4.15) is not applicable. While it is possible to derive

[2] This bound, however, does not include the effect of rounding in the computation of $p_n(x)$. For simplicity, this is the way bounds are usually stated. If rounding errors need to be taken into consideration, a separate analysis has to be made.

Table 4.2 Maximum absolute error for interpolation in Example 4.6.

n	$\alpha = 1$	$\alpha = 5$
5	1.41×10^{-2}	4.33×10^{-1}
10	7.91×10^{-4}	1.92×10^{0}
20	5.85×10^{-6}	5.98×10^{1}

other bounds for this case, we will not do so here. Intuitively, we should expect that less smoothness results in less accuracy.

The bound (4.15) does not immediately guarantee convergence of the approximation. If the higher derivatives of f remain bounded (as they do for simple functions such as sine and cosine), then (4.15) indicates fast convergence, but this is the exception rather than the rule. For many functions, the magnitude of the higher derivatives increases, sometimes so much that the right side of (4.15) goes to infinity. Since the expression is only a bound, this does not necessarily mean that the approximation does not converge, but it should be taken as a cautionary sign. The fact is that for many relatively simple functions approximation by interpolation does not work very well. The situation is fairly complicated so we will not go into it, except to illustrate it with a simple example.

Example 4.6 The function

$$f(x) = \frac{1}{1 + \alpha^2 x^2}$$

was approximated in the interval $[-1, 1]$ by simple interpolation at $n + 1$ equally spaced points. The observed error, shown in Table 4.2, indicates apparent convergence for $\alpha = 1$. However, the results for $\alpha = 5$ are useless. ■

The interpolating approximation and its error depend on the choice of the interpolation points. The error is zero at the interpolation points and generally largest at points distant from any of these points. If there is no clear reason for the choice of the interpolation points, it may be better to find an approximation that keeps the overall error small.

EXERCISES

1. Let x_i be n equally spaced points in the interval $[0, 1]$. Examine conditioning of the matrix in (4.9) as n increases.

2. Why can you expect the matrix in (4.9) to be nonsingular?

3. Use the Lagrange form to find the third degree polynomial approximation to e^x by simple interpolation on the points $\{0,\ 0.25,\ 0.5,\ 1\}$.

4. Verify by direct computation that for $n = 1$ and $n = 2$ the divided difference formulas (4.12) and (4.13) yield a polynomial that satisfies the stated interpolation conditions.

5. Add the interpolation point $x = 0.5$ to Example 4.4 and find the resulting interpolating polynomial of degree four.

6. In deriving the bound (4.15) we used

$$|(t - t_0)(t - t_1)\ \ldots\ (t - t_n)| \leq |b - a|^{n+1},$$

which is obvious but not very precise. Examine the term $|(t - t_0)(t - t_1)\ \ldots\ (t - t_n)|$ to get a better bound for Example 4.4.

7. Compare the accuracy of the approximation in Example 4.4 with the four-term Taylor expansion about $x = 0$.

8. Find a third degree polynomial that agrees with $f(x) = \sin(x)$ and its first and second derivatives at $x = 0$, and that passes through the point $(\pi/2,\ 1)$, using the method of undetermined coefficients.

9. Simple interpolation is a special case of a general scheme in which the approximating polynomial is defined by imposing a variety of conditions. For example, we can ask that the approximating polynomial not only agree with the given function f at selected points, but that there is also agreement of the derivatives. Given points $x_0,\ x_1,\ \ldots,\ x_n$ we look for a polynomial $p_{2n+1}(x)$ of degree $2n + 1$, such that

$$p_{2n+1}(x_i) = f(x_i),$$
$$p'_{2n+1}(x_i) = f'(x_i),$$

for $i = 0,\ 1,\ \ldots,\ n$. This is the *Hermite interpolation* problem.

Use the method of undetermined coefficients to find the third degree Hermite interpolating polynomial for e^x using the interpolating points $x = 0$ and $x = 1$.

10. Exercise 9 can be generalized even further. Since a polynomial of degree n has $n+1$ coefficients, we can impose any well-chosen $n+1$ conditions to select them, most easily by the method of undetermined coefficients.

Use this observation to find a third degree polynomial that agrees with $e^{-x}\cos(x)$ at $x = 0$ and $x = 1$, and that has the same first and second derivatives at $x = 0$.

11. Using $x_k = \cos(\frac{2k-1}{42}\pi)$, for $k = 1,\ 2,\ \ldots,\ 21$ as the interpolation points to approximate the function $f(x) = \frac{1}{1 + 25x^2}$ by simple interpolation, estimate the maximum absolute error for the interpolation and compare it with Table 4.2 in Example 4.6.

4.3 Polynomial Approximation by Least Squares

Instead of asking that an approximation agree with the given function f at certain selected points, we can require instead that it agrees with the function closely on the whole interval, say by minimizing the square of 2-norm[3] of the difference between the function and its approximation

$$||f - p_n||_2^2 = \int_a^b \{f(x) - p_n(x)\}^2 dx \tag{4.16}$$

with respect to all polynomials of degree n. This gives us the *least squares* approximation of the function by a polynomial.

We can find a solution of the least squares problem by writing the approximation as

$$p_n(x) = \sum_{i=0}^{n} a_i x^i, \tag{4.17}$$

and use the same reasoning as in Section 3.3. That is, we set the derivative of (4.16) with respect to each coefficient a_i to zero. This leads to the system for determining the coefficients,

$$\begin{bmatrix} c_{11} & c_{12} & \cdots & c_{1,n+1} \\ c_{21} & c_{22} & \cdots & c_{2,n+1} \\ \vdots & \vdots & & \vdots \\ c_{n+1,1} & c_{n+1,2} & \cdots & c_{n+1,n+1} \end{bmatrix} \begin{bmatrix} a_0 \\ a_1 \\ \vdots \\ a_n \end{bmatrix} = \begin{bmatrix} b_0 \\ b_1 \\ \vdots \\ b_n \end{bmatrix}, \tag{4.18}$$

where

$$\begin{aligned} c_{ij} &= \int_a^b x^{i+j-2} dx \\ &= \frac{1}{i+j-1}(b^{i+j-1} - a^{i+j-1}), \end{aligned}$$

and

$$b_i = \int_a^b x^i f(x) dx.$$

---

[3]The 2-norm of an integrable function f over $[a, b]$ is $||f||_2 = \sqrt{\int_a^b |f(x)|^2 dx}$. This is a generalization of the vector norm to integrable functions. It can be shown that this norm satisfies all properties for vector norms given in Chapter 3.

While this looks like an easy and straightforward solution to the problem, there are some issues of concern. One point is that the integral for b_i may not have a closed form answer, so that its value would have to be determined numerically. Even though this is only a minor complication, it can be avoided by using a *discrete least squares* approach. For this we use m sample points $t_1, t_2, \ldots, t_m$ and try to make the discrepancy,

$$\rho_i = p_n(t_i) - f(t_i),$$

as small as possible for all $i = 1, 2, \ldots, m$. If $m > n + 1$ we cannot of course make all residuals vanish, but we can minimize the sum of the squares of the ρ_i. When we do this, we see that we need to find the least squares solution of the system

$$\begin{bmatrix} c_{11} & c_{12} & \cdots & c_{1,n+1} \\ c_{21} & c_{22} & \cdots & c_{2,n+1} \\ \vdots & \vdots & & \vdots \\ c_{m,1} & c_{m,2} & \cdots & c_{m,n+1} \end{bmatrix} \begin{bmatrix} a_0 \\ a_1 \\ \vdots \\ a_n \end{bmatrix} = \begin{bmatrix} b_0 \\ b_1 \\ \vdots \\ b_m \end{bmatrix}, \tag{4.19}$$

where

$$c_{ij} = t_i^{j-1}$$

and

$$b_i = f(t_i).$$

In this version, there is no need to evaluate integrals. From Chapter 3 we know how to solve the matrix least squares problem, so we have a straightforward computational process. When m is much larger than n, the results of the discrete and the continuous least squares methods are not much different.

There is, however, a much more serious problem in the conditioning of the matrices that arise from this approach.

Example 4.7 If we take $a = 0,\ b = 1$, then in (4.18)

$$c_{ij} = \frac{1}{i + j - 1},$$

and the matrix is a Hilbert matrix, which is very ill-conditioned. We should therefore expect that any attempt to solve the full least squares problem

(4.18) or the discrete version via (4.19) is likely to yield disappointing results.

■

The situation in this example is typical of similar cases. Least squares polynomial approximations, using form (4.1), are invariably very ill-conditioned and the method should be avoided except for very small n (say $n \leq 4$). If higher degree least squares polynomials are needed, we have to take a different path.

EXERCISES

1. Use the method (4.18) to approximate $\sin(x)$ in (0, 1) by a polynomial of degree three.
2. Use the method (4.18) to approximate $f(x) = e^{-x}\cos(x)$ in (0, 1) by a linear function.
3. Repeat Exercise 1, this time using a discrete least squares method with 100 equally spaced sample points in (0, 1).
4. What are the condition numbers of the matrices involved in the solution of the linear systems that arise in Exercises 1 and 3?
5. If p_n and q_n are the least-squares polynomial approximation of degree n to the functions f and g, respectively, is $p_n + q_n$ necessarily the least squares approximation of $f + g$?

4.4 Least Squares Approximation and Orthogonal Polynomials*

Some of the difficulties in polynomial approximation come from the unsuitability of the monomial form (4.1). More effective ways can be found by relying on the concept of orthogonal polynomials.

Definition 4.1

A sequence of polynomials $p_0, p_1, \ldots, p_n, \ldots$ is said to be *orthogonal* over the interval (a, b) with respect to the *weight function* $w(x)$ if

$$\int_a^b w(x)p_i(x)p_j(x)dx = 0, \tag{4.20}$$

for all $i \neq j$. The sequence is said to be *orthonormal* if

$$\int_a^b w(x)p_i(x)p_i(x)dx = 1, \tag{4.21}$$

for all i.

Orthogonal polynomials are a major topic in approximation theory and many important results have been developed. Here we list only a few of the many types of orthogonal polynomials that have been studied, with recipes for computing the individual terms. As we will see, orthogonal polynomials are quite important in numerical analysis. Constructing algorithms with orthogonal polynomials often requires a knowledge of their special properties; these will be covered as needed. For more information, refer to books on approximation theory or to Abramovitz and Stegun [1], where a variety of results are tabulated.

Legendre Polynomials

The Legendre polynomials $P_n(x)$ are orthogonal with respect to $w(x) = 1$ over $(-1,\ 1)$. The first few Legendre polynomials are

$$\begin{aligned} P_0(x) &= 1, \\ P_1(x) &= x, \\ P_2(x) &= \frac{(3x^2-1)}{2}, \\ P_3(x) &= \frac{(5x^3-3x)}{2}. \end{aligned}$$

Successive Legendre polynomials can be generated by the three-term recurrence

$$P_{n+1}(x) = \frac{2n+1}{n+1}xP_n(x) - \frac{n}{n+1}P_{n-1}(x). \tag{4.22}$$

These polynomials are orthogonal, but not orthonormal. They can be normalized by multiplying with a suitable constant.

Chebyshev Polynomials of the First Kind

The Chebyshev polynomials of the first kind $T_n(x)$ are orthogonal with respect to $w(x) = \dfrac{1}{\sqrt{1-x^2}}$ over $(-1,\ 1)$. The first few Chebyshev polynomials of the first kind are

$$
\begin{aligned}
T_0(x) &= 1, \\
T_1(x) &= x, \\
T_2(x) &= 2x^2 - 1, \\
T_3(x) &= 4x^3 - 3x.
\end{aligned}
$$

Successive Chebyshev polynomials of the first kind can be generated by the three-term recurrence

$$T_{n+1}(x) = 2xT_n(x) - T_{n-1}(x). \tag{4.23}$$

Chebyshev polynomials of the first kind have many properties that make them attractive for numerical work. One of these is the identity

$$T_n(x) = \cos(n\cos^{-1}(x)). \tag{4.24}$$

In spite of its appearance, the right side of (4.24) is a polynomial. From (4.24) follows the important fact that the roots of T_n are explicitly given as

$$z_i = \cos\left\{\frac{(2i+1)\pi}{2n}\right\}, \quad i = 0,\ 1,\ \ldots,\ n-1. \tag{4.25}$$

Chebyshev Polynomials of the Second Kind

The Chebyshev polynomials of the second kind $U_n(x)$ are orthogonal with respect to $w(x) = \sqrt{1-x^2}$ over $(-1,\ 1)$. The Chebyshev polynomials of the first kind are related to those of the second kind by

$$T_n'(x) = nU_{n-1}(x).$$

An explicit form for Chebyshev polynomials of the second kind is

$$U_n(x) = \frac{\sin((n+1)\cos^{-1}(x))}{\sqrt{1-x^2}}.$$

Laguerre Polynomials

These polynomials are orthogonal with respect to e^{-x} over the interval $(0, \infty)$. Laguerre polynomials can be generated from the three-term recurrence

$$L_{n+1}(x) = \frac{1}{n+1}\{(2n+1-x)L_n(x) - nL_{n-1}(x)\} \tag{4.26}$$

starting with

$$
\begin{aligned}
L_0(x) &= 1, \\
L_1(x) &= 1 - x.
\end{aligned}
$$

Hermite Polynomials

The weight e^{-x^2} over $(-\infty, \infty)$ leads to the Hermite polynomials. With

$$H_0(x) = 1,$$

$$H_1(x) = 2x,$$

successive Hermite polynomials can be generated by

$$H_{n+1}(x) = 2xH_n(x) - 2nH_{n-1}(x). \tag{4.27}$$

Orthogonal polynomials have many uses, but here we are interested only in orthogonal polynomials for the solution of the least squares problem. For simplicity, assume that $a = -1$ and $b = 1$. Instead of the traditional monomial form, we write the approximation as an expansion in Legendre polynomials

$$p_n(x) = \sum_{i=0}^{n} a_i P_i(x).$$

We substitute this into (4.16) and repeat the process leading to (4.18) to find

$$c_{ij} = \int_{-1}^{1} P_i(x)P_j(x)dx.$$

Because of the orthogonality of the Legendre polynomials, the matrix in (4.18) is diagonal, making it well-conditioned and leading to the immediate solution

$$a_i = \frac{\int_{-1}^{1} f(x)P_i(x)dx}{\int_{-1}^{1} P_i^2(x)dx}. \tag{4.28}$$

Example 4.8 Approximate $f(x) = e^{x^2}$ on $[-1, 1]$ by a polynomial of degree two, using a least squares method.

Writing the solution as

$$p_2(x) = a_0P_0(x) + a_1P_1(x) + a_2P_2(x),$$

we can determine the coefficients by (4.28) as

$$a_0 = \frac{\int_{-1}^{1} e^{x^2}dx}{\int_{-1}^{1} dx},$$

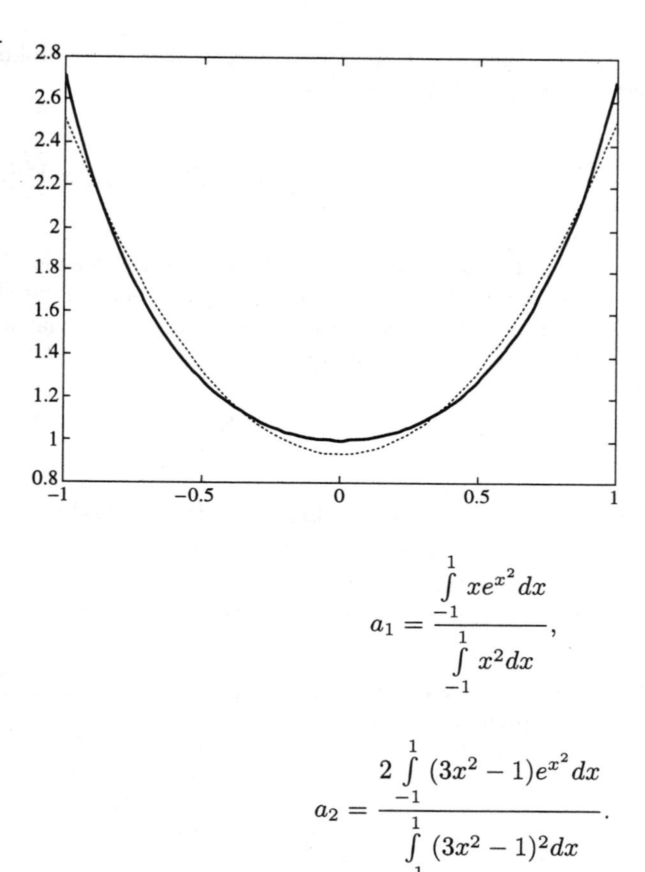

Figure 4.2 Least squares quadratic approximation to e^{x^2}. The solid line is the true function, the dashed line its approximation.

$$a_1 = \frac{\int_{-1}^{1} x e^{x^2} dx}{\int_{-1}^{1} x^2 dx},$$

$$a_2 = \frac{2\int_{-1}^{1} (3x^2-1)e^{x^2} dx}{\int_{-1}^{1} (3x^2-1)^2 dx}.$$

Some of the integrals in the numerator cannot be done analytically, but we can evaluate them approximately.[4] The result obtained this way is

$$p_2(x) = 0.9367 + 1.5780x^2.$$

Figure 4.2 shows the function and its least squares approximation. ■

It is worthwhile to note that the approximation by orthogonal polynomials gives in theory the same result as (4.18). But because the coefficients of the Legendre polynomials are all integers, there is no rounding in this step and the whole process is much more stable. Least squares approximation by polynomials is a good example of the stabilizing effect of a rearrangement of the computation.

[4]We will study efficient methods for computing such integrals a little later. For the moment you can think of this as having been done by the rectangular method described in Chapter 1.

We can generalize the least squares process by looking at the weighted error of the approximation

$$||f - p_n||_w^2 = \int_a^b w(x)\{f(x) - p_n(x)\}^2 dx, \tag{4.29}$$

where the weight function $w(x)$ is positive throughout (a, b). This weighs the error more heavily where $w(x)$ is large, something that is often practically justified. If we know the polynomials $\phi_i(x)$ that are orthogonal with respect to this weight function, we write

$$p_n(x) = \sum_{i=1}^{n} a_i \phi_i(x),$$

and, repeating the steps leading to (4.18), we find that

$$c_{ij} = \int_a^b w(x)\phi_i(x)\phi_j(x)dx.$$

The orthogonality conditions then make the matrix in (4.18) diagonal and we get immediately that

$$a_i = \frac{\int_a^b w(x)\phi_i(x)f(x)dx}{\int_a^b w(x)\phi_i^2(x)dx}. \tag{4.30}$$

Example 4.9 Find a second degree polynomial approximation to e^{x^2} on $(-1, 1)$, using a weighted least squares approximation with $w(x) = \dfrac{1}{\sqrt{1-x^2}}$.

Here we use Chebyshev polynomials of the first kind. Using symmetry and (4.30) we find that the approximation is

$$p_2(x) = a_0 + a_2(2x^2 - 1),$$

where

$$a_0 = \frac{\int_{-1}^{1} \frac{e^{x^2}}{\sqrt{1-x^2}}dx}{\int_{-1}^{1} \frac{1}{\sqrt{1-x^2}}dx},$$

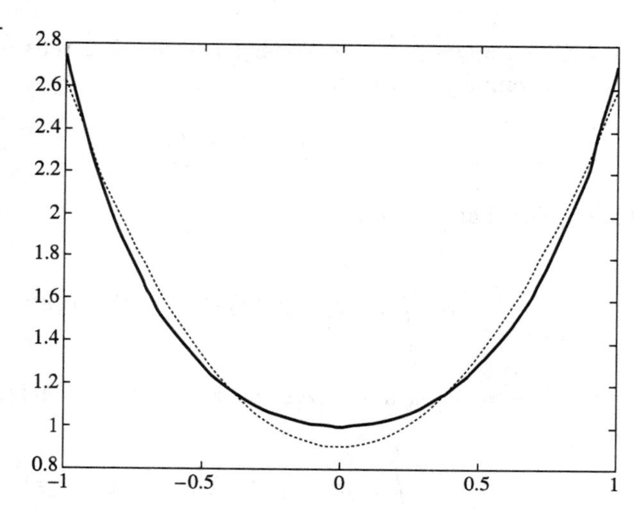

Figure 4.3 Weighted least squares quadratic approximation to e^{x^2}. The solid line is the true function, the dashed line its approximation.

$$a_2 = \frac{\int_{-1}^{1} \frac{e^{x^2}(2x^2-1)}{\sqrt{1-x^2}}dx}{\int_{-1}^{1} \frac{(2x^2-1)^2}{\sqrt{1-x^2}}dx}.$$

These integrals will have to be approximated numerically, but this is quite a bit harder than in Example 4.8 since the integrands are not bounded. We will see later how one deals with such difficulties. When the coefficients are evaluated, we get the approximation

$$p_2(x) = 0.9030 + 1.7008x^2.$$

The graph in Figure 4.3 shows the expected behavior. Compared to the solution of Example 4.8, the errors near the end points are much smaller. To compensate for this, we get larger errors in the center of the interval. ■

An advantage of least squares methods over interpolation is that they converge under fairly general conditions. The proof of this is straightforward.

Theorem 4.4 Let $p_n(x)$ denote the nth degree least squares polynomial approximation to a function f, defined on a finite interval (a, b). Then

$$\lim_{n\to\infty} ||f - p_n||_2 = 0. \tag{4.31}$$

We call this *convergence in the mean.*

Proof: Given any $\varepsilon > 0$, the Weierstrass theorem tells us that there exists some polynomial $\bar{p}_n$ such that

$$\max_{a \le x \le b} |\bar{p}_n(x) - f(x)| \le \varepsilon.$$

This implies that

$$\int_a^b \{f(x) - \bar{p}_n(x)\}^2 dx \le |b-a| \, \varepsilon^2.$$

But then, because p_n minimizes the square of the deviation, we must have

$$\int_a^b \{f(x) - p_n(x)\}^2 dx \le |b-a| \, \varepsilon^2$$

and convergence follows. ■

Under suitable assumptions, almost identical arguments lead to the convergence of the weighted least squares method. But note that the arguments only prove convergence in the mean and do not imply that the error is small at every point of the interval.

EXERCISES

1. Use a Legendre polynomial approximation to compute the least squares polynomial approximation of degree three to $f(x) = \sin(x)$ in the interval $[-1, 1]$.
2. Find the first three normalized Legendre polynomials.
3. Verify that the first four Legendre polynomials form an orthogonal set.
4. Find the first three Chebyshev polynomials of the second kind.
5. In Example 4.8, what is the next term in the least squares approximation?
6. Show that the expressions of $T_2(x)$ and $T_3(x)$ satisfy the recurrence relation stated in (4.23).
7. Show that if $w(x)$ is absolutely integrable, then Theorem 4.4 also holds for weighted least squares approximations.
8. Approximate the function $f(x) = \dfrac{1}{1+\alpha^2 x^2}$ in Example 4.6 by the least-squares method using Legendre polynomial of degree 3 for $\alpha = 1$ and 5. Observe the maximum absolute error for the approximation in each case.

9. Suppose $p_n(x) = \sum_{i=0}^{n} \alpha_i \varphi_i(x)$ is the least squares approximation of the function $f(x)$ in $[a, b]$ and $\varphi_1, \ldots, \varphi_n$ are orthonormal polynomials. Show that

(a) $\alpha_i = \int_a^b f(x)\varphi_i(x)dx$.

(b) $\int_a^b \{f(x) - p_n(x)\}\varphi_i(x)dx = 0$.

4.5 Fourier Approximations*

The concept of orthogonality of polynomials can be generalized to other functions. Trigonometric functions, for example, have important and useful orthogonality properties. The term *Fourier analysis* is used to denote a variety of issues connected with the approximation of functions by trigonometric sums.

Example 4.10 The functions $\cos(nx)$, $n = 0, 1, 2, \ldots$ are orthogonal with respect to $w(x) = 1$ over the interval $(-\pi, \pi)$. The sequence can be normalized by dividing each function by a suitable constant. Suppose that $f(x)$ is a function that can be suitably approximated in the interval $(-\pi, \pi)$ by a sum of cosines

$$f_n(x) = \sum_{i=0}^{n} a_i \cos(ix). \tag{4.32}$$

The techniques used in the polynomial case for choosing the coefficients a_i can be carried over to (4.32) without any difficulty. For example, to get a least squares solution, we choose the coefficients so that

$$||f - f_n||_2^2 = \int_{-\pi}^{\pi} \{f(x) - f_n(x)\}^2 dx$$

is minimized. The computation is greatly simplified by the known orthogonality results

$$\int_{-\pi}^{\pi} \cos(ix)\cos(jx)dx = \begin{cases} 0, & \text{if } i \neq j \\ 2\pi, & \text{if } i = j = 0 \\ \pi, & \text{if } i = j \neq 0 \end{cases} \cdot \tag{4.33}$$

With this, repeating the manipulations in Section 4.3, we find immediately that

$$\begin{aligned} a_0 &= \tfrac{1}{2\pi} \int_{-\pi}^{\pi} f(x)dx, \\ a_i &= \tfrac{1}{\pi} \int_{-\pi}^{\pi} f(x)\cos(ix)dx, \quad i = 1,\ 2,\ \ldots,\ n. \end{aligned} \tag{4.34}$$

This solution involves integrals whose integrands contain trigonometric functions.

■

To avoid the computation of the occasionally difficult integrals in (4.34) we can resort to a discrete least squares method in which we minimize

$$\rho = \sum_i \{f(x_i) - f_n(x_i)\}^2$$

where the x_i are certain sample points in $(-\pi,\ \pi)$. For carefully chosen x_i there is a simplification, utilizing a *summation-orthogonality*, that avoids the need for solving a linear system.

Let m be a positive integer and set

$$x_i = \frac{(2i+1)\pi}{2m}, \quad i = -m,\ -m+1,\ \ldots,\ m-1. \tag{4.35}$$

Then it can be shown that

$$\sum_{i=-m}^{m-1} \cos(jx_i)\cos(kx_i) = \begin{cases} 0, \text{ if } j \neq k \\ m, \text{ if } j = k \neq 0 \\ 2m, \text{ if } j = k = 0 \end{cases} . \tag{4.36}$$

From this and (4.32) we find the coefficients

$$\begin{aligned} a_0 &= \frac{1}{2m} \sum_{i=-m}^{m-1} f(x_i), \\ a_j &= \frac{1}{m} \sum_{i=-m}^{m-1} \cos(jx_i) f(x_i), \quad j = 1,\ 2,\ \ldots,\ n. \end{aligned} \tag{4.37}$$

Expansions with cosines alone are of limited use. Since the approximations are periodic and even, we cannot expect to get good results unless f

has the same properties. To generalize, we develop analogous expansions with sines for odd functions, then use the fact that any function can be decomposed into the sum of an even and an odd function.

Theorem 4.5 The sequence of functions $1, \sin x, \cos x, \sin 2x, \cos 2x, \ldots$ is orthogonal over the interval $(-\pi, \pi)$ with respect to the weight function $w(x) = 1$. This sequence of functions is also summation orthogonal on the set of points (4.35).

Proof: This orthogonality can be shown directly by taking the definite integral over the interval $(-\pi, \pi)$ with the following trigonometric identities:

$$\cos(jx)\cos(kx) = \frac{1}{2}[\cos(j+k)x + \cos(j-k)x],$$

$$\sin(jx)\cos(kx) = \frac{1}{2}[\sin(j+k)x + \sin(j-k)x],$$

and

$$\sin(jx)\sin(kx) = \frac{1}{2}[\cos(j-k)x - \cos(j+k)x].$$

For the summation orthogonality, we outline the main steps of the proof as follows and leave the details as an exercise.

Using Euler's formula, $e^{i\theta} = \cos\theta + i\sin\theta$, we find that for any integer r,

$$\begin{aligned}\sum_{j=-m}^{m-1}\cos(rx_i) + i\sum_{j=-m}^{m-1}\sin(rx_i) &= \sum_{j=-m}^{m-1}[\cos(rx_j) + i\sin(rx_j)] \\ &= \sum_{j=-m}^{m-1} e^{irx_j}.\end{aligned}$$

From this and (4.35), we have

$$\begin{aligned}\sum_{j=-m}^{m-1} e^{irx_j} &= \sum_{j=0}^{2m-1} e^{i[\frac{(2j+1)}{2m}-1]r\pi} = e^{-ir\pi}\sum_{j=0}^{2m-1} e^{i\frac{(2j+1)r\pi}{2m}} \\ &= e^{-ir\pi}\frac{e^{ir\pi/2m}(1-e^{ir\pi})}{1-e^{ir\pi/m}} = e^{-ir\pi}\frac{e^{ir\pi/2m}(1-1)}{1-e^{ir\pi/m}} = 0.\end{aligned}$$

So

$$\sum_{j=-m}^{m-1}\cos(rx_i) + i\sum_{j=-m}^{m-1}\sin(rx_i) = \sum_{j=-m}^{m-1} e^{irx_j} = 0,$$

implies

$$\sum_{j=-m}^{m-1} \cos(rx_i) = \sum_{j=-m}^{m-1} \sin(rx_i) = 0.$$

From this point, it is now straightforward to prove the summation orthogonality (Exercise 1). ■

With this result, we can now look for approximations of the form

$$f_n(x) = \sum_{i=0}^{n} a_i \cos(ix) + \sum_{i=1}^{n+1} b_i \sin(ix). \tag{4.38}$$

The coefficients a_i are given by (4.37) and

$$\begin{aligned} b_j &= \frac{1}{m} \sum_{i=-m}^{m-1} \sin(jx_i) f(x_i), \quad j = 1, 2, \ldots, n, \quad \text{if } n \leq m-1 \\ b_m &= \frac{1}{2m} \sum_{i=-m}^{m-1} \sin(mx_i) f(x_i). \end{aligned} \tag{4.39}$$

Slightly different formulas exist for other data point arrangements.

The expansion (4.38) has $2n+2$ undetermined coefficients from $2m$ data points. Therefore, for $n = m - 1$ every data point can be matched and we have an interpolation problem. For $n < m-1$, the answer is a least squares fit. As for all orthogonal expansions, changing values of n does not require a complete recomputation. To increase n from k to $k+1$, we simply compute the coefficients a_k and b_{k+1}, then add two terms to (4.38).

Example 4.11 To find the discrete least-squares Fourier approximate for twenty equally spaced data points$\{(x_i, y_i)\}_{i=-10}^{9}$, where $x_i = \frac{2i+1}{20}$ and $y_i = (x_i + 3)e^{-x_i^2}$, we first map the problem into $(-\pi, \pi)$ by $z = \pi x$.

Suppose that a five-term Fourier approximation $S(x)$ is used for the least-squares solution. Then

$$\begin{aligned} S(x) &= \sum_{i=0}^{2} a_i \cos(iz) + \sum_{i=1}^{2} b_i \sin(iz) \\ &= \sum_{i=0}^{2} a_i \cos(i\pi x) + \sum_{i=1}^{2} b_i \sin(i\pi x) \end{aligned}$$

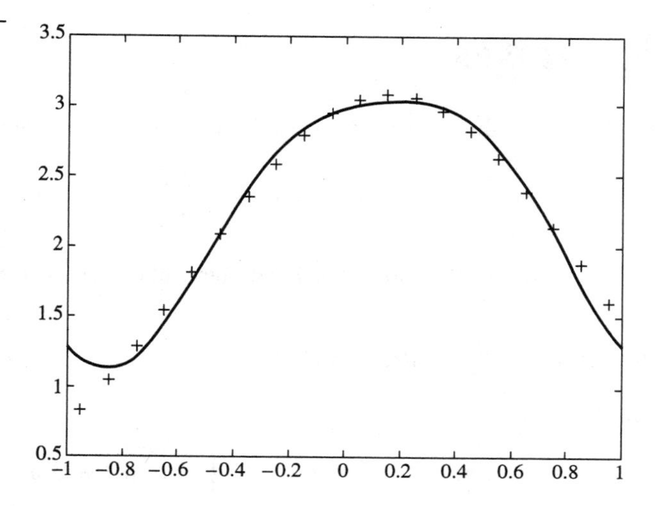

Figure 4.4
A least squares Fourier approximation for $(x+3)\,e^{-x^2}$.

where

$$a_0 = \frac{1}{20} \sum_{j=-10}^{9} y_j,$$

$$a_i = \frac{1}{10} \sum_{j=-10}^{9} \cos(iz_j) y_j,$$

$$b_i = \frac{1}{10} \sum_{j=-10}^{9} \sin(iz_j) y_j, \quad i = 1,\ 2.$$

Evaluating the coefficients, we find

$$\begin{aligned} S(x) = 2.2414 &+ 0.8405\cos(\pi x) - 0.1142\cos(2\pi x) \\ &+ 0.4420\sin(\pi x) - 0.1235\sin(2\pi x). \end{aligned}$$

The approximation and the values at the sample points are plotted in Figure 4.4. Note that the approximation is rather poor, particularly near $x = -1$. This reflects the fact that Fourier approximations do not work well if the function to be approximated is not periodic. This is the case here, so the results are not unexpected.

■

A great deal has been written on the mathematics and the applications of Fourier transforms and approximation. Computing with Fourier series has been greatly aided by the discovery of fast methods (the *Fast Fourier transform*) for computing the sums in (4.37) and (4.39). For an introduction to the voluminous literature on the subject, consult Weaver [27].

EXERCISES

1. Prove the summation orthogonality relations (4.36), using the equality

$$\sum_{j=-m}^{m-1} \cos(rx_i) = \sum_{j=-m}^{m-1} \sin(rx_i) = 0$$

in the proof of Theorem 4.5. Hint: Remember that $\cos a \cos b = \frac{1}{2}\{\cos(a+b) + \cos(a-b)\}$.

2. Show that $\sum_{i=-m}^{m-1} \sin(jx_i)\cos(kx_i) = 0$.

3. Show that

$$\sum_{i=-m}^{m-1} \sin(jx_i)\sin(kx_i) = \begin{cases} 0, \text{ if } j \neq k \text{ or } j = k = 0 \\ m, \text{ if } j = k \leq n-1 \\ 2m, \text{ if } j = k = m \end{cases}$$

4. Use Exercise 3 and (4.38) to derive the coefficients in (4.39).

5. Find 6-term approximation for Example 4.11.

6. Compute the discrete least-squares Fourier approximation for the sixteen points $\{(x_i, y_i)\}_{i=0}^{15}$ using $n = 7$ in (4.38), where $x_i = \frac{i}{8}$ and $y_i = e^{-x_i}$. Estimate the error of the approximation at 100 equally spaced points and plot the approximation and $y = e^{-x}$ for the interval [0, 2]. Note that this approximation is in fact a trigonometric interpolation problem.

7. Find a five-term discrete least squares Fourier approximation on (−1, 1), using 30 uniformly spaced sample points, for the following functions

   (a) $f(x) = x$,

   (b) $f(x) = x^2$,

   (c) $f(x) = \dfrac{\sin(\pi x)}{2 + \cos(\pi x)}$.

   Compare the effectiveness of the approximations and explain any significant differences.

4.6 Best Approximations in the Maximum Norm*

The least squares method gives a solution that is best in the 2-norm; that is, in some average sense. However, a good least squares solution does not preclude the possibility of large errors at individual points. In some applications it is important that the error be small at all points of the interval; in such instances we may want to get an approximation for which the error

$$e = \max_{a \le x \le b} |f(x) - p_n(x)| \tag{4.40}$$

is minimized. This is a best approximation in the maximum norm or a best *uniform* approximation.

Best uniform approximations are more difficult to compute than approximations by interpolation or least squares. To get a handle on this problem, let us return to simple interpolation, in particular the error formula (4.14), from which we know that

$$e \le \max_{a \le x \le b} \frac{|(x - x_0)(x - x_1) \ \dots \ (x - x_n)|}{(n+1)!} \max_{a \le t \le b} |f^{(n+1)}(t)|. \tag{4.41}$$

Since this bound depends on the quantity $(x - x_0)(x - x_1) \ \dots \ (x - x_n)$, a question arises: Is there a clever way of choosing unequally spaced interpolation points so that this bound and the actual error are as small as possible? The answer has some surprising connection with Chebyshev polynomials.

For simplicity, let us look at the interval $[-1, \ 1]$; other cases can be reduced to this by a linear transformation. Since the coefficient of the highest power in the polynomial in (4.41) is one, we consider all polynomials of the form

$$q_n(x) = x^n + \text{lower power terms}. \tag{4.42}$$

In the set of all polynomials of this form, is there one whose maximum absolute value is smallest? Such a polynomial will be called a *min-max* polynomial. Min-max polynomials are characterized by a rather special property.

Definition 4.2

Suppose that f is a function on $[-1, \ 1]$ that has at least n distinct extrema of equal magnitude and alternating sign. More precisely, we require that there are n points $z_1 < z_2 < \ \dots \ < z_n$, such that

$$f(z_{i+1}) = -f(z_i), \quad i = 1, \ 2, \ \dots \ , \ n-1 \tag{4.43}$$

and

$$\max_{-1 \le x \le 1} |f(x)| = |f(z_1)|. \tag{4.44}$$

Then f is said to have an n-point *equi-oscillation property.*

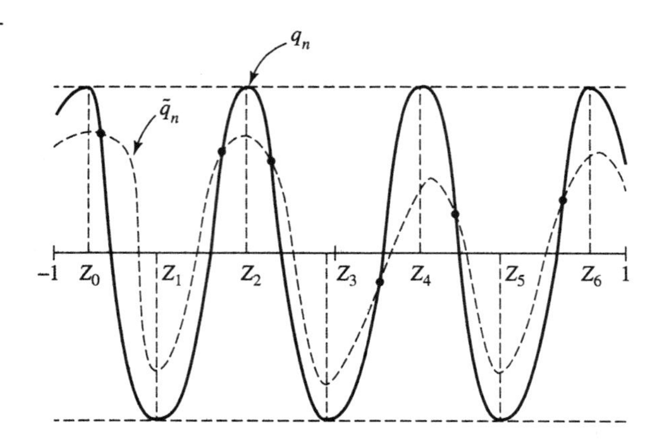

Figure 4.5 Polynomials with the equi-oscillation property. The value of $q_n - \tilde{q}_n$ is zero at the points marked with •.

Theorem 4.6 Suppose that $q_n(x)$ is an nth degree polynomial of the form (4.42) that has an $(n+1)$-point equi-oscillation property. Then $q_n(x)$ is a min-max polynomial.

Proof: Let $z_0, z_1, \ldots, z_n$ denote the extreme points of $q_n(x)$, as illustrated in Figure 4.5. Assume now that that $q_n(x)$ is not a min-max polynomial. Then there must exist some other polynomial $\tilde{q}_n(x)$ of the form (4.42), such that

$$\max_{-1\leq x\leq 1} |\tilde{q}_n(x)| < \max_{-1\leq x\leq 1} |q_n(x)|.$$

Since $q_n(x)$ has its extrema at z_i, this means that

$$\tilde{q}_n(z_i) \leq q_n(z_i), \quad \text{if } q_n(z_i) > 0$$

and

$$\tilde{q}_n(z_i) \geq q_n(z_i), \quad \text{if } q_n(z_i) < 0$$

for all i. Since polynomials are continuous functions, there are at least n points in the interval at which q_n agrees with $\tilde{q}_n$, so that the polynomial $q_n - \tilde{q}_n$ has at least n zeros. However, because $q_n - \tilde{q}_n$ is a polynomial of degree $n-1$ it must be identical to zero.[5] This contradiction proves that q_n is a min-max polynomial. ■

We have already encountered a min-max polynomial; from (4.24) we know that the Chebyshev polynomials T_n have an equi-oscillation property. From the recurrence for the Chebyshev polynomials we also know that

$$T_n(x) = 2^{n-1}x^n + \text{lower power terms.}$$

[5] This follows from the well-known and elementary result that a polynomial of degree n can have at most n zeros.

Therefore,

$$q_n(x) = \frac{1}{2^{n-1}} T_n(x) \tag{4.45}$$

is a min-max polynomial.

Suppose now that we approximate a function f by interpolating at the roots of the Chebyshev polynomial T_{n+1}. The error bound (4.41) then contains the polynomial

$$q_{n+1}(x) = (x - x_0)(x - x_1) \ \ldots \ (x - x_n),$$

where the x_i are the roots of T_{n+1}. Appealing to the unique factorization property of polynomials, we see that this polynomial is identical to the one in (4.45), so that it is a min-max polynomial.

Theorem 4.7 Let p_n be the polynomial approximation to f obtained by simple interpolation at the zeros of T_{n+1}. Then

$$\max_{-1 \leq x \leq 1} |f(x) - p_n(x)| \leq \frac{1}{2^n (n+1)!} \max_{-1 \leq t \leq 1} |f^{(n+1)}(t)|. \tag{4.46}$$

Proof: This follows directly from (4.45) and the observation that $|T_i(x)| \leq 1$ for all i and x.

Note that the bound (4.46) is much better than the bound (4.15). This leads us to believe that interpolation at the Chebyshev roots may give better results than interpolation at a set of uniformly spaced points. Since we are dealing with bounds we cannot be absolutely sure that this is always so, but in practice our expectations are usually met.

Example 4.12 Approximations to

$$f(x) = \sin(2x)e^x$$

on $[-1, \ 1]$ were computed using an n-point simple interpolation. In one case, we used equidistant points, in the other the roots of the Chebyshev polynomials. The results in Table 4.3 show that the second approach gives much better accuracy.

Table 4.3 Observed maximum errors for Example 4.12.

n	Equidistant	Chebyshev
5	3.15×10^{-2}	1.73×10^{-2}
10	1.06×10^{-5}	1.62×10^{-6}

Interpolating at the roots of Chebyshev polynomials minimizes the polynomial term in (4.14). This suggests that if we want to approximate a function by simple interpolation, and we are free to choose the interpolation points, using the zeros of the Chebyshev polynomials of the first kind is likely to give better results than using equally spaced points. In fact, what we get this way is often very close to the best approximation. However, interpolation at the roots of T_{n+1} usually does not give the best uniform approximation. This is because (4.14) also contains the term $f^{(n+1)}(t)$, where t depends on the interpolation points in some complicated way. If we want to find the best approximation we have to abandon interpolation altogether.

The theory of uniform approximation has been studied thoroughly and there do exist algorithms for finding the best solution. The methods are based on the following characterization of the best uniform approximation.

Theorem 4.8 Suppose that f is a continuous function. Then p_n is the best uniform approximating polynomial to f if and only if the error

$$e(x) = f(x) - p_n(x)$$

has an $(n+2)$-point equi-oscillation property.

Proof: The proof is involved and we refer the reader to the literature, e.g. Isaacson and Keller [12]. ■

One of the most effective methods for finding best uniform approximations is the *Remez algorithm*. To find the best polynomial approximation p_n to a given function f, we carry out the following steps:

(i) Compute a good approximation $p_n^{(1)}$ to f by interpolating at the roots of T_{n+1}.

(ii) The error $f - p_n^{(1)}$ has $n+1$ zeros, so we can look at the $n+2$ subintervals created by these zeros and the ends of the entire interval. In each subinterval, we find a point where the error $f - p_n^{(1)}$ achieves an extremum. This will result in $n+2$ local extrema, say, $z_0, z_1, \ldots, z_{n+1}$.

(iii) Find a new polynomial $p_n^{(2)}$ such that $f - p_n^{(2)}$ has alternating values of equal magnitude at the points z_i. That is

$$f(z_i) - p_n^{(2)}(z_i) = (-1)^i e. \tag{4.47}$$

The $n+1$ coefficients of $p_n^{(2)}$ and the unknown e are uniquely determined by the $n+2$ interpolating conditions.

(iv) The new polynomial $p_n^{(2)}$ does not necessarily have the required equioscillation property either, as its extrema may not be at z_i and so may differ in magnitude. In general, though, $p_n^{(2)}$ is a better guess of the best approximation than $p_n^{(1)}$. To improve further, steps (ii) to (iii) are repeated with $p_n^{(2)}$ and subsequent guesses until sufficient accuracy is obtained. We can see from this that there is quite a bit of work involved in getting a best approximation.

Best uniform approximations are widely used in the construction of libraries of elementary functions. Because of the heavy use of these library programs and the high accuracy required, run-time efficiency is essential and the initial work in getting uniform approximations is of little concern. In other areas of numerical analysis, though, uniform approximations are of less importance.

Example 4.13 Use a Remez algorithm to approximate

$$f(x) = e^{-x}\cos(x)$$

on $[-1,\ 1]$ by a third degree polynomial. We first compute the polynomial approximation

$$\begin{aligned} p_3^{(1)}(x) &= 1.02078994125418 - 1.00396350741001x \\ &\quad - 0.16636912068282x^2 + 0.36524522813762x^3 \end{aligned}$$

to f by interpolating at the roots of T_4. The error $f - p_3^{(1)}$ has four real zeros (see Figure 4.6), and five local extrema: $z_0 = -1$, $z_1 = -0.71172089315902$, $z_2 = -0.01187458251072$, $z_3 = 0.70044731906578$, and $z_5 = 1$.

If we now apply the third step of the Remez algorithm, we get the new approximation

$$\begin{aligned} p_3^{(2)}(x) &= 1.02066422108664 - 1.01539642874286x \\ &\quad - 0.16611064679550x^2 + 0.38043251395812x^3. \end{aligned}$$

As is apparent from Figure 4.6, this new approximation has smaller error than $p_3^{(1)}$. Another iteration of the Remez algorithms gives

$$\begin{aligned} p_3^{(3)}(x) &= 1.02056809440733 - 1.01535681088710x \\ &\quad - 0.16591844407434x^2 + 0.38039289610237x^3, \end{aligned}$$

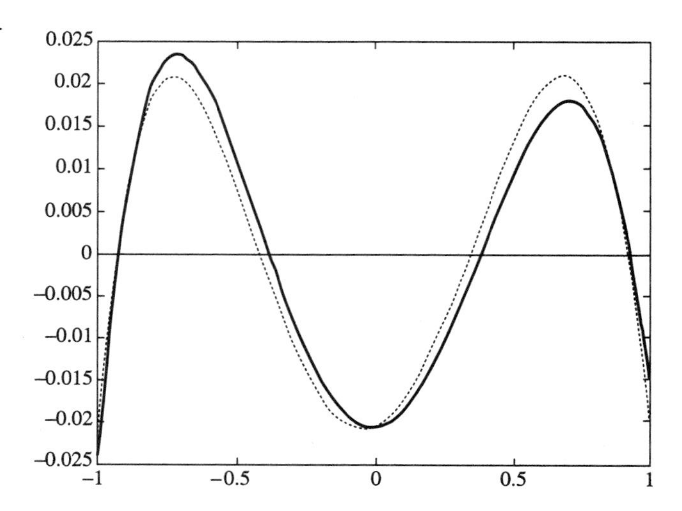

Figure 4.6
Errors of $f - p_3^{(i)}$ for the best approximation in maximum norm. The solid line represents $i = 1$ and the dashed line represents $i = 2, 3$.

Table 4.4
Observed maximum errors for Example 4.13.

i	Observed maximum $\lvert f - p_3^{(i)} \rvert$
1	0.02444515992786
2	0.02101715614576
3	0.02091968558056

which is only a slight improvement over $p_3^{(2)}$. Table 4.4 shows the observed errors for these approximations.

EXERCISES

1. Show that T_n has an $(n+1)$-point equi-oscillation property.
2. Prove Theorem 4.8 for $n = 1$. Hint: If the equi-oscillation condition does not hold, you can adjust the approximation to get a better answer.
3. Consider the approximation on $[-1, 1]$ of a polynomial of degree n by a polynomial of degree $n-1$. Write

$$p_n(x) = a_n x^n + \text{lower power terms}.$$

Then use

$$x^n = \frac{1}{2^{n-1}} T_n(x) + \ \ldots$$

and rearrange to get

$$p_n(x) = \frac{a_n}{2^{n-1}} T_n(x) + p_{n-1}(x).$$

This process is called the *economization* of a power series. Show that p_{n-1} is the best uniform approximation to p_n.

4. Consider the following suggestion for finding best uniform approximations: To find the best polynomial approximation $p_n(x)$ to a function $f(x)$, first approximate $f(x)$ very accurately by a polynomial $p_N(x)$, with $N >> n$. Then use the economization process in the previous exercise to find the best approximating polynomial $p_{N-1}(x)$ to $p_N(x)$. Use economization again to find the best approximation $p_{N-2}(x)$to $p_{N-1}(x)$. Continue this way until you arrive at the polynomial of desired degree. If N is chosen so large that $|f(x) - p_N(x)|$ is negligible, then p_n is in effect the best nth degree polynomial approximation to f. What is wrong with this idea?

5. For $-1 \leq x \leq 1$, find the best uniform approximation to

$$p_4(x) = 1 - \frac{1}{2}x + \frac{1}{4}x^2 - \frac{1}{8}x^3 + \frac{1}{16}x^4$$

by a third degree polynomial p_3. Find a bound for $||p_4 - p_3||_\infty$. Compare this with the error in the third degree polynomial that results from simply dropping the term $x^4/16$.

6. For $n = 1, 2, \ldots,$ let p_n denote the best uniform polynomial approximation to a continuous function f. Show that the sequence of approximating polynomials converges to f in the maximum norm.

Chapter 5

Piecewise Polynomials and Data Fitting

Polynomial approximations are important in the design of many numerical algorithms. We will see shortly several instances of this in connection with numerical integration and the solution of differential equations. In most of these instances polynomials of fairly low degree are used. High degree polynomial approximations, while occasionally useful, are limited by convergence and stability difficulties. These concerns often rule out approximations by polynomials of degree ten or more.

Example 5.1 A common practical problem is the need to graph data; that is, to plot data points and connect them with a smooth curve. Often we do this in an ad hoc fashion, but sometimes we need to be more careful. Before the advent of computers engineers used physical devices, such as French curves, to produce good-looking graphs. Today we rely much more on computer software.

Table 5.1 represents a typical data set, in this case the mean daily temperature in Chicago. We want to use this data to construct a nice, smooth graph.

Table 5.1 Mean daily temperature in Chicago, IL in degrees F.

January	26
February	27
March	36
April	46
May	57
June	69
July	72
August	71
September	67
October	53
November	41
December	30

For twelve data points there is a unique polynomial of degree eleven that passes though all the points. The interpolation formulas that were developed in Chapter 4 can be used to find this polynomial, which can then be used to plot the graph. Unfortunately here, as in many similar cases, the results are quite disappointing. The interpolating polynomial (shown in Figure 5.1) has features that are neither physically plausible, nor warranted by the data.

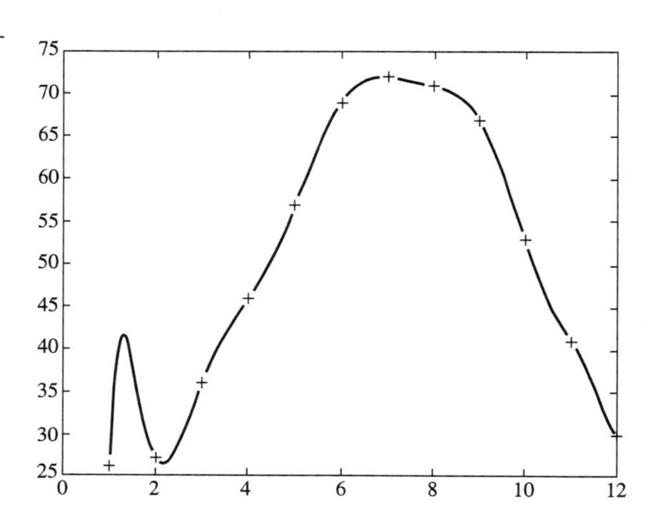

Figure 5.1 Interpolating polynomial of degree eleven for the data in Table 5.1.

To avoid such unpleasantness, numerical analysts commonly use a different approach, one that retains the convenience of polynomials but increases the range of applicability considerably. For an approximation, a larger interval is split into smaller subintervals, and a polynomial of low degree is used in each subinterval. The resulting functions are called *piecewise polynomials*.

5.1 Piecewise Polynomial Functions

To construct piecewise polynomials on some interval $[a, b]$, the interval is partitioned into subintervals $a = x_0 < x_1 < \dots < x_n = b$. The partition is called a *mesh*, and the points of division are the *mesh points*, or *knots*. A piecewise polynomial function is a polynomial in each subinterval, but not necessarily the same polynomial or even of the same degree in all subintervals (Figure 5.2).

When piecewise polynomials are used in practice, the application usually dictates how the pieces are to be connected at the knots. It is possible to have discontinuities in the actual function value at the knots, but this is a rare occurrence, so we will ignore this possibility. Most of the time we need at least continuity, but sometimes we also require a certain amount of differentiability. If a function is continuous but not differentiable at the knots, we say that it has *zero order* continuity; if the function and its first derivative are continuous, it has *first order* continuity, and so on.

Interpolation with piecewise polynomial functions is straightforward. The simplest case involves functions that are composed of straight-line segments or piecewise linear functions. Suppose that we are given function values $y_0, y_1, \dots, y_n$ at the knots $x_0, x_1, \dots, x_n$. In each subinterval $[x_i, x_{i+1}]$ we can then construct a linear function by interpolation at the endpoints. The ith piece is given by the expression

$$p_i(x) = \frac{x_{i+1} - x}{x_{i+1} - x_i} y_i + \frac{x - x_i}{x_{i+1} - x_i} y_{i+1}. \tag{5.1}$$

By putting together the pieces in all the subintervals, we get a piecewise linear function that agrees with the data at all the knots. The function is continuous, but generally has derivative jumps at the knots.

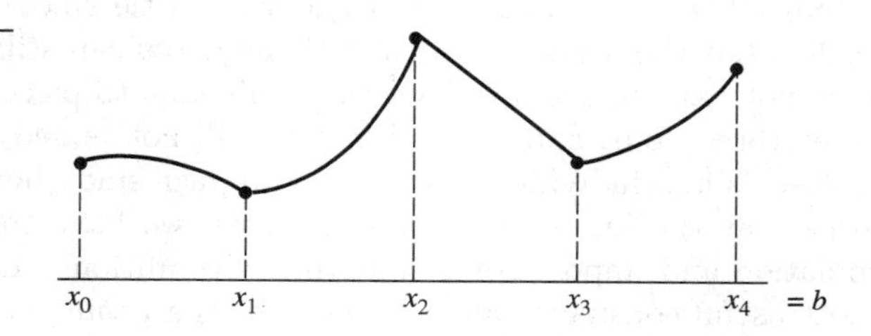

Figure 5.2
A piecewise polynomial function.

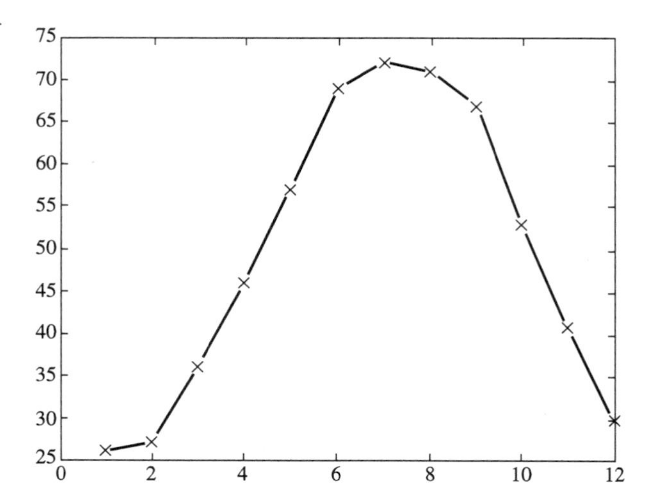

Figure 5.3 Piecewise linear interpolation for the data in Table 5.1.

If we apply this to graph the data in Table 5.1 we get Figure 5.3. This may be adequate for some applications, but frequently we want a smoother result. We can then try piecewise polynomials of degree two, or piecewise quadratics.

To construct a quadratic we need three pieces of information; if we have the value at the knots x_i and x_{i+1}, we will need one extra point in the subinterval. Suppose that we call this point z, with a corresponding data value w. Using the Lagrange form (4.10) we find the quadratic

$$\begin{aligned} p_i(x) = & \frac{(x-z)(x-x_{i+1})}{(x_i-z)(x_i-x_{i+1})} y_i + \frac{(x-x_i)(x-x_{i+1})}{(z-x_i)(z-x_{i+1})} w \\ & + \frac{(x-x_i)(x-z)}{(x_{i+1}-x_i)(x_{i+1}-z)} y_{i+1} \end{aligned} \tag{5.2}$$

interpolates at the points x_i, x_{i+1}, and z. A complete piecewise quadratic can be determined by piecing together each subinterval part. As in the linear case, the resulting function is continuous, but may have a discontinuous first derivative.

To apply this to the data in Table 5.1, we use the odd-numbered data points as knots and the even-numbered points as the intermediate points. Since this can only be done with an odd number of points, we omit the last data item. The graph in Figure 5.4 looks a little smoother than that in Figure 5.3, but the derivative breaks at the knots can still be seen.

It is not hard to see how to extend this idea to piecewise polynomials of degree three, four, and so on, but this will not remedy the difficulty at the knots. While individual pieces may appear smoother, there may still be corners at the knots. To overcome this, we have to abandon simple interpolation and impose some smoothness conditions. In practice, one of the most useful constructs along these lines is a *cubic spline.*

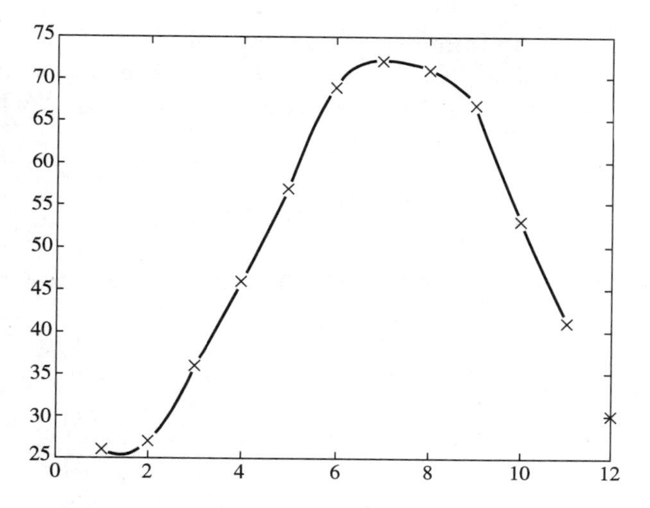

Figure 5.4 Piecewise quadratic interpolation for the data in Table 5.1.

Definition 5.1

Let $a = x_0 < x_1 < \ldots < x_n = b$ denote a mesh on the interval $[a,\ b]$. Then $C(x)$ is said to be a cubic spline on this mesh if it satisfies the following conditions:

(a) In each subinterval $(x_i,\ x_{i+1})$

$$C(x) = C_i(x), \tag{5.3}$$

where $C_i(x)$ is a polynomial of degree at most three.

(b) $C(x)$ is continuous and has continuous first and second derivatives on the whole interval. This implies that

$$\begin{aligned} C_i(x_{i+1}) &= C_{i+1}(x_{i+1}), \\ C_i'(x_{i+1}) &= C_{i+1}'(x_{i+1}), \\ C_i''(x_{i+1}) &= C_{i+1}''(x_{i+1}), \end{aligned} \tag{5.4}$$

for $i = 0,\ 1,\ \ldots,\ n-2$. A cubic spline is a piecewise polynomial of degree three with second order continuity.

In principle, interpolating cubic splines can be constructed in an elementary fashion. For simplicity, take the case of a mesh with three knots $x_0,\ x_1,\ x_2$, with corresponding y values $y_0,\ y_1,\ y_2$. We first write

$$\begin{aligned} C_0(x) &= a_0 + a_1 x + a_2 x^2 + a_3 x^3, \\ C_1(x) &= b_0 + b_1 x + b_2 x^2 + b_3 x^3, \end{aligned}$$

then impose the desired conditions. The interpolating conditions at $x_0,\ x_1,\ x_2$ give

$$\begin{aligned} a_0 + a_1 x_0 + a_2 x_0^2 + a_3 x_0^3 &= y_0, \\ a_0 + a_1 x_1 + a_2 x_1^2 + a_3 x_1^3 &= y_1, \\ b_0 + b_1 x_1 + b_2 x_1^2 + b_3 x_1^3 &= y_1, \\ b_0 + b_1 x_2 + b_2 x_2^2 + b_3 x_2^3 &= y_2, \end{aligned} \tag{5.5}$$

while the continuity requirements yield two more equations,

$$\begin{aligned} a_1 + 2a_2 x_1 + 3a_3 x_1^2 &= b_1 + 2b_2 x_1 + 3b_3 x_1^2, \\ 2a_2 + 6a_3 x_1 &= 2b_2 + 6b_3 x_1. \end{aligned} \tag{5.6}$$

Since we have only six equations for eight unknowns, the answer cannot be unique unless we impose two additional conditions. These conditions can be chosen in many ways; for example, we can require that the second derivatives vanish at the leftmost and rightmost knots. This gives the two extra conditions

$$\begin{aligned} 2a_2 + 6a_3 x_0 &= 0, \\ 2b_2 + 6b_3 x_2 &= 0. \end{aligned} \tag{5.7}$$

Equations (5.5), (5.6), and (5.7) constitute a system of eight equations in eight unknowns whose solution is a cubic interpolating spline.

The idea in this simple case can be extended to the case with an arbitrary set of knots, but it gets a little lengthy to write down all the equations explicitly. In the next section we will study a systematic way of creating splines. When these methods are used to graph the data in Table 5.1, we get Figure 5.5. The results, at least in this case, appear much more pleasing than in previous attempts.

There are other ways of creating piecewise polynomials and splines such as, for example, by the least squares method. But the simple-minded approach used in this section quickly becomes cumbersome. To design automatic algorithms for approximation and data fitting with piecewise polynomials we need more powerful tools.

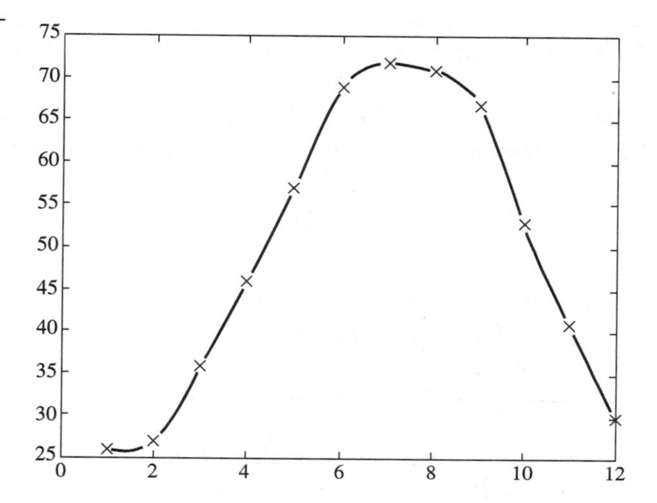

Figure 5.5 Cubic spline interpolation for the data in Table 5.1.

EXERCISES

1. Develop a formula for creating a piecewise polynomial of degree three, using y values at the knots x_i, x_{i+1}, and two additional points in the subinterval. Use this to graph the data in Table 5.1 with piecewise polynomials of degree three.

2. Write down the equations analogous to (5.5) to (5.7) for constructing a cubic interpolating spline on four knots x_0, x_1, x_2, x_3.

3. Write the equations for creating a cubic spline on three knots, if the first derivatives of the spline at the leftmost and rightmost knots are to have pre-assigned values.

4. Find a cubic spline, satisfying condition (5.7), that passes through the data points (0, 1), (1, 6), (2, 5).

5. Find a cubic spline, satisfying condition (5.7), that passes through the data points (0, 1), (1, 6), (2, 5), (3, 4).

6. A kth degree piecewise polynomial with $k > 0$, which has continuous derivatives up to order k-1 on a mesh $a = x_0 < x_1 < \ldots < x_n = b$, is called a spline of degree k. Can a spline of degree k be uniquely determined if only values at knots are given? If not, how many additional pieces of information are needed? Justify your answer.

5.2 Bases for Piecewise Polynomials

Working with piecewise polynomials can be greatly simplified by introducing the concept of a *basis*. At first this may seem unnecessarily formal and abstract, but there are many practical uses for the idea.

Definition 5.2

Let F be a set of functions. Then the set $\{\phi_1, \phi_2, \ldots, \phi_n\}$ is a *basis* for F if every element f in F can be written uniquely as a linear combination

$$f = \sum_{i=1}^{n} a_i \phi_i. \tag{5.8}$$

We say that (5.8) is the *expansion* of f in terms of the basis elements ϕ_i. The totality of all functions that can be written in the form (5.8) is the *span* of the set $\{\phi_1, \phi_2, \ldots, \phi_n\}$.

The qualifier "uniquely" in this definition is important and places a restriction on the expansion set $\{\phi_1, \phi_2, \ldots, \phi_n\}$.

Definition 5.3

The set $\{\phi_1, \phi_2, \ldots, \phi_n\}$ is said to be *linearly independent* if

$$\sum_{i=1}^{n} a_i \phi_i = 0 \tag{5.9}$$

implies that all the a_i are zero.[1] If there exists a set of coefficients a_i not all zero, for which the linear combination (5.9) is identically zero, then the set is said to be *linearly dependent.*

It is not difficult to show that if a set is linearly independent, then any element in its span has a unique expansion of the form (5.8); that is, any linearly independent set is a basis for its span.

Example 5.2 The monomials $1, x, x^2$ are linearly independent on the interval $[0, 1]$. To see this, assume that

$$a_0 + a_1 x + a_2 x^2 = 0.$$

[1] Equalities involving functions imply equality at all points and equation (5.9) has to be interpreted this way. For example, if the ϕ_i are functions of a single variable x, then the linear combination has to be zero for all x. Thus, linear dependence and independence are defined with respect to some interval for x.

Figure 5.6
A hat function.

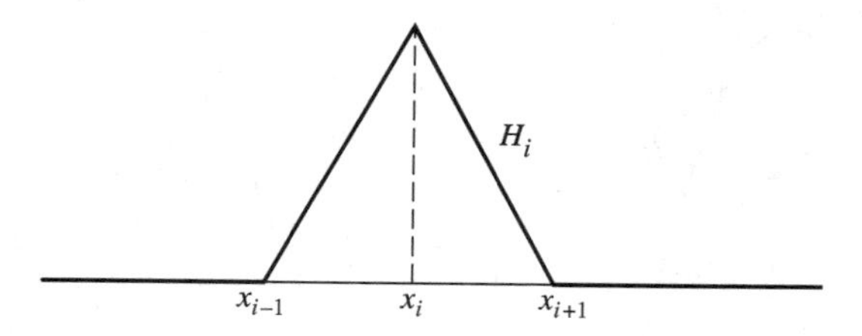

Then, putting $x = 0$ gives $a_0 = 0$. Choosing any other two points, say $x = \frac{1}{2}$ and $x = 1$, and solving the resulting linear system shows that we must also have $a_1 = a_2 = 0$. Therefore, the monomials are linearly independent.

In general, it is known that for any n, the monomials $1,\ x,\ x^2,\ \dots,\ x^n$ are linearly independent[2] and so form a basis for the set of all polynomials of degree n.

To construct a basis for piecewise linear functions on a given mesh we need a set of elements that are piecewise linear, linearly independent, and are such that any linear combination of the form (5.8) is continuous. As we will see, it is also desirable that these basis elements vanish outside a small interval. The *hat* functions, defined for $i = 0,\ 1,\ \dots,\ n$, by

$$H_i(x) = \begin{cases} (x - x_{i-1})/(x_i - x_{i-1}), & x_{i-1} \le x \le x_i, \\ (x_{i+1} - x)/(x_{i+1} - x_i), & x_i \le x \le x_{i+1}, \\ 0, & \text{otherwise,} \end{cases} \tag{5.10}$$

fulfill all these requirements. Figure 5.6 shows that each H_i is a linear, hat-shaped function that vanishes everywhere except in the interval $(x_{i-1},\ x_{i+1})$. It is obvious that any linear combination of hat functions is piecewise linear and continuous. It is also not hard to see that any piecewise linear function can be written as a linear combination of the hat functions. The hat functions are therefore a basis for the set of all piecewise linear functions and suitable for numerical work.

In general, we can choose any set of mesh points, but in practice we often use a uniform partition so that the sizes of all the subintervals are identical. In this case, we will denote the *mesh width* by

$$h = x_{i+1} - x_i. \tag{5.11}$$

Clearly $h = (b - a)/n$ and $x_i = a + ih$. For simplicity, we restrict the rest of the discussion in this chapter to approximations on such uniform meshes, but it is not too difficult to extend the results to nonuniform meshes.

[2]However, it is also known that as n increases, the basis becomes nearly linearly dependent; that is, x^n can be approximated by a polynomial of degree $n - 1$ with a very small error. This is the reason for the instability that we observed in some algorithms such as the one in Example 4.7.

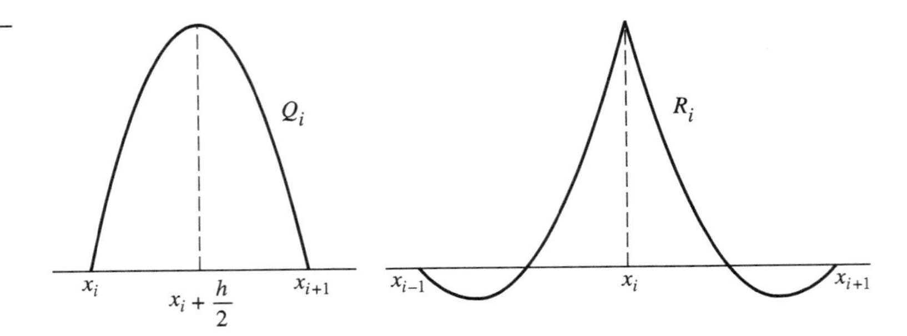

Figure 5.7
Basis elements for piecewise quadratics.

For piecewise quadratics, two types of basis elements are used, constructed by introducing a point halfway between the knots (Figure 5.7). If the mesh is uniform with spacing h, expressions for the basis elements are as follows:

For $i = 0,\ 1,\ \ldots,\ n-1$

$$Q_i(x) = \begin{cases} -\dfrac{4(x - x_i)(x - x_{i+1})}{h^2}, & x_i \leq x \leq x_{i+1}, \\ 0, & \text{otherwise.} \end{cases} \tag{5.12}$$

For $i = 0,\ 1,\ \ldots,\ n$

$$R_i(x) = \begin{cases} \dfrac{2(x - x_{i-1})(x - x_i + \frac{h}{2})}{h^2}, & x_{i-1} \leq x \leq x_i, \\ \dfrac{2(x - x_{i+1})(x - x_i - \frac{h}{2})}{h^2}, & x_i \leq x \leq x_{i+1}, \\ 0, & \text{otherwise.} \end{cases} \tag{5.13}$$

We will leave it as an exercise to show that these functions form a basis for the set of piecewise quadratics on a uniform mesh, and to derive the somewhat more complicated expressions for a nonuniform mesh.

Bases for cubic spline approximations can be constructed in a variety of ways. One of the easiest kinds to visualize are the so-called *B-splines*, the name deriving from their characteristic bell-shaped form, shown in Figure 5.8.

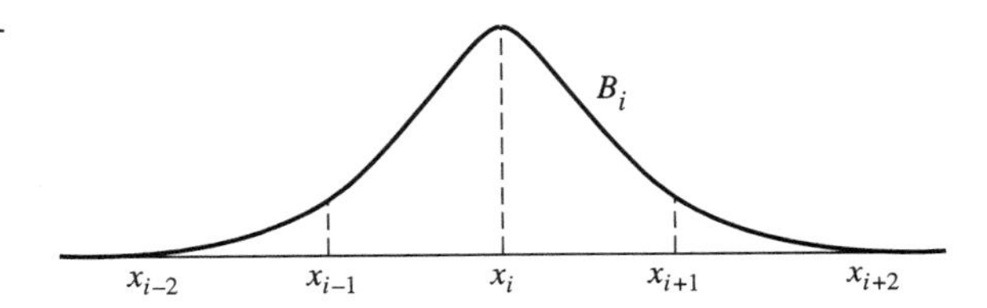

Figure 5.8
A cubic B-spline, centered at x_i.

For a uniform partition, the formulas for the cubic B-spline centered at x_i are, for $i = 0,\ 1,\ \dots,\ n$,

$$B_i(x) = \begin{cases} \dfrac{(x - x_{i-2})^3}{4h^3}, \quad x_{i-2} \le x \le x_{i-1}, \\[2mm] \dfrac{h^3 + 3h^2(x - x_{i-1}) + 3h(x - x_{i-1})^2 - 3(x - x_{i-1})^3}{4h^3}, \\ \qquad x_{i-1} \le x \le x_i, \\[2mm] \dfrac{h^3 + 3h^2(x_{i+1} - x) + 3h(x_{i+1} - x)^2 - 3(x_{i+1} - x)^3}{4h^3}, \\ \qquad x_i \le x \le x_{i+1}, \\[2mm] \dfrac{(x_{i+2} - x)^3}{4h^3}, \quad x_{i+1} \le x \le x_{i+2}, \\[2mm] 0, \text{ otherwise.} \end{cases} \tag{5.14}$$

This definition requires knots outside the interval $[a,\ b]$. They are defined as $x_{-1} = a - h,\ x_{-2} = a - 2h,\ x_{n+1} = b + h,\ x_{n+2} = b + 2h$. Formulas for a nonuniform mesh are explored in the exercises.

An important point is that all the bases for piecewise polynomials we have discussed have the property that its elements are nonzero only over a small part of the whole interval $[a,\ b]$. Such bases are called *local.* This distinguishes them from the *global* bases such as $\{1,\ x,\ \dots,\ x^n\}$. As we will see in several instances, local bases have many advantages over global ones.

EXERCISES

1. Show that the set $\{1,\ \sin x,\ \cos x,\ \sin^2 x,\ \cos^2 x\}$ is linearly dependent over the interval $(0,\ \pi)$.

2. Show that if the set $\{\phi_1,\ \phi_2,\ \dots,\ \phi_n\}$ is linearly independent, then any element in its span has a unique expansion of the form (5.8).

3. Show that the functions in (5.12) and (5.13) are a basis for the set of piecewise quadratics with zero order continuity.

4. Derive formulas corresponding to (5.12) and (5.13) for a nonuniform mesh. To derive the expression for a nonuniform mesh, we use simple interpolation. For example, to get the left part of R_i we find the second degree polynomial that is zero at x_{i-1} and $\frac{1}{2}(x_{i-1} + x_i)$ and has a value one at x_i.

5. Construct a basis for the set of piecewise cubics with zero order continuity. Use the following scheme:

(a) In the interval $[x_i,\ x_{i+1}]$ introduce two new points x_{i1} and x_{i2}.

(b) Find a cubic that has value one at x_{i1} and values zero at $x_i,\ x_{i2}$, and x_{i+1}. The first basis element has the values of this cubic in $[x_i,\ x_{i+1}]$ and is zero outside.

(c) A second basis element is found by repeating a similar process at x_{i2}.

(d) Find a cubic that has value one at x_i and values zero at x_{i1}, x_{i2}, and x_{i+1}. The third basis element has the value of this cubic in $[x_i, x_{i+1}]$. In $[x_{i-1}, x_i]$ the basis element has values that are symmetric about x_i to values in $[x_i, x_{i+1}]$. Outside $[x_{i-1}, x_{i+1}]$ the value of this basis element is zero.

Find explicit expressions for the basis elements and show that the collection of these is a basis for a set of piecewise cubics with zero order continuity.

6. Show that the formulas

$$\overline{B}_i(x) = \begin{cases} A_i + C_i + D_i + E_i, & \text{if } x_{i-2} \le x < x_{i-1}, \\ A_i + C_i + D_i\,, & \text{if } x_{i-1} \le x < x_i\,, \\ A_i + C_i\,, & \text{if } x_i \le x < x_{i+1}\,, \\ A_i\,, & \text{if } x_{i+1} \le x < x_{i+2}\,, \\ 0\,, & \text{otherwise,} \end{cases}$$

with

$$A_i = \frac{(x_{i+2} - x)^3}{(x_{i+2} - x_{i-1})(x_{i+2} - x_i)(x_{i+2} - x_{i+1})},$$

$$C_i = \frac{-(x_{i+2} - x_{i-2})(x_{i+1} - x)^3}{(x_{i+1} - x_{i-2})(x_{i+1} - x_{i-1})(x_{i=1} - x_i)(x_{i+2} - x_{i+1})},$$

$$D_i = \frac{(x_{i+2} - x_{i-2})(x_i - x)^3}{(x_{i+1} - x_i)(x_i - x_{i-2})(x_i - x_{i-1})(x_{i+2} - x_i)},$$

$$E_i = \frac{-(x_{i+2} - x_{i-2})(x_{i-1} - x)^3}{(x_{i+1} - x_{i-1})(x_i - x_{i-1})(x_{i-1} - x_{i-2})(x_{i+2} - x_{i-1})},$$

define B-splines on a non-uniform mesh. What is the relation for a uniform mesh between $\bar{B}$ and B in (5.14)?

7. Show that the hat functions are linearly independent.

8. Show that the B-splines on a uniform mesh are linearly independent.

9. Other types of piecewise polynomials that have found use in some types of numerical work are cubic approximations with first order continuity. These are sometimes called *Hermite splines.*

Consider

$$\Phi_i(x) = \begin{cases} \dfrac{(x - x_{i-1})^2(2x_i - 2x + h)}{h^3}, & x_{i-1} \le x \le x_i, \\ \dfrac{(x - x_{i+1})^2(2x - 2x_i + h)}{h^3}, & x_i \le x \le x_{i+1}, \\ 0, & \text{otherwise,} \end{cases}$$

$$X_i(t) = \begin{cases} \dfrac{(x - x_{i-1})^2(x - x_i)}{h^3}, & x_{i-1} \le x \le x_i, \\ \dfrac{(x - x_{i+1})^2(x - x_i)}{h^3}, & x_i \le x \le x_{i+1}, \\ 0, & \text{otherwise.} \end{cases}$$

Sketch these functions and show that they form a basis for the Hermite splines.

5.3 Approximation of Functions by Piecewise Polynomials

The bases that were developed in the last section have many applications in numerical work. A typical case is the approximation of functions by interpolation or by least squares.

Suppose that $f(x)$ is a function defined on some interval $[a,\ b]$. The function is to be approximated by a piecewise polynomial, using interpolation on a uniform mesh $x_0,\ x_1,\ \ldots,\ x_n$, with mesh width h. The function

$$\tilde{f}(x) = \sum_{i=0}^{n} H_i(x) f(x_i) \tag{5.15}$$

is a piecewise linear function that satisfies the interpolation conditions at the knots, as is easily verified. Given the simple nature of the approximation we should expect low accuracy, something that can be checked by an application of the error bound (4.15).

Theorem 5.1 Let f be a function that is at least twice continuously differentiable on an interval $[a,\ b]$ and let $\tilde{f}$ denote the piecewise linear approximation to f of the form (5.15) on a uniform mesh of width h. Then

$$\max_{a\le x\le b} |f(x) - \tilde{f}(x)| \le \frac{h^2}{2} \max_{a\le t\le b} |f''(t)|. \tag{5.16}$$

Proof: This follows immediately from (4.15) by setting $n = 1$ and $b - a = h$. ■

We see from this that $\tilde{f}$ converges to f with order[3] two in h. Note that, unlike the polynomial case, convergence does not require the boundedness

[3] Reminder: We say $\tilde{f}$ converges to f with order n in h if the error $\max_{a\le x\le b} |f(x) - \tilde{f}(x)|$ decreases at the rate proportional to h^n as h decreases.

of higher derivatives. We can therefore expect that piecewise polynomials will converge for a much broader range of functions than do polynomials.

To get higher convergence rates, we increase the order of the piecewise polynomial. For interpolation with a second degree piecewise polynomial we use function values at the knots as well as the halfway points $x_i + \frac{h}{2}$ as data. This yields the interpolation formula

$$\tilde{f}(x) = \sum_{i=0}^{n} f(x_i)R_i(x) + \sum_{i=0}^{n-1} f\left(x_i + \frac{h}{2}\right) Q_i(x), \tag{5.17}$$

where R_i and Q_i are the basis functions defined in (5.12) and (5.13). Again an appeal to (4.15) gives an error bound.

Theorem 5.2 Let f be a function that is three times continuously differentiable on an interval $[a,\ b]$ and let $\tilde{f}$ denote the piecewise quadratic approximation to f of the form (5.17) on a uniform mesh of size h. Then

$$\max_{a \le x \le b} |f(x) - \tilde{f}(x)| \ \le \ \frac{h^3}{6} \max_{a \le t \le b} |f'''(t)|. \tag{5.18}$$

Proof: This follows again from (4.15). ■

This theorem shows that piecewise quadratic interpolation has order of convergence three. The decrease in the error when halving the mesh size is approximately eight, in contrast to piecewise linear interpolation, where the error reduction is by a factor of four. Approximations by piecewise quadratic interpolation tend to converge more quickly and normally have a much smaller error.

The situation for cubic splines is quite a bit more complicated. Suppose we want to find a cubic spline that satisfies simple interpolation conditions at a set of uniformly spaced knots. If we write

$$\tilde{f}(x) = \sum_{i=0}^{n} a_i B_i(x), \tag{5.19}$$

and take into account the values of the B-splines at the knots, we see that the coefficient vector $(a_0, a_1, \ldots, a_n)$ is the solution of the linear system

$$\begin{bmatrix} 1 & \frac{1}{4} & 0 & & \cdots & 0 \\ \frac{1}{4} & 1 & \frac{1}{4} & 0 & & \\ 0 & \frac{1}{4} & 1 & \frac{1}{4} & & \vdots \\ & & & & & \\ \vdots & & \ddots & \ddots & \ddots & 0 \\ & & & & & \frac{1}{4} \\ 0 & \cdots & & 0 & \frac{1}{4} & 1 \end{bmatrix} \begin{bmatrix} a_0 \\ a_1 \\ \\ \vdots \\ \\ \\ a_n \end{bmatrix} = \begin{bmatrix} f(x_0) \\ f(x_1) \\ \\ \vdots \\ \\ \\ f(x_n) \end{bmatrix}. \tag{5.20}$$

The matrix in this equation is diagonally dominant, and from matrix theory we know that it has an inverse. Therefore the expansion coefficients are uniquely determined. But in spite of its plausibility, the method does not always work well, and the solutions produced by it sometimes show pronounced oscillations.

Example 5.3 The poor performance of (5.20) for a cubic spline can be shown in the approximation for the function

$$f(x) = e^{-0.1x^2}$$

at 10 equally spaced points in $[-1, 1]$. The result is shown in Figure 5.9. The high oscillations near the endpoints are obviously unacceptable. ■

One way to understand the poor results in this example is to realize that the spline produced by (5.20) is not the only spline that satisfies the interpolation conditions. As it turns out, the B-splines $B_0, B_1, \ldots, B_n$ do not form a basis for general cubic splines, and simple interpolation at the knots does not have a unique solution. We can see this by a counting argument: There are n intervals. In each interval the function is a cubic defined by four parameters. Therefore, we have $4n$ free parameters. The interpolation at the knots takes care of $n+1$ conditions. The continuity requirements at the interior knots gives another $3(n-1)$ conditions, for a total of $4n-2$. We therefore have two free parameters, allowing for many solutions. This non-uniqueness must be taken as a warning of potential difficulty in the method.

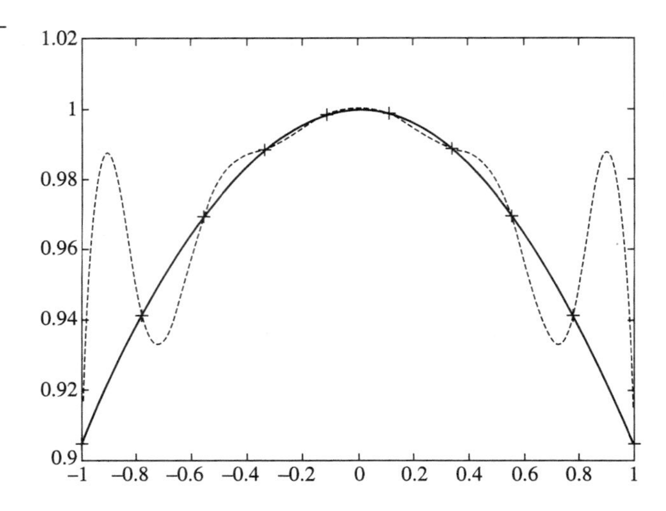

Figure 5.9 Cubic spline approximation of (5.20) for $f(x) = e^{-0.1x^2}$.

To construct a complete basis we introduce the additional B-splines B_{-1} and B_{n+1} centered at $x_{-1} = a - h$ and $x_{n+1} = b + h$, respectively, and write the approximation as

$$\tilde{f}(x) = \sum_{i=-1}^{n+1} a_i B_i(x). \tag{5.21}$$

The set $\{B_{-1}, B_0, B_1, \ldots, B_n, B_{n+1}\}$ is a basis for cubic spline approximations on a uniform mesh, and simple interpolation at the knots plus two additional end conditions will make the solution unique.

A frequent choice for the end conditions is to require that

$$f_n''(x_0) = f_n''(x_n) = 0. \tag{5.22}$$

The result is called the *natural* cubic spline. If we use the fact that

$$B_i''(x_{i-1}) = B_i''(x_{i+1}) = \frac{3}{2h^2},$$

and

$$B_i''(x_i) = \frac{3}{h^2},$$

we get the natural spline interpolant by

$$\begin{bmatrix} 1 & -2 & 1 & & \cdots & 0 \\ \frac{1}{4} & 1 & \frac{1}{4} & 0 & & \\ 0 & \frac{1}{4} & 1 & \frac{1}{4} & & \vdots \\ & & & & \ddots & \\ \vdots & & & \ddots & \ddots & 0 \\ & & & \frac{1}{4} & 1 & \frac{1}{4} \\ 0 & & \cdots & 1 & -2 & 1 \end{bmatrix} \begin{bmatrix} a_{-1} \\ a_0 \\ \\ \vdots \\ \\ a_n \\ a_{n+1} \end{bmatrix} = \begin{bmatrix} 0 \\ f(x_0) \\ \\ \vdots \\ \\ f(x_n) \\ 0 \end{bmatrix}. \tag{5.23}$$

Example 5.4 The same 10 equally spaced points will be used to approximate the same function as that in Example 5.3 using (5.23). The result in Figure 5.10 is obviously superior. ■

While natural splines are useful in data fitting and similar situations, they are not always suitable. For example, when we use a cubic spline to approximate a given function $f(x)$, it makes more sense to choose the two free parameters by matching the derivatives of the cubic spline and f at some points in the interval. The formulas for such splines are easy to derive and will be left as an exercise.

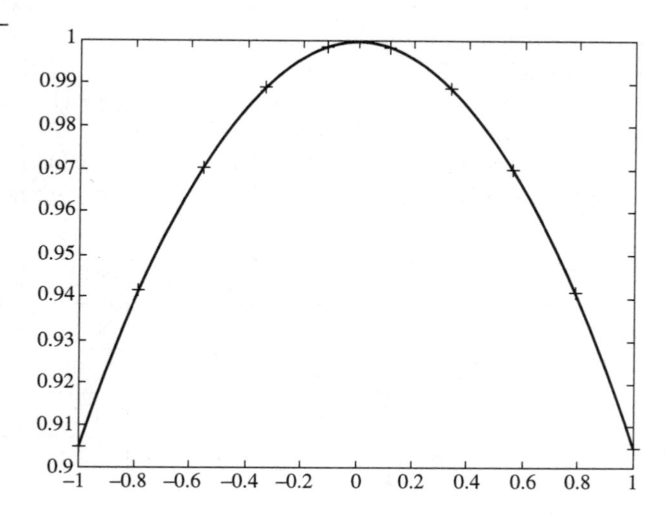

Figure 5.10 Natural cubic spline approximation by (5.23) for $f(x) = e^{-0.1x^2}$. The function $f(x)$ and the approximation are indistinguishable on the graph.

Error bounds for spline approximations are quite difficult to derive and the best we can do here is to quote a representative result. One thing to note in the next theorem is that splines approximate not only the values of a function but also its derivatives.

Theorem 5.3

Assume that f is four times continuously differentiable on an interval $[a, b]$. Then there exists a cubic spline $\tilde{f}$ on a uniform mesh of width h such that

$$\max_{a \leq x \leq b} |\tilde{f}^{(p)}(x) - f^{(p)}(x)| = O(h^{4-p}), \quad \text{for } p = 0,\ 1,\ 2,\ 3.$$

Proof: For this result you will have to consult a book on spline approximation. See, for example, Ahlberg, Nilsen, and Walsh [2]. ■

This theorem shows that the values of the interpolating spline converge to the true function values with order four. This suggests that the spline approximation (if properly chosen) can be considerably better than the approximation by piecewise quadratics, a fact that is easily demonstrated with specific examples.

Example 5.5

The function

$$f(x) = e^x$$

is sampled at 81 equally spaced points in $[0, 1]$. This data is then approximated by interpolation, using a piecewise linear and a piecewise quadratic, each with $n + 1$ uniformly spaced mesh points. For the cubic spline, we use the same set of knots, with an approximation of the form (5.21), and subsidiary conditions $\tilde{f}'(0) = f'(0)$, $\tilde{f}'(1) = f'(1)$. The maximum observed error at the data points is shown in Table 5.2.

The results in Table 5.2 show the expected behavior, with higher accuracy for the higher order approximations. For the linear approximation

Table 5.2 Maximum absolute error for different piecewise polynomial interpolations.

n	Piecewise linear	Piecewise quadratic	Cubic spline
5	1.23×10^{-2}	1.60×10^{-4}	1.10×10^{-5}
10	3.23×10^{-3}	2.08×10^{-5}	6.96×10^{-7}
20	8.29×10^{-4}	2.66×10^{-6}	4.39×10^{-8}
40	2.10×10^{-4}	3.36×10^{-7}	2.54×10^{-9}
80	5.07×10^{-5}	4.22×10^{-8}	1.59×10^{-10}

we expect the error to drop off by a factor of four with each halving of the mesh size, while for the piecewise quadratic the factor should be eight. A factor of 16 drop-off rate for the cubic spline is clearly revealed in the table. ■

From the theorems and the practical observations, we can draw some conclusions and make some comparisons of the relative merits of different approximations:

- Piecewise quadratic interpolation has higher order convergence than piecewise linear interpolation, so it is normally more accurate. Cubic spline interpolation is likely to be even more accurate.
- Although higher-order approximations tend to be more accurate, they are usually more difficult to apply.
- Higher-order approximations have higher smoothness requirements. If these requirements are not satisfied, lower order approximation may do better.

Such trade-offs are characteristic not only of piecewise polynomial approximations, but of all numerical methods. Note, though, that the results in Theorems 5.1 to 5.3 are only bounds. In some cases, any of the methods may perform better than indicated by these bounds.

Using the established bases we can also approximate functions in other ways, for example, by least squares.

Example 5.6 If we approximate a function f by least squares, using the hat function basis on a uniform mesh, we get the system

$$\mathbf{Aa} = \mathbf{b}, \tag{5.24}$$

where

$$[\mathbf{A}]_{ij} = \int_a^b H_{i-1}(x) H_{j-1}(x) dx, \tag{5.25}$$

and

$$[\mathbf{b}]_i = \int_a^b f(x) H_{i-1}(x) dx. \tag{5.26}$$

Because the basis functions are piecewise polynomials, the elements of the matrix $\mathbf{A}$ can be evaluated explicitly. The elements for the right side may have to be done numerically. Working out the integrals, we see that **A** has the form

$$\mathbf{A} = h \begin{bmatrix} \frac{1}{3} & \frac{1}{6} & 0 & \cdots & 0 \\ \frac{1}{6} & \frac{2}{3} & \frac{1}{6} & 0 & \vdots \\ 0 & \ddots & \ddots & \ddots & 0 \\ \vdots & 0 & \frac{1}{6} & \frac{2}{3} & \frac{1}{6} \\ 0 & \dots & 0 & \frac{1}{6} & \frac{1}{3} \end{bmatrix}.$$

This matrix has two nice properties. First, it is *sparse*; that is, it has mostly zero elements. The sparseness arises from the fact that the hat function basis is local. Since sparse systems can be solved much more efficiently than nonsparse ones, local bases offer great computational advantages. The second desirable feature of **A** is that it is diagonally dominant, so it is well-conditioned. This makes the least squares approximation stable. ■

Another use for the bases for piecewise polynomials is in the approximate representations of functions of several variables, such as surfaces in three dimensions. This topic is one of considerable complication, but simplifies quite a bit in special circumstances. Suppose we have one set of mesh points $\{x_0, x_1, \dots, x_n\}$ in the first dimension and another set of mesh points $\{y_0, y_1, \dots, y_m\}$ in the second dimension. The pairs (x_i, y_j) then define a rectangular, two-dimensional mesh we call a *grid* (see Figure 5.11).

Approximations on this grid can be generated by the products of the one-dimensional bases on the two meshes. If we denote these by $\{\phi_0, \phi_1, \dots, \phi_n\}$ and $\{\psi_0, \psi_1, \dots, \psi_m\}$, respectively, then the approximations

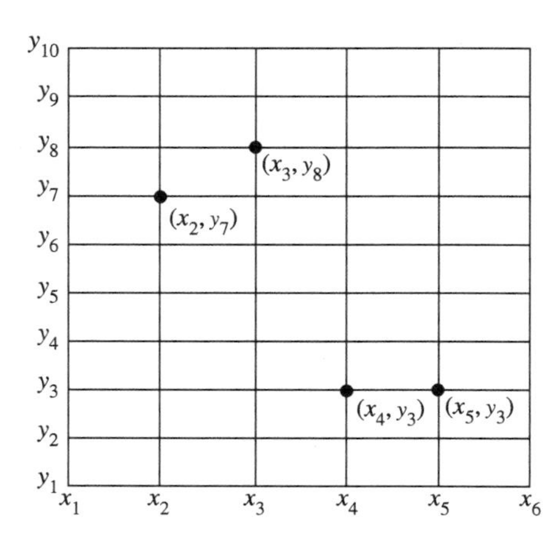

Figure 5.11
A rectangular grid.

will look like

$$\hat{f}(x,y) = \sum_{i=0}^{n}\sum_{j=0}^{m} a_{ij}\phi_i(x)\psi_j(y). \tag{5.27}$$

The coefficients a_{ij}, $i = 0, 1, \ldots, n$, $j = 0, 1, \ldots, m$ can be chosen by interpolation, least squares, or any other suitable set of conditions in essentially the same way as for the one-dimensional case. While the amount of work to get the approximations rapidly increases with the dimension, there is little that is conceptually new here.

Example 5.7 If we have a grid, uniform in both dimensions, and use the corresponding hat functions as bases, we can create an approximation by simple linear interpolation

$$\hat{f}(x,y) = \sum_{i=0}^{n}\sum_{j=0}^{m} f(x_i, y_j)H_i(x)H_j(y). \tag{5.28}$$

The result is a surface that is piecewise linear in each dimension, but it is not a plane. The approximating surface agrees with the function f at all the grid points (Figure 5.12).

■

Similarly, we can use higher degree polynomials or splines to get better approximations to the surface defined by a function of two variables. The extension to more dimensions is conceptually simple. This way of creating higher dimensional approximating elements is known as the *tensor-product* method and is useful in many circumstances. The error analysis of such approximation also tends to follow the one-dimensional case with few surprises, only some technical complications.

Figure 5.12 A bilinear approximation to a surface.

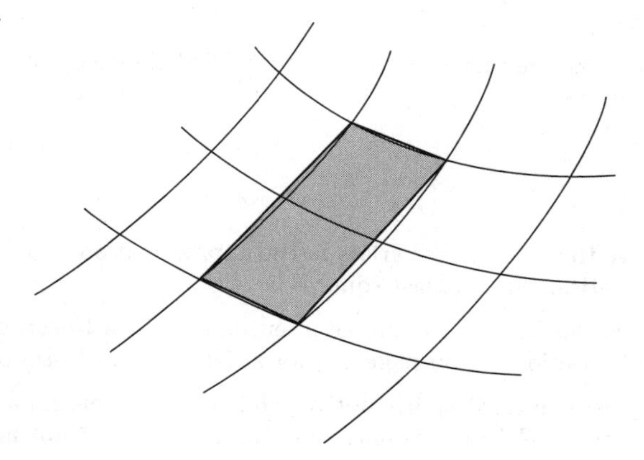

The difficulty in multidimensional approximations comes when we cannot work on a nice rectangular grid, which happens often in applications. Multidimensional approximation occurs in connection with the solution of partial differential equations, modeling of physical bodies in computer graphics, and other such applications. In all of these areas we usually work with objects that have complicated boundaries and other geometric features that make rectangular grids unsuitable. How one deals with these difficulties depends very much on the specifics of a situation, and any discussion of more sophisticated methods is well beyond the scope of this text. We will, however, look at some of the simpler issues in connection with partial differential equations in Chapters 12 and 13.

EXERCISES

1. Show that in (5.15) $\tilde{f}(x_i) = f(x_i)$ for all i.

2. Show that in (5.17) $\tilde{f}(x_i) = f(x_i)$ and $\tilde{f}(x_i + \frac{h}{2}) = f(x_i + \frac{h}{2})$ for all i.

3. Suppose the function

$$f(x) = \frac{e^x}{1+x}$$

is to be approximated in the interval $[0,\ 1]$ by a piecewise interpolating polynomial on a uniform partition to an accuracy of 10^{-5}. What is the smallest value of n that will guarantee this accuracy if

(a) a piecewise linear approximation is used, and

(b) a piecewise quadratic is used?

4. Prove Theorem 5.2.

5. Show that the spline derived by (5.20) satisfies the stated interpolation conditions.

6. Show that the spline derived by (5.23) satisfies the stated interpolation conditions.

7. Show how the matrix in (5.23) is modified if instead of (5.22) we impose the conditions

$$\tilde{f}'(x_0) = g_1,$$
$$\tilde{f}'(x_n) = g_2.$$

8. Derive the matrix that arises in the approximation of a function by piecewise quadratics, using a least squares fit.

9. Derive the formula for simple interpolation with Hermite splines. (See Exercise 9, Section 5.2 on page 104 for a definition of Hermite splines.)

10. Why is a natural spline not a good way of approximating the function in Example 5.3? In particular, why can Theorem 5.3 not hold in this case?

11. Find the specific form of the approximation (5.28) when the interpolation points are $(0,0),(1,0),(0,1),(1,1)$ with values $f(0,0) = 1, f(1,0) = 2$, $f(0,1) = 2$, and $f(1,1) = 3$.

5.4 Data Fitting*

The approximation methods that we have developed so far were primarily aimed at replacing complicated functions with simpler ones, such as polynomials and piecewise polynomials. This is often the first step in the design of other numerical methods, so the emphasis was on the approximation errors and the speed of convergence. There are, however, situations where the underlying assumptions and the purposes of approximation are quite different.

Example 5.1 is a case of such a different need, where we are given a finite set of data points that we want to represent or fit by a continuous, smooth curve. In *data fitting* we start with a *data set*, consisting of N distinct pairs $(x_1,\ y_1),\ (x_2,\ y_2),\ \ldots\ ,\ (x_N,\ y_N)$. Typically, the data set represents experimental observations in which we measure the value of some variable at distinct points x_i. Because experimental data inevitably contains some measurement error, we assume that

$$y_i = f(x_i) + \varepsilon_i, \tag{5.29}$$

where $f(x)$ is the function that represents the underlying phenomenon (the signal) and ε_i is the error (or noise) in the ith observation. We want a function $f_n(x)$ that represents the signal $f(x)$ closely and, at the same time, filters out the undesirable noise. In data fitting we have to make two immediate choices: the type of function that we use, e.g., polynomials, piecewise polynomials, trigonometric approximation, and so on, and the discretization parameter n, that represents the degree of the polynomial or the number of points in the mesh. These choices could affect the answer significantly.

To solve the data fitting problem in a plausible way we need to make some assumptions. One assumption is that the underlying function f is well-behaved, and that it can be approximated with a full polynomial with just a few terms or by a piecewise polynomial with a small number of knots. As a consequence, we deal only with cases where n is considerably less than N. The second assumption is that experimental errors, although significant, are small compared with the size of f and that they are random. Most frequently, one assumes that the error is a random variable with mean zero, variance σ^2, normally distributed.[4]

When N is large, interpolation is usually not suitable for data fitting. Not only does it violate the condition that n should be small, but it also

---

[4]If this assumption is violated, the subsequent discussion in this section has to be modified suitably. However, such modifications do not affect the essential conclusions.

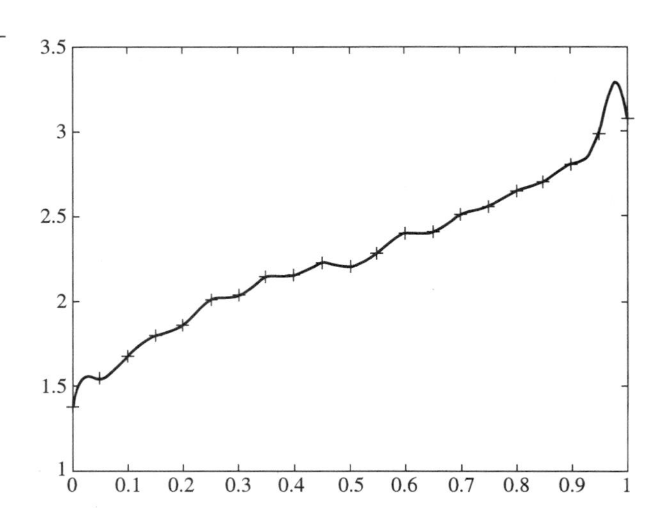

Figure 5.13 Data of Example 5.8 fitted with an interpolating spline.

tends to give rapidly oscillating results that violate physical plausibility. If we make n too large we run the risk of *overfitting*; that is, producing a representation that not only fits the function, but also the undesirable error.

Example 5.8 For this and the next two examples, we use

$$f(x) = 2 + x^2 - 3e^{-x}(0.6 - x)^3,$$

evaluated at 21 equally spaced points in the interval [0, 1]. The data is generated by (5.29), using as error a normally distributed random variable with zero mean and $\sigma = 0.03$. In the plots the data is represented by the discrete points marked with +.

The result of a cubic spline interpolation on this data is shown in Figure 5.13. It is obvious that the approximation is flawed. While the interpolating curve fits the data very well, the many small oscillations are an indication that we are fitting the noise as well as the signal. This is clearly a case of overfitting and the result would not be acceptable in most situations. ■

An alternative to interpolation is discrete least squares; such approximations are very common in data fitting. Here, the condition that n be small lets us reconsider full polynomial approximations. Quite frequently a low degree polynomial does very well. The primary question is how to choose this degree.

Example 5.9 Using the data in Example 5.8, we compute the discrete least squares polynomial approximations with degrees $n = 2$, 3, and 4. The results are shown in Figures 5.14 through 5.16. In Figure 5.17 we show the least squares cubic spline approximation, computed with five equally spaced knots.

The second degree polynomial does not fit very well. We suspect a case of *underfitting*, in which the approximation is not accurate enough to represent the signal function. Underfitting can be solved by increasing n. The approximation for $n = 3$ in Figure 5.15 looks much better and little change is noticed when we increase n to four. The spline approximation in Figure 5.17 is also almost identical to the polynomial cases. Any of the three fits in Figures 5.15 through 5.17 seems reasonable. When several options are present we tend to accept the simplest one. By this rule of thumb, we would pick the least squares solution by a polynomial of degree three as the best answer.

■

Trying different fits and selecting the one that seems best is sometimes a reasonable way of solving a data fitting problem. In general, though, we prefer a less intuitive and more algorithmic approach. To do this we must introduce some ideas from statistics. The assumption we have made is that the errors are normally distributed with mean zero and variance σ^2. Because any specific case is only a finite sample, we cannot expect the ideal values to be attained. However, if N is large we expect the sample mean of random errors

$$m_N = \frac{1}{N}\sum_{i=1}^{N} \varepsilon_i$$

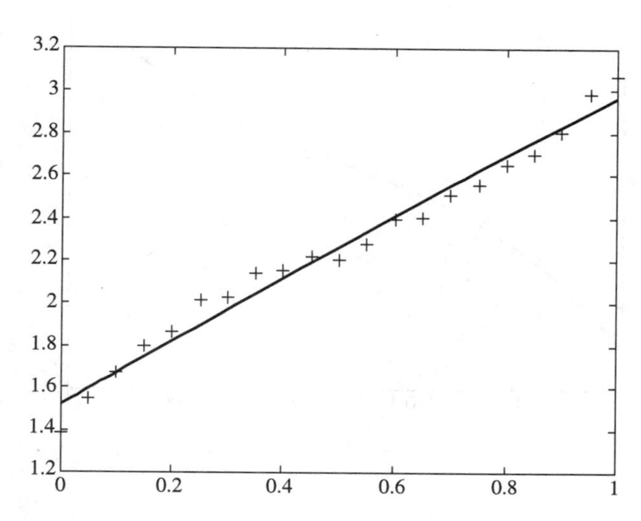

Figure 5.14 Data fitted with second degree polynomial.

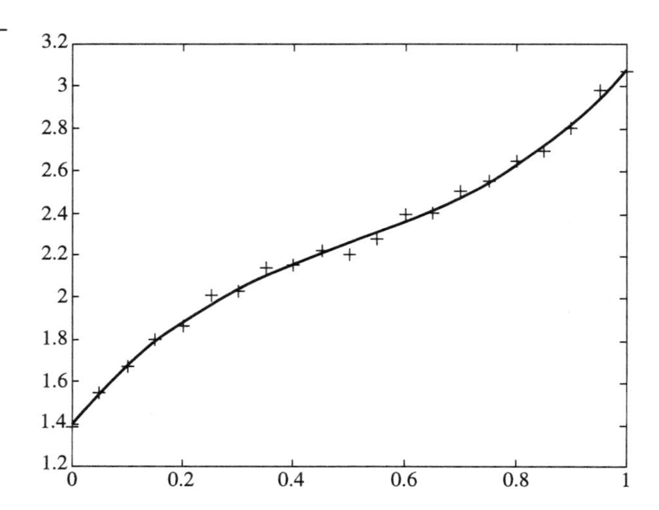

Figure 5.15 Data fitted with third degree polynomial.

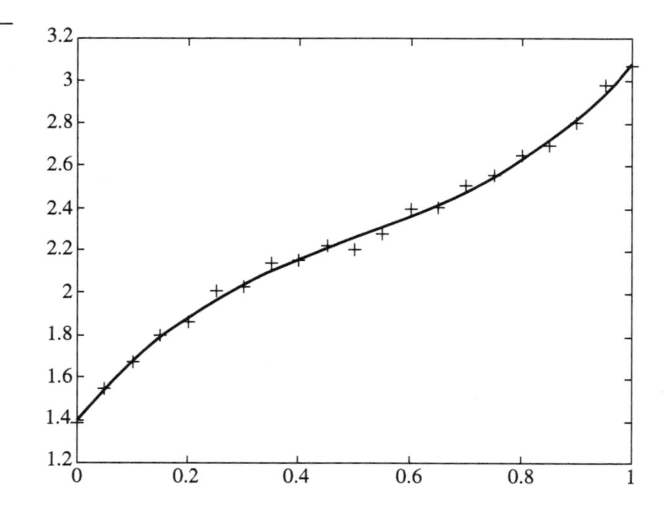

Figure 5.16 Data fitted with fourth degree polynomial.

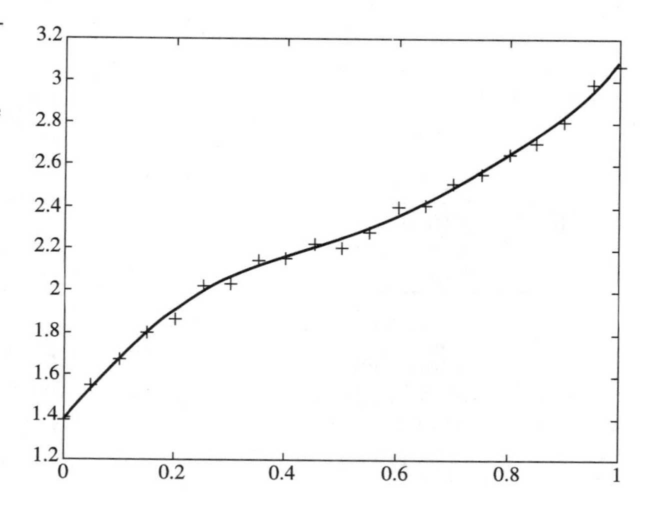

Figure 5.17
Data fitted with a cubic natural spline on knots (0, 0.25, 0.5, 0.75, 1).

to be nearly zero and the corresponding sample variance

$$\sigma_N^2 = \frac{1}{N}\sum_{i=1}^{N}\varepsilon_i^2$$

to be close to σ^2. If the approximation fits the signal with a deviation much less than the observational error, we expect that the mean of $y_i - f_n(x_i)$ should also be close to zero and that the observed variance

$$\bar{\sigma}_N^2 = \frac{1}{N}\sum_{i=1}^{N}(y_i - f_n(x_i))^2 \tag{5.30}$$

should be close to σ_N^2 and hence close to the true σ^2. We normally must expect differences from the ideal values, but there are statistical tests that tell us whether or not they are sufficiently close. Thus, if we know σ we can start with a low value of n, compute the mean and variance of the approximation, and then increase n until these are within acceptable limits of the expected values.

If we do not have a good value for σ, we can use the known fact that if $f = f_n$, then

$$\hat{\sigma}_N^2 = \frac{1}{N-n}\sum_{i=1}^{N}(y_i - f_n(x_i))^2 \tag{5.31}$$

is an unbiased estimator of σ^2. What this means is that when n is so low that we are underfitting, what we get from (5.31) is not necessarily a good estimate of σ and is likely to change with n. However, once we reach the stage where we have fitted f, (5.31) will give a value of $\hat{\sigma}_N^2$ that is close

Table 5.3 Estimated variance for least squares polynomials of different degrees.

n	$\hat{\sigma}_N^2$
1	4.8×10^{-3}
2	5.0×10^{-3}
3	7.7×10^{-4}
4	7.6×10^{-4}
5	8.1×10^{-4}

to σ^2 and so will not change very much with n. The strategy, then, is to compute least squares approximations with increasing n until $\hat{\sigma}_N^2$ becomes nearly constant, showing a plateau in the graph of $\hat{\sigma}_N^2$ versus n. The first n that gets us to the plateau is then used as the most appropriate value.

Example 5.10 For the data in Example 5.8 we computed $\hat{\sigma}_N^2$ for the least squares polynomial approximations with $n = 1,\ 2,\ 3,\ 4,\ 5$. The results are in Table 5.3. Since there is a considerable drop in $\hat{\sigma}_N^2$ going from $n = 2$ to $n = 3$, we expect that a polynomial of at least degree three is needed to represent the signal. Since after $n = 3$ there is little further drop in $\hat{\sigma}_N^2$, we might accept this as the best value. But the choice is subjective and one could not argue very much with someone who preferred $n = 4$. Note also that there is only a small change between $n = 1$ and $n = 2$. Such instances should warn us that there may be other reasons for a lack of change in $\hat{\sigma}_N^2$ (such as a small second power term in the approximation). Usually we need at least three or four values of n to establish a plateau.

■

While simple polynomials are suitable for data fitting in which the signal is smooth, they are less appropriate under more severe conditions. Data fitting with piecewise polynomials, particularly splines, is much more adaptable.

Example 5.11 The raw data for this example is plotted in Figure 5.18. This data, sometimes called the *titanium data*, is used in several books on numerical analysis or data fitting. (See for example Rice [22], p. 106.) Because of the apparent sharp peak, this data is more difficult to fit than the data in the previous examples.

When we examine Figure 5.18 we can see easily why a polynomial approximation works poorly. The sharp changes of the data in the central range would require a fairly high degree polynomial , but this opens up the potential for instability and overfitting. A spline looks more promising.

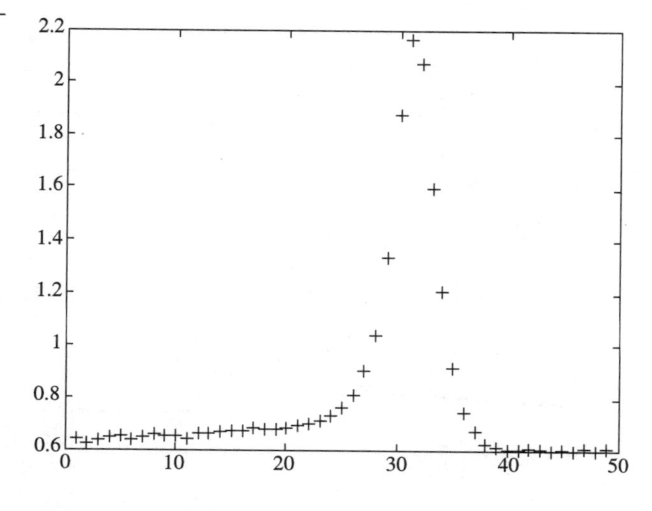

Figure 5.18
The titanium data.

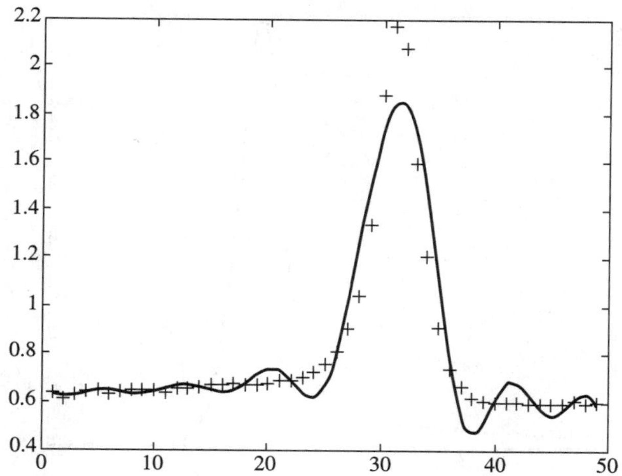

Figure 5.19
Fitting the titanium data with a spline on 13 equally spaced knots.

Our first attempt in Figure 5.19 is not too successful. Using 13 equally spaced knots we get a fit that is not very good, but at the same time has oscillations that look like overfitting. Actually, here the wiggles arise more from the continuity requirements of the spline and are not uncommon in spline approximations. The situation improves when we increase the number of knots.

With 25 equally spaced knots we get the approximation in Figure 5.20. The fit looks quite good; there is only a slight hint of overfitting at the left edge. In this case we can do even better by using knots that are not equally spaced. The results in Figure 5.21 give slightly better results with fewer knots. Unfortunately, good placement of the knots is not always easy. There has been some work done on best knot placement, but the resulting

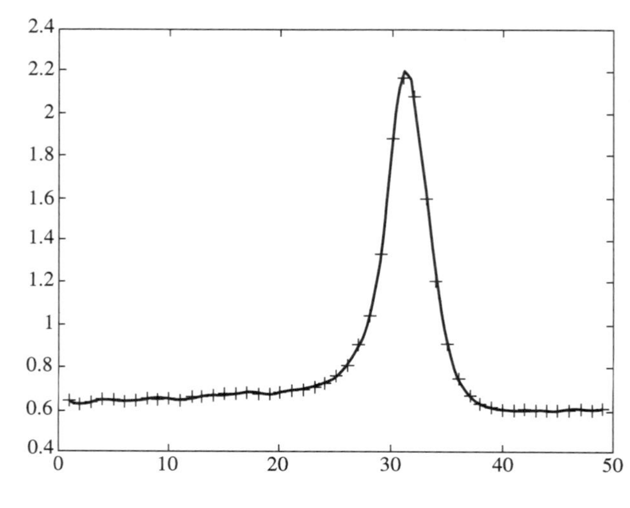

Figure 5.20 Fitting the titanium data with a spline of 25 equally spaced knots.

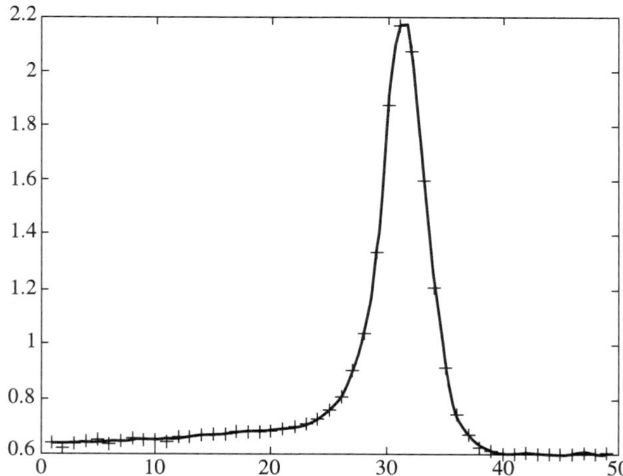

Figure 5.21 Fitting the titanium data with knots at 1, 7, 13, 19, 25, 26, ..., 36, 37, 43, 49.

algorithms are not easy to understand. For our purposes, the best rule of thumb is to place closely spaced knots where the data values change rapidly and put only a few knots where the data is relatively constant. ■

The issues in data fitting are different from the issues in approximation. In data fitting we are not concerned with convergence and high accuracy, as obviously the data can be fitted exactly by interpolation. The challenge lies in finding a simple representation of the signal and filtering out the noise. Because the noise is usually random, statistics comes into most data fitting questions, and much of what has been written on this subject is in statistics-oriented literature. The book by Lancaster and Salkauskas [16] is a very nice introduction that avoids the sometimes heavy-duty statistical machinery.

EXERCISES

1. According to the statement in Example 5.8 the value of σ^2 is 9×10^{-4}, but the estimated value from Table 5.3 is slightly larger than 8×10^{-4}. How do you account for this difference?

2. Fit the data in Example 5.8 with a polynomial of degree 6. Plot the results. Use (5.31) to estimate σ^2. What conclusions can you draw from the results?

3. Find least squares spline fits, using equally spaced knots, to the data

x	y	x	y
0.0	0.10	0.6	0.83
0.1	0.21	0.7	0.84
0.2	0.33	0.8	0.85
0.3	0.44	0.9	0.85
0.4	0.57	1.0	0.86
0.5	0.70		

What is the best choice for the number of knots? Explain why you consider your choice reasonable.

Chapter 6

Numerical Differentiation and Integration

Even the fundamental operations of calculus, differentiation, and integration, often have to be attacked numerically. In Chapter 1 we encountered one method for numerical integration, the rectangular rule, but this method is very inefficient. As we will see, the idea can be extended to get much better results. Differentiation, although more amenable to analytic treatment, also needs a numerical counterpart. Numerical differentiation is the key to the approximate solution of differential equations, and numerical differentiation formulas are used in many circumstances.

Throughout this chapter we develop formulas that require that the underlying functions have certain differentiability properties; strictly speaking the results hold only under those conditions. To avoid frequent disclaimers, we will simply assume that the functions we deal with have all the differentiability required by the manipulations involved. We also make liberal use of order of magnitude arguments and O-notation. This makes the discussion more intuitive and results easier to derive, but carries with it the danger of imprecision. It is possible to make all arguments rigorous, but for the sake of clarity, tying up all the loose ends will be left to the reader.

6.1 Finite Differences and Numerical Evaluation of Derivatives

The rules of differentiation are simple and widely applicable, so that even if a function is complicated we may be able to get an analytic expression for its derivative. Still, there are instances where, for one reason or another, we need a discretized approximation to a derivative. The most commonly used approximations are the *finite differences.*

One way to get numerical differentiation formulas is to approximate the function f by a polynomial, then take the derivative of this approximation. It is hoped that the derivative of the approximation is an approximation to the derivative. A simple way, which at the same time tells us something about the error, is to start from Taylor's Theorem. From (4.4) we have that

$$f(x+h) = f(x) + hf'(x) + O(h^2),$$

so that

$$f'(x) = \frac{f(x+h) - f(x)}{h} + O(h).$$

Ignoring the $O(h)$ term, we get

$$f'(x) \cong \frac{f(x+h) - f(x)}{h}. \tag{6.1}$$

The expression on the right side of (6.1) is the *forward difference* approximation to f' at x. It is computed using only values of f and converges to the derivative with order one. In an almost identical way we get the *backward difference*

$$f'(x) \cong \frac{f(x) - f(x-h)}{h}. \tag{6.2}$$

The forward and backward difference rules have relatively low accuracy, but it is not hard to get better approximations. Again, from Taylor's Theorem we know that

$$f(x+h) = f(x) + hf'(x) + \frac{h^2}{2}f''(x) + O(h^3),$$

$$f(x-h) = f(x) - hf'(x) + \frac{h^2}{2}f''(x) + O(h^3),$$

so that, neglecting the higher order terms,

$$f'(x) \cong \frac{f(x+h) - f(x-h)}{2h}. \tag{6.3}$$

This is the *centered difference* approximation to the first derivative. It has second order convergence and normally gives a much more accurate answer than either (6.1) or (6.2).

Many other formulas can be developed along these lines. For example, the expression

$$f'(x) \cong \frac{-f(x+2h) + 4f(x+h) - 3f(x)}{2h} \tag{6.4}$$

is another forward difference approximation for the first derivative. We will leave it as an exercise to show that it has second order convergence.

We can also derive approximations for higher derivatives this way. Taking more terms in the Taylor expansions

$$f(x+h) = f(x) + hf'(x) + \frac{h^2}{2}f''(x) + \frac{h^3}{6}f'''(x) + O(h^4),$$

$$f(x-h) = f(x) - hf'(x) + \frac{h^2}{2}f''(x) - \frac{h^3}{6}f'''(x) + O(h^4),$$

we find that

$$f''(x) = \frac{f(x+h) - 2f(x) + f(x-h)}{h^2} + O(h^2). \tag{6.5}$$

Thus,

$$f''(x) \cong \frac{f(x+h) - 2f(x) + f(x-h)}{h^2}.$$

This is a very useful centered difference approximation to the second derivative that has second order accuracy.

These difference formulas have a major application in the approximate solution of differential equations, and the effect of the discretization on this will be discussed in the appropriate place. Here we examine what happens when these formulas are used to actually approximate the derivatives of a given function. Because all the finite differences are convergent as $h \to 0$, in an ideal setting we can get arbitrarily good accuracy by simply making h small enough. But in practice we work with finite precision computations, which puts a limit on what we can hope to achieve. Intuitively, we expect that the accuracy of an approximation will improve as h decreases until rounding and other computational errors become significant. As h decreases, one cannot expect unlimited increases in accuracy and at some point the accuracy may actually decrease as shown in Figure 6.1.

If the situation in Figure 6.1 holds, there is some optimal value of h at which the maximum attainable accuracy is achieved. A straightforward analysis can tell us at least roughly what the maximum accuracy is. As an illustration, we consider the computation of the first derivative by means of the centered difference (6.3).

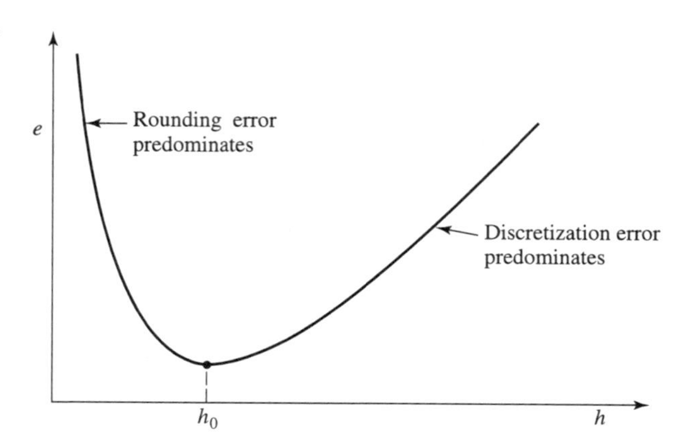

Figure 6.1
Error in computing a derivative numerically.

Computational errors arise from inaccuracies involved in computing f as well as from rounding errors in forming the difference expressions. Let us denote the combined error in computing the numerator in (6.3) by η. Then the error e in the approximation of the derivative has two components, the error due to the discretization from (6.3), and the effect of η. Putting these two errors together, we see that

$$e \cong ch^2 + \frac{\eta}{h}, \tag{6.6}$$

where c is the constant implied in the $O(h^2)$ term in (6.3). The function at the right of (6.6) behaves exactly like the function in Figure 6.1. Its minimal value is attained at h_0 which can be obtained by differentiating (6.6), giving

$$2ch_0 - \frac{\eta}{h_0^2} = 0$$

and

$$h_0 = \left(\frac{\eta}{2c}\right)^{\frac{1}{3}}.$$

Without going into details about the values of c and η, we can at least claim that the best result is likely to occur for

$$h = O(\eta^{1/3})$$

with an error

$$e = O(\eta^{2/3}). \tag{6.7}$$

Roughly, if η is a few ulps in IEEE double precision, say about 10^{-15}, the optimal value of h is around 10^{-5} and the maximum attainable accuracy is of the order of magnitude 10^{-10}.

Table 6.1
The error in the approximation of a first derivative as a function of h.

h	Error
1×10^{-3}	8.0×10^{-8}
1×10^{-4}	8.0×10^{-10}
1×10^{-5}	7.7×10^{-12}
1×10^{-6}	2.2×10^{-12}
1×10^{-7}	1.1×10^{-10}
1×10^{-8}	5.1×10^{-9}
1×10^{-9}	1.6×10^{-8}
1×10^{-10}	4.0×10^{-7}
1×10^{-11}	3.5×10^{-6}

While estimates of this kind are not precise, they are of value in many instances. Superficial reasoning such as this is often employed not to avoid rigorous mathematics, but because the necessary information, such as c in (6.6), is not available. Despite the imprecision in the arguments, the conclusions we draw can be useful.

Example 6.1

To check out the prediction (6.7), we computed the derivative of $\cos(x)$ at $x = 0.5$, using the centered difference formula (6.3). The magnitudes of the observed errors, for a range of h, are shown in Table 6.1.

Table 6.1 is in reasonably good agreement with the prediction, given the imprecision of the arguments leading to (6.7). The predicted best h is about 10^{-5}, with a corresponding error of 10^{-10}; the actual results are within one or two orders of magnitude of this. From a practical standpoint, this is often quite adequate.

■

Numerical computation of the first derivative is slightly unstable because the rounding errors can be magnified by a factor of $1/h$. For very small h , this may result in the loss of several significant digits. If the error in f is of the order of machine accuracy, the results are still sufficiently accurate for most practical purposes. However, when the accuracy in f is much less, as it might be if f comes from some experimental observations, it can be quite difficult to get an accurate value of the derivative numerically. For the computation of higher derivatives the situation is even worse, and it is generally not advisable to compute high order derivatives numerically.

EXERCISES

1. Find bounds for the order of magnitude terms in (6.1) and (6.3).

2. Show that the approximation in (6.4) has an $O(h^2)$ accuracy.

3. Derive difference formulas for the first, second, and third derivatives of a function f at a point x, using interpolation at $f(x-h)$, $f(x)$, $f(x+h)$, and $f(x+2h)$. Find the order of convergence for all the approximations.

4. Derive the coefficients in the approximation

$$f'(x) \cong a_0 f(x-h) + a_1 f(x+h/3) + a_2 f(x+h).$$

   Choose coefficients so that the approximation is exact for all polynomials of degree two. Establish the order of convergence for this approximation.

5. Find the orders of magnitude of the optimal value of h and the maximum attainable accuracy for the computation of the second derivative by the centered difference formula (6.5).

6. Devise a rough rule that relates the magnitude of the computational errors to the maximum attainable accuracy for numerically computing derivatives $f^{(k)}$, as a function of k.

7. Consider the two-dimensional grid below.

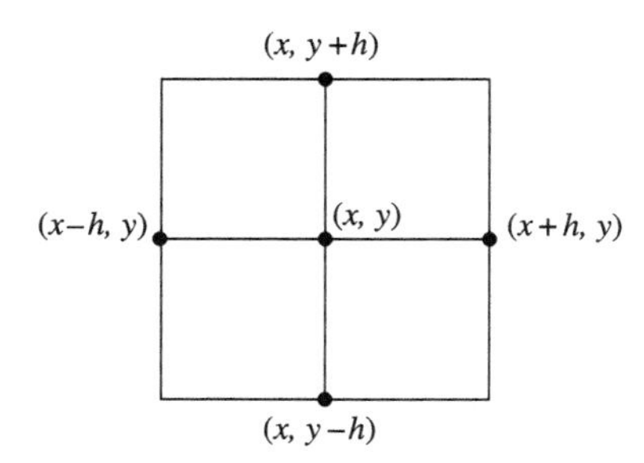

   Develop a formula for the approximation of the partial derivative $\dfrac{\partial^2 f(x, y)}{\partial x \partial y}$ at (x, y) in terms of the functional values at the five points (x, y), $(x+h, y)$, $(x-h, y)$, $(x, y+h)$, $(x, y-h)$.

8. Consider the following scenario. The form of a function $f(x)$ is not explicitly known, but we have a computer program that gives its value for all x with an error less that 10^{-15}. We want to estimate the value $f'(1)$ numerically.

   When we do this by the forward difference formula (6.1) using different values of h, we get the results on the next page.

h	Forward difference
0.1	0.0000
0.01	0.0000
0.001	0.0000

But when we use the centered difference formula (6.3), we get quite different answers.

h	Centered difference
0.1	0.5000
0.01	0.5000
0.001	0.5000

The first set of results suggests that the derivative is zero; the second sequence of results seems to converge to one half. What is a possible reason for this discrepancy?

6.2 Numerical Integration by Interpolation

Numerical integration involves replacing an integral by a sum. Suppose that f is a function defined on some interval $[a, b]$ and let $\{x_0, x_1, \ldots, x_n\}$ be a set of distinct points.[1] Then the approximation

$$\int_a^b f(x)dx \cong \sum_{i=0}^{n} w_i f(x_i), \tag{6.8}$$

with properly chosen w_i, is a *numerical integration* rule. For historical reasons, the term *quadrature* is often used as a synonym for numerical integration in one dimension. The w_i are the *quadrature weights* and the x_i the *quadrature points.* All the numerical integration rules that we discuss here will have the form (6.8); the only difference between them is how the quadrature weights and points are chosen.

An entire class of integration rules can be constructed as follows. First, the integrand is approximated by a polynomial, using simple interpolation. This approximation is then integrated and, because all we have to do is integrate powers of x, the quadrature weights can be evaluated explicitly. Depending on what approximation we start with, we get different quadrature rules. Methods of this type are termed *interpolatory quadrature* rules.

[1] In most cases, the points $x_0, x_1, \ldots, x_n$ are in the interval $[a, b]$, but there are some exceptions. We assume that f is defined at all points where it is needed.

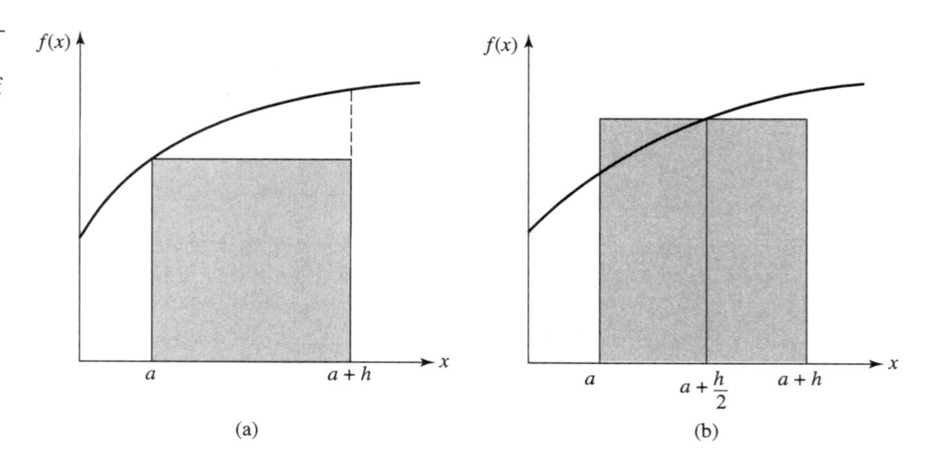

Figure 6.2 A graphical view of two simple quadrature methods: (a) rectangular rule, and (b) midpoint method. The shaded area is the approximation of the area under the curve.

The Rectangular and Midpoint Rules

The simplest approximation we can make to a function is to replace it by a constant. If, in the interval $[a,\ a+h]$, we replace $f(x)$ by $f(a)$ we get the quadrature rule

$$\int_{a}^{a+h} f(x)dx \cong hf(a). \tag{6.9}$$

This is the rectangular rule that we encountered in Chapter 1. Since it is based on a poor approximation of the integrand, its accuracy is low.

If we approximate $f(x)$ by its value at the center of the interval, we get the *midpoint* rule

$$\int_{a}^{a+h} f(x)dx \cong hf(a+h/2). \tag{6.10}$$

An examination of the graphical interpretation of these two simple rules, Figures 6.2a and 6.2b, suggests that the midpoint method is superior to the rectangular rule.

The Trapezoidal Rule

We approximate f by a linear function, interpolating at the points a and $a+h$. Then

$$\int_{a}^{a+h} f(x)dx \cong \frac{h}{2}\left\{f(a)+f(a+h)\right\}. \tag{6.11}$$

The name *trapezoidal rule* arises because here we approximate the area under a curve by a trapezoid (Figure 6.3).

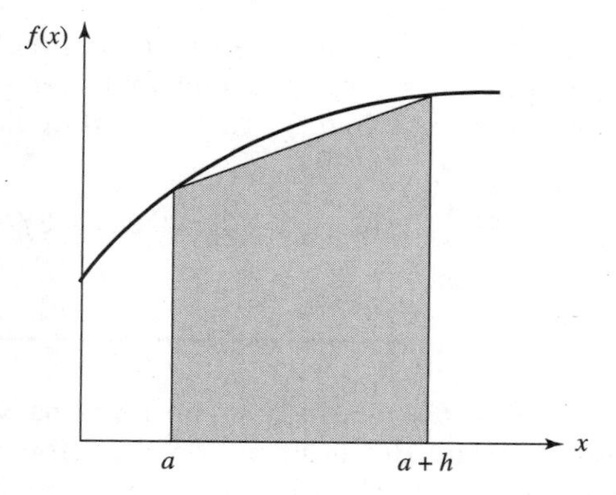

Figure 6.3 The trapezoidal method.

The Simpson Method

We approximate f by a quadratic interpolation at the points a, $a+h$, and $a+2h$. Then

$$\int_a^{a+2h} f(x)dx \cong \frac{h}{3}\{f(a)+4f(a+h)+f(a+2h)\}. \qquad (6.12)$$

The Three-Eights Rule

Using four uniformly spaced points a, $a+h$, $a+2h$, and $a+3h$, and approximating f by a cubic interpolation on these points gives

$$\int_a^{a+3h} f(t)dt \cong \frac{3h}{8}\{f(a)+3f(a+h)+3f(a+2h)+f(a+3h)\}. \qquad (6.13)$$

All of these formulas are derived by interpolating on a set of uniformly spaced mesh points. Such methods are said to be of the *Newton-Cotes* type. The trapezoidal, Simpson, and three-eights rules are called *closed* because both endpoints are used. The midpoint rule, which does not involve function values at the endpoints of the interval, is an *open* rule, while the rectangular rule is sometimes called *semi-closed*. Any number of closed and open Newton-Cotes quadrature formulas can be derived by simply taking different interpolation points.

Example 6.2 The open Newton-Cotes formula

$$\int_a^{a+3h} f(x)dx \cong \frac{3}{2}\{f(a+h)+f(a+2h)\}$$

is derived by using $f(a+h)$ and $f(a+2h)$ as interpolation points, then integrating the resulting first degree polynomial over $[a,\ a+3h]$. By an analogous process we can find higher order open formulas, such as

$$\int_a^{a+4h} f(x)dx \cong \frac{4}{3}\left\{2f(a+h) - f(a+2h) + 2f(a+3h)\right\}.$$

■

Newton-Cotes rules have relatively simple patterns of points and weights and are therefore popular. But it should be clear from the discussion that we can work out any number of other quadrature formulas by simply using different, not necessarily equally spaced, interpolation points. If we take the points $\{x_0,\ x_1,\ \ldots,\ x_n\}$, the simple interpolation polynomial is given as

$$P_n(x) = \sum_{i=0}^{n} l_i(x) f(x_i),$$

where the l_i are the Lagrange polynomials defined in (4.11). It follows immediately that the quadrature weights are given by

$$w_i = \int_a^b l_i(x)dx. \tag{6.14}$$

Since this involves only an integration of polynomials, the weights can be evaluated explicitly.

While (6.14) always gives the quadrature weights, using it may involve some tedious manipulations. Sometimes it is easier to use the method of undetermined coefficients. Since every quadrature formula is of the form (6.8), once we have chosen the quadrature points the only remaining issue is to compute the weights w_i. Since there are $n+1$ weights, we can select them so that the rule is exact for all polynomials up to degree n. Imposing these conditions, we are led to the system

$$\begin{aligned}
w_0 + w_1 + \ldots + w_n &= b - a,\\
w_0x_0 + w_1x_1 + \ldots + w_nx_n &= \frac{b^2 - a^2}{2},\\
&\vdots\\
w_0x_0^n + w_1x_1^n + \ldots + w_nx_n^n &= \frac{b^{n+1} - a^{n+1}}{n+1},
\end{aligned} \tag{6.15}$$

where the first equation comes from the requirement that $f(x) = 1$ is to be integrated exactly, the second equation comes from integrating $f(x) = x$, and so on. Solving the system gives the quadrature weights.

Example 6.3 Derive the weights for the five-point closed Newton-Cotes formula. It should be clear that the weights are independent of a and proportional to h, so for simplicity we choose $a = 0$ and $x_0 = 0,\ x_1 = 1,\ x_2 = 2,\ x_3 = 3,\ x_4 = 4$. Substituting into (6.15) we get the system

$$\begin{aligned} w_0 + w_1 + w_2 + w_3 + w_4 &= 4, \\ w_1 + 2w_2 + 3w_3 + 4w_4 &= 8, \\ w_1 + 4w_2 + 9w_3 + 16w_4 &= 64/3, \\ w_1 + 8w_2 + 27w_3 + 64w_4 &= 64, \\ w_1 + 16w_2 + 81w_3 + 256w_4 &= 1024/5. \end{aligned}$$

Solving this and scaling to the given interval, we arrive at the quadrature formula

$$\begin{aligned} \int_a^{a+4h} f(x)dx \cong \frac{2h}{45} \{7f(a) + 32f(a+h) + 12f(a+2h) \\ + 32f(a+3h) + 7f(a+4h)\}. \end{aligned} \tag{6.16}$$

Example 6.4 Find the weights for the numerical quadrature

$$\int_{-1}^{1} f(x)dx \cong w_0 f(x_0) + w_1 f(x_1) + w_2 f(x_2) + w_3 f(x_3), \tag{6.17}$$

where the x_i are the roots of the Chebyshev polynomial[2] T_4.

The roots of the Chebyshev polynomial T_4 are $x_0 = -0.9238795,\ x_1 = -0.3826834$, and $x_3 = -x_0,\ x_2 = -x_1$. When we use this in (6.15), we get

$$\begin{aligned} w_0 = w_3 &= 0.2642977, \\ w_1 = w_2 &= 0.7357023. \end{aligned}$$

To compare this with the four-point closed Newton-Cotes rule; that is, the three-eights rule, we tested it on several simple functions. The results are in Table 6.2. Since the results for the Chebyshev roots are always better, we might suspect that there is some advantage in unequally spaced quadrature points. This issue will be explored more in Section 6.5.

[2] A definition of the Chebyshev polynomials and an expression for their roots is in Section 4.4.

Table 6.2 Maximum observed errors for Example 6.4.

$f(x)$	Newton-Cotes	Chebyshev
$\cos(x)$	4.6×10^{-3}	6.3×10^{-4}
e^x	5.2×10^{-3}	7.6×10^{-4}
$1/(2+x)$	6.1×10^{-3}	1.0×10^{-3}

Example 6.5 It is also possible to design quadrature rules in which the quadrature points are outside the interval of integration. For example, using the method of undetermined coefficients, we find the rule

$$\int_a^{a+h} f(x)dx \cong \frac{h}{12}\{-f(a-h)+8f(a)+5f(a+h)\}.$$

We leave it as an exercise to show that this rule is exact for all polynomials of degree two. Formulas like these are rarely used in numerical integration, but the idea has some application in the numerical solution of differential equations.

■

When the interval $[a,\ b]$ is not small, a single application of a quadrature rule may not give adequate accuracy. In this case, we split the interval $[a,\ b]$ into several smaller subintervals, called *panels*, and then apply the quadrature rule in *composite form*; that is, separately over each panel. The result is the same as if we had approximated the function f by a piecewise polynomial and then integrated this approximation. The weights of a composite formula depend not only on the underlying quadrature rule but also on the subdivisions used. For a uniform partition, the weight patterns are often quite simple. If the panel endpoints are $a = x_0 < x_1 < \ldots < x_n = b$ and $h = x_{i+1} - x_i$, then the composite trapezoidal rule is

$$\int_a^b f(x)dx \cong \frac{h}{2}\{f(x_0)+2f(x_1)+\ \ldots\ +2f(x_{n-1})+f(x_n)\}. \tag{6.18}$$

For the composite Simpson rule, we apply (6.12) over the panels $[x_{2i},\ x_{2i+2}]$, so that n must be even. The resulting formula is

$$\begin{aligned}\int_a^b f(x)dx =& \frac{h}{3}\{f(x_0)+4f(x_1)+2f(x_2)+4f(x_3)+\ \ldots \\ &+2f(x_{n-2})+4f(x_{n-1})+f(x_n)\}.\end{aligned} \tag{6.19}$$

The efficiency of a quadrature rule is largely governed by how many times we have to evaluate the integrand to achieve a desired accuracy. To construct efficient composite quadrature rules it is often best to use closed formulas. With closed formulas, adjacent subintervals share points so that programs can be written that take advantage of this, computing the function values at the common points only once. Even though a single interval open rule may be more accurate than a closed rule with the same number of points, the advantage of sharing values is lost and composite open methods may end up being more expensive.

EXERCISES

1. Find the weights for the six-point closed Newton-Cotes method.
2. Find the weights for the three-point semi-open Newton-Cotes method

$$\int_a^{a+3h} f(x)dx \cong w_0 f(a) + w_1 f(a+h) + w_3 f(a+2h).$$

3. Derive the formulas in Example 6.2.
4. Show how the weights and points in Example 6.4 can be changed to get a quadrature rule over an arbitrary interval $[a,\ b]$.
5. Describe the pattern of weights for the composite three-eights rule on a set of uniform panels.
6. Derive the quadrature rule in Example 6.5 and show that it is exact for all polynomials of degree two.
7. Derive the weights of the quadrature over $[-1,\ 1]$ in which the quadrature points are the roots of the Chebyshev polynomial T_5.
8. Why do we not use spline interpolation to develop quadrature rules?
9. It is possible to establish quadrature rules that are not of the form (6.8), for example, rules that also use the derivatives of a function. Find the weights for the quadrature

$$\int_{-1}^{1} f(x)dx \cong w_1 f(-1) + w_2 f(0) + w_3 f(1) + w_4 f'(-1) + w_5 f'(1),$$

chosen so that the rule is exact for all polynomials of degree four.

10. Use the quadrature in Exercise 2 in composite form to approximate the centroid of the region $(\bar{x},\ \bar{y})$, bounded by $y = 2^x$ and $y = x^2$ for $0 \le x \le 2$, where

$$\bar{x} = \frac{1}{A}\int_0^2 x(2^x - x^2)dx$$

and

$$\bar{y} = \frac{1}{A} \int_0^2 \frac{1}{2}(2^{2x} - x^4)dx$$

with A the area of the region. Choose a uniform panel size small enough to obtain an accuracy of 10^{-4}.

6.3 Error Analysis for Interpolatory Quadratures

A primary concern in the design and use of numerical methods is accuracy. In some applications only two or three significant digits are needed, but other cases call for much more accuracy. This is why convergence is an essential property of any numerical method; it allows us to get arbitrarily good answers by simply increasing the amount of work we do. To judge the efficiency and applicability of a method, we rely on convergence theorems, but when we apply the algorithms we are concerned with the accuracy of the specific results that are delivered.

There are two distinct ways to deal with these concerns. The first is to use rigorous mathematics to get error bounds and orders of convergence. When such results can be applied effectively we are in the safest possible position, but unfortunately this is a rare situation. Often error bounds cannot be found or, if found, are very hard to apply. Even for the simple case of numerical integration, for example, error bounds involve higher derivatives of the integrand that may not be easy to get. Finally, even if these obstacles can be overcome, the bounds may be very pessimistic and so of limited use. The other alternative is to use less precise, more heuristic and ad hoc ways of estimating errors. While such methods are not entirely without justification, they usually involve some reasonable, but unverifiable assumptions. But they tend to be more flexible and effective, so they are often used in practice.

In this section, we discuss rigorous error bounds for numerical integration. In the next section, we will see how less precise, but practically more useful ways can be devised.

An analysis of the error in interpolatory quadrature is a fairly straightforward matter. The bounds from polynomial approximations can be substituted in the quadrature expressions to get bounds for the quadrature errors. Depending on the level of detail, some of this can be fairly technical, but most of it involves little more than calculus. It is quite easy to establish some general rough results.

Theorem 6.1 Let I_n be an approximation of the form (6.8) to

$$I = \int_a^b f(x)dx,$$

generated by a composite quadrature. Assume that the interval $[a,\ b]$ is divided into a set of panels of uniform width h, and that in each panel the quadrature is based on simple interpolation with a polynomial of degree k. Then, provided f is sufficiently smooth,

$$|I - I_n| = O(h^{k+1}), \tag{6.20}$$

so the order of convergence of I_n to I is at least $k+1$.

Proof: Let $\hat{f}$ denote the piecewise polynomial approximation that results in the composite quadrature rule. Then

$$\begin{aligned} |I - I_n| &= \left| \int_a^b \left\{ f(x) - \hat{f}(x) \right\} dx \right| \\ &\le |b-a| \max_{a \le x \le b} |f(x) - \hat{f}(x)|. \end{aligned}$$

From (4.15) we know that

$$\max_{a \le x \le b} |f(x) - \hat{f}(x)| = O(h^{k+1})$$

and the desired result follows. ■

The conclusion we can draw from this is that the rectangular and midpoint rules have an order of convergence at least one, the trapezoidal method an order two, the Simpson method an order three, and the three-eights rule an order four. But there are two reasons why this is not completely satisfactory. The first is that it just tells us the order of convergence, but gives no explicit error bounds. The second is that it sometimes predicts too low an order of convergence. Because of the inequalities involved we cannot rule out that the order of convergence may in fact be higher than that guaranteed by the theorem. With a little more work, we can get better results.

Example 6.6 The composite trapezoidal method is based on integrating a piecewise linear approximation $\hat{f}$ on a uniform panel of size h. From the expression for the error in this approximation, we know that for $x_i \le x \le x_{i+1}$,

$$f(x) - \hat{f}(x) = \frac{(x_i - x)(x_{i+1} - x)}{2} f''(\xi_i),$$

where $x_i < \xi_i < x_{i+1}$, from which it follows that

$$\int_a^b \left\{ f(x) - \hat{f}(x) \right\} dx = \sum_{i=0}^{n-1} \int_{x_i}^{x_{i+1}} \frac{(x_i - x)(x_{i+1} - x)}{2} f''(\xi_i) dx.$$

Then

$$\left| \int_a^b \left\{ f(x) - \hat{f}(x) \right\} dx \right| \leq \max_{a \leq t \leq b} |f''(t)| \sum_{i=0}^{n-1} \int_{x_i}^{x_{i+1}} \frac{|(x_i - x)(x_{i+1} - x)|}{2} dx.$$

But

$$\int_{x_i}^{x_{i+1}} |(x_i - x)(x_{i+1} - x)| dx = \int_{x_i}^{x_{i+1}} (x - x_i)(x_{i+1} - x) dx$$

$$= \frac{h^3}{6}$$

so that

$$\begin{aligned} \left| \int_a^b \left\{ f(x) - \hat{f}(x) \right\} dx \right| &\leq \frac{nh^3}{12} \max_{a \leq x \leq b} |f''(x)| \\ &\leq \frac{|b-a|h^2}{12} \max_{a \leq x \leq b} |f''(x)|. \end{aligned} \tag{6.21}$$

This bound is realistic in the sense that there are functions for which the error is exactly $|b-a|\, h^2 \max_{a \leq x \leq b} |f''(x)| \,/\, 12$.

■

For some of the other rules, such as the midpoint method and Simpson's rule, Theorem 6.1 underestimates the order of convergence.

Example 6.7 The composite midpoint rule on uniform panels is

$$\int_a^b f(x) dx = h \sum_{i=0}^{n-1} f(x_i + h/2).$$

If we use Taylor's theorem to expand $f(x)$ about $x_i + h/2$, we get

$$\int_{x_i}^{x_{i+1}} f(x) dx = \int_{x_i}^{x_{i+1}} f(x_i + h/2) dx + \int_{x_i}^{x_{i+1}} (x - x_i - h/2) f'(x_i + h/2) dx +$$

$$\int_{x_i}^{x_{i+1}} \frac{(x - x_i - h/2)^2}{2} f''(\xi_i) dx.$$

Because of the symmetry of the integrand, the second term on the right cancels, leaving us with

$$\int_{x_i}^{x_{i+1}} f(x)dx - hf(x_i + h/2) = O(h^3).$$

This is a bound on the error in one panel. Since there are $O(1/h)$ subdivisions, we see that the error in the composite midpoint method is of order two. This is better than what we expect from Theorem 6.1.

By a similar but more complicated argument, one can show that the composite Simpson method on a uniform subdivision has order of convergence four and that an error bound is

$$|I - I_n| \le \frac{|b-a|h^4}{180} \max_{a \le x \le b} |f^{(4)}(x)|. \tag{6.22}$$

It is known that all closed Newton-Cotes methods with an odd number of quadrature points have an order of convergence one higher than that of the underlying approximation. For closed Newton-Cotes rules with an even number of points, including the three-eights rule, the order of convergence from Theorem 6.1 is correct. An explicit error bound for the three-eights method is

$$|I - I_n| \le \frac{|b-a|h^4}{80} \max_{a \le x \le b} |f^{(4)}(x)|. \tag{6.23}$$

Comparing (6.22) and (6.23), we conclude that Simpson's method tends to be more accurate than the three-eights rule, even though the latter is derived from a more accurate approximation.

The order of convergence of a composite quadrature can often be most easily obtained from the following result.

Theorem 6.2 Consider an $n+1$-point interpolatory quadrature that is exact for all polynomials of degree $m > n$. When this rule is used on a finite interval $[a, b]$, in a composite form with uniform panels of width h, the order of convergence is at least $m+1$.

Proof: Suppose that the quadrature, when applied to one panel $[z, z+h]$, has quadrature points $x_0, x_1, \ldots, x_n$ and corresponding weights $w_0, w_1, \ldots, w_n$. Let p_m be an mth degree polynomial that interpolates f at the points $x_0, x_1, \ldots, x_n$ as well as at $m-n$ additional points in $[z, z+h]$. Then

$$\int_z^{z+h} f(x)dx = \int_z^{z+h} p_m(x)dx + \int_z^{z+h} e_m(x)dx,$$

where $e_m(x)$ is the error in approximating $f(x)$ by the interpolating polynomial. From (4.17)

$$||e_m||_\infty = O(h^{m+1}).$$

Since the rule is exact for polynomials of degree m,

$$\int_z^{z+h} p_m(x)dx = \sum_{i=0}^{n} w_i p_m(x_i).$$

Also, from the interpolating conditions,

$$f(x_i) = p_m(x_i), \quad i = 0,\ 1,\ \ldots,\ n,$$

so that

$$\int_z^{z+h} f(x)dx = \sum_{i=0}^{n} w_i f(x_i) + O(h^{m+2}).$$

Since there are $O(1/h)$ panels in $[a,\ b]$, it follows by summing over all the panels that the error in the composite quadrature is $O(h^{m+1})$. ■

Example 6.8 It is an easy matter to show that Simpson's rule is exact for all polynomials of degree three. Theorem 6.2 then guarantees the fourth order convergence of the composite Simpson method. ■

EXERCISES

1. Show that the rectangular rule has actual order of convergence one.
2. In the approximate computation of

$$I = \int_0^1 \frac{\sin(x)}{1+x}dx$$

by the composite trapezoidal rule, how small must h be in order to guarantee an accuracy of 10^{-5}?

3. Find a function f for which the error bound (6.21) is actually attained.

4. What order of convergence would you expect when the quadrature rule in Example 6.3 is applied in composite form?

5. Show that a realistic error bound for the composite midpoint method is

$$|I - I_n| \leq \frac{(b-a)h^2}{24} \max_{a \leq x \leq b} |f''(x)|.$$

6. Can we conclude from (6.22) and (6.23) that Simpson's rule is always better than the three-eights rule?

7. Show that the quadrature rule in Example 6.5 is exact for all polynomials of degree two, but is not exact for polynomials of degree three. If this rule were to be used in a composite form on a uniform panel, what would you expect its order of convergence to be?

8. Use Theorem 6.2 to justify the statement that all composite closed Newton-Cotes rules with an odd number of points have an order of convergence one greater than the accuracy of the underlying approximation of f.

9. Show that a conclusion similar to Exercise 8 holds for open Newton-Cotes rules, but not for semi-closed ones.

10. Suppose we want to find the value of

$$\int_0^2 \frac{e^x}{1+x^2} dx$$

to an accuracy of 10^{-8}, using a composite quadrature on a set of uniform panels of width h. What is the largest value of h that will guarantee this accuracy, if

(a) we use a composite trapezoidal method, and

(b) we use a composite Simpson's method?

6.4 Adaptive Quadrature

General purpose software that is used in many different circumstances and that is expected to receive wide distribution should be as automatic as possible. It has to deliver results efficiently, accurately, and require little user invention. For quadrature software this means that the user should only have to specify the integrand, the endpoints, and the desired accuracy, with the software doing the rest. One major decision the software then has to make is how to choose the quadrature points to achieve this efficiently. While the various error bounds are useful in establishing convergence rates and give us some idea of the applicability of the method, they are not quite as suitable for establishing the error in a specific computation. Bounds on the higher derivatives are hard to get and even if they are available, the results are likely to be too pessimistic. The situation is further complicated because integrands may behave differently in the region of integration, so that a good strategy must depend on the integrand. For numerical software that can handle a variety of integrands without much human intervention, we utilize a technique called *adaptive* quadrature.

Adaptive quadrature is based on a simple and widely used idea for dealing with the discretization error in a numerical computation: perform the computation twice, once with a rough discretization, then with a finer discretization and use the difference between the two results to estimate the error. Although this process is sometimes hard to justify, it is frequently effective and is without doubt the most widely used heuristic for accuracy estimation. We illustrate the process with Simpson's rule. Suppose we want to evaluate the integral I over some panel $[x_0,\ x_2]$. An application of Simpson's rule gives the approximation

$$I_1 = \frac{h}{3}\left\{f(x_0) + 4f(x_1) + f(x_2)\right\}. \tag{6.24}$$

Next, we introduce the additional points $s_1 = (x_0 + x_1)/2$ and $s_2 = (x_1 + x_2)/2$ and use a composite Simpson's rule to get a second approximation

$$I_2 = \frac{h}{6}\left\{f(x_0) + 4f(s_1) + 2f(x_1) + 4f(s_2) + f(x_2)\right\}. \tag{6.25}$$

We do not know the errors in I_1 and I_2, but what we know about Simpson's method gives us some idea how they are related. If we write

$$I = I_1 + \eta, \tag{6.26}$$

we expect, from the fact that Simpson's rule converges with order four, that the error should drop by about a factor of sixteen when the interval size is halved. Therefore,

$$I \cong I_2 + \frac{\eta}{16}. \tag{6.27}$$

From (6.26) and (6.27) we can then estimate the magnitude of error in I_1 by

$$|\eta| \cong \frac{16}{15}|I_2 - I_1|. \tag{6.28}$$

To compute the integral over $[x_0,\ x_2]$ we apply (6.24) and (6.25) and use (6.28) to estimate the error. If this estimate is smaller than the desired accuracy, we are done. If not, we subdivide the interval into two panels, and repeat the process for each panel. If the estimated error in some panel is deemed small enough, we accept its contribution to I and proceed to the next panel. If the estimated error is too large, we again subdivide the panel into two parts and add it to a list of panels to be processed.

The algorithm continues until the contributions of all panels have been successfully evaluated. The overall effect is that the algorithm adapts itself to the characteristics of the integrand. In regions where the integrand is well-behaved, a large panel size will be found adequate to get the required accuracy, while in regions where the integrand shows rapid changes a very fine discretization will be chosen. This gives a method that is efficient but at the same time can handle integrands with very diverse behaviors.

Adaptive integration algorithms need a parameter, say ε, that defines the accuracy we wish to obtain. The accuracy required for each panel will depend on this as well as the panel width, in such a way that the combined errors do not exceed the specified tolerance. For an integration over $[a, b]$ we might, for instance, ask that the error in a panel is no larger than

$$\varepsilon \times \frac{\text{width of panel}}{b-a}, \tag{6.29}$$

but other choices are possible. This strategy works reasonably well if ε specifies an absolute error tolerance. To compute an integral with a pre-assigned relative error is much more difficult. The adaptive process estimates the absolute error, so it needs a rough estimate of the absolute value of the integral to translate the relative error to an absolute one. Since the estimate of the value of the integral can only come as the computation proceeds, it sometimes is a very poor estimate. If an integrand has regions where it is small and computations start there, it may use many points to try to get an accuracy that is too high for the integral as a whole. The result can be a very inefficient computation. On the whole though, adaptive quadrature works well and gives the required accuracy with no intervention by the user.

Example 6.9 As a test of performance, we computed

$$I(\alpha) = \int_{-1}^{1} e^{-\alpha x^2}\, dx$$

to an accuracy of 10^{-6}, for $\alpha = 0.01, 0.1, 1, 10,$ and 100, using the adaptive Simpson's routine described in this section. The results, with their actual approximate errors and number of function evaluations, are shown in Table 6.3.

There are two things to notice that are typical for adaptive quadratures. The observed errors are considerably smaller than the specified tolerance. This reflects the conservative implementation described shortly. The second is the fact that for slowly varying integrands (small α), the method uses only a few function evaluations. As α increases and the integrand becomes more peaked, the adaptive method places more quadrature points in the region

Table 6.3 Results from the adaptive Simpson method.

α	Computed $I(\alpha)$	Observed error	Number of function evaluations
0.01	1.99335411	8.2×10^{-7}	5
0.1	1.93528690	2.8×10^{-7}	17
1	1.49364836	9.5×10^{-8}	49
10	0.56049496	1.7×10^{-7}	105
100	0.17724546	7.1×10^{-8}	137

where the integrand is not well-behaved and consequently the number of function evaluations rises.

In spite of the success of such typical examples, it must be understood that adaptive methods do not guarantee that the actual error is within the specified tolerance. The reason for this is that expressions like (6.27) are not exact, but only hold asymptotically as $h \to 0$. For finite h the improvement from halving the panel size may be much less than expected, leading to an underestimation of the error, and consequently an incorrect final answer. Even worse, one can construct counterexamples for which the adaptive technique fails altogether.

Example 6.10 Suppose we want to find an approximation to

$$I = \int_0^1 e^{-2x} \sin(100\pi x) dx$$

to an accuracy of 10^{-6}, using an adaptive trapezoidal method.

If we start with a panel width of 0.1, all values of the integrand will be zero and the estimated value will be $I = 0$. If we then halve the panel size and repeat the computation, we still get $I = 0$. This might lead us to accept this as the correct result in spite of its incorrectness.

Admittedly, this example is artificial and it is very naive to apply a method designed for well-behaved cases to a difficult situation, but it does illustrate the theoretical possibility that adaptive quadrature can fail completely.

Less spectacularly, there are practical instances where an adaptive method will underestimate the error. To reduce the chances for this we often take a conservative approach, using (6.28) as the estimated error, but the normally more accurate I_2 as the value of the integral. Actually, the estimated error can be used to improve the result by computing

$$I_3 = I_2 + \frac{\eta}{16}. \tag{6.30}$$

For sufficiently small h we expect that I_3 will be a better approximation than I_2. The process of using the estimated error to improve the result is known as *Richardson's extrapolation.* Extrapolation techniques are attractive in many circumstances. The most conservative strategy is to use the extrapolated value (6.30) as the answer, but the estimated error as an indication of the error in the answer. This normally overestimates the error, but this overestimation tends to compensate for occasional failures in estimating the error in the first place.

The basic ideas of adaptive integration are not hard to understand, but actual implementations involve subtle issues, such as how to deal with rounding or the effect of the presence of singularities in the integrand. For a deeper discussion of this and a number of nicely implemented adaptive quadratures see Piessens et al [20].

EXERCISES

1. Describe how to estimate the error in an adaptive method based on the trapezoidal rule.

2. Can you foresee possible difficulties in evaluating

$$I = \int_0^{2\pi} \frac{\sin(20x)}{1+x^2} dx$$

with an adaptive method?

3. Why is (6.29) a reasonable way of setting the error level for a panel?

4. How do you think rounding affects the numerical quadrature? What are the implications for adaptive methods?

5. Why should you expect the evaluation of

$$I = \int_0^{2\pi} \cos(1000x + \sqrt{x}) dx$$

to a relative accuracy of 10^{-15} to be a lengthy and difficult problem?

6.5 Gaussian Quadrature*

It is clear from the preceding discussion that we can develop many different numerical integration methods and achieve high orders of convergence by simply increasing the degree of the approximating polynomial. In practice though, relatively low order composite methods are often preferred. Higher order quadratures not only have more complicated weight patterns, but may also show increasing instability as the degree of the approximating polynomial gets larger. But there is one kind of quadrature where rapid convergence can be achieved while maintaining stability. The derivation of the formulas is not obviously based on interpolation, so we need a different approach.

In (6.15) we showed how, once the quadrature points are fixed, the $n+1$ weights can be determined by making the rule exact for polynomials of degree n. Example 6.4 suggested that there may be some advantage in using unequally spaced quadrature points, so let us suppose that we are completely free to choose these points. In that case, we have an extra $n+1$ parameters which we could use to try to make the rule exact for polynomials up to degree $2n+1$. In principle, all we have to do is to extend (6.15) by treating $x_0, x_1, \ldots, x_n$ as unknowns, writing down the $n+1$ extra equations that assure exact integration for $x^{n+1}, x^{n+2}, \ldots, x^{2n+1}$, and solving the resulting system. Unfortunately, now the system is nonlinear and quite difficult to solve. But the idea is still useful, we just have to look at the problem from a different angle. The way we can solve it shows a surprising connection with orthogonal polynomials. For the interval $[-1, 1]$ it turns out that the quadrature points we need are the roots of Legendre polynomials.

Theorem 6.3 Let $x_0, x_1, \ldots, x_n$ be the roots of the Legendre polynomial L_{n+1}. Then there exists a set of quadrature weights so that the integration rule with $x_0, x_1, \ldots, x_n$ as quadrature points is exact for all polynomials of degree $2n+1$.

Proof: We want to show that there exist weights such that

$$\int_{-1}^{1} p_{2n+1}(x)dx = \sum_{i=0}^{n} w_i p_{2n+1}(x_i)$$

for all polynomials p_{2n+1} of degree $2n+1$. Any polynomial p_{2n+1} can be written as

$$p_{2n+1}(x) = q_n(x) + r_{2n+1}(x),$$

where r_{2n+1} is a polynomial of degree $2n+1$ and q_n is the nth degree interpolation polynomial to p_{2n+1} at the points $x_0, x_1, \ldots, x_n$. Because of the interpolation

$$p_{2n+1}(x_i) = q_n(x_i),$$

so that

$$r_{2n+1}(x_i) = 0, \quad \text{for } i = 0,\ 1,\ \ldots,\ n.$$

The theory of polynomials tells us that every polynomial has a unique factorization, so r_{2n+1} must have the form

$$r_{2n+1}(x) = (x - x_0)(x - x_1)\ \ldots\ (x - x_n)u_n(x),$$

where u_n is a polynomial of degree n. Then

$$\int_{-1}^{1} p_{2n+1}(x)dx = \int_{-1}^{1} q_n(x)dx + \int_{-1}^{1} (x - x_0)(x - x_1)\ \ldots\ (x - x_n)u_n(x)dx. \tag{6.31}$$

For the next step, we draw on Exercise 1 at the end of this section, that states that

$$\int_{-1}^{1} L_{n+1}(x)x^i dx = 0,$$

for all $i = 0,\ 1,\ \ldots,\ n$. Since $(x-x_0)(x-x_1)\ \ldots\ (x-x_n)$ is just a multiple of L_{n+1}, the second term on the right of (6.31) vanishes and

$$\int_{-1}^{1} p_{2n+1}(x)dx = \int_{-1}^{1} q_n(x)dx.$$

If we select the quadrature weights so that with points $x_0,\ x_1,\ \ldots,\ x_n$ every polynomial of degree n is integrated exactly, then

$$\begin{aligned}
\int_{-1}^{1} p_{2n+1}(x)dx &= \int_{-1}^{1} q_n(x)dx \\
&= \sum_{i=0}^{n} w_i q_n(x_i) \\
&= \sum_{i=0}^{n} w_i p_{2n+1}(x_i)
\end{aligned}$$

which is the desired result. ■

To establish quadratures of this type, we first must find the roots of the Legendre polynomials. For n greater than three this has to be done numerically and we will show in the next chapter how to do it. Once the quadrature points have been found, we can use either (6.14) or the method of undetermined coefficients to get the weights.

Example 6.11 Find the points and weights for a three-point quadrature on $[-1,\ 1]$ that is exact for all polynomials up to degree five.

We know that

$$L_3(x) = \frac{1}{2}(5x^3 - 3x),$$

so its roots are

$$x_0 = -\sqrt{\frac{3}{5}}, \quad x_1 = 0, \quad x_2 = \sqrt{\frac{3}{5}}.$$

When we use these values in (6.15), we obtain

$$w_0 = \frac{5}{9}, \quad w_1 = \frac{8}{9}, \quad w_2 = \frac{5}{9}.$$

It is easy to verify that the rule integrates all polynomials of degree five exactly.

■

Numerical integration rules in which the quadrature points are selected for extra accuracy are called *Gaussian quadratures.* Because of the association with the Legendre polynomials, the methods addressed by Theorem 6.3 are referred to as *Gauss-Legendre* rules. Quadrature weights and points for different values of n have been tabulated extensively, for example, in Stroud and Secrest [24]. In Table 6.4 we give the values (to the accuracy of 10^{-15}) for $n = 3,\ 4,\ 5,\ 6$.

There are other kinds of Gaussian quadrature rules; in fact, for every set of orthogonal polynomials we can develop a corresponding Gaussian rule.

Theorem 6.4 Let $w(x) > 0$ and let $\{q_0,\ q_1,\ \ldots\}$ be a set of polynomials orthogonal over the interval $[a,\ b]$ with respect to the function w. Suppose that each orthogonal polynomial q_i of degree i has i real zeros in the interval $[a,\ b]$.[3] Then the quadrature rule

$$\int_a^b w(x)f(x)dx \cong \sum_{i=0}^{n} w_i f(x_i), \tag{6.32}$$

[3]It is known that this assumption holds in all cases.

Table 6.4 Quadrature points and weights for Gauss-Legendre rule.

n	Quadrature points ($\pm x_i$)	Weights (w_i)
3	0.000000000000000	0.888888888888889
	0.774596669241483	0.555555555555556
4	0.339981043584856	0.652145154862546
	0.861136311594053	0.347854845137454
5	0.000000000000000	0.568888888888889
	0.538469310105683	0.478628670499366
	0.906179845938664	0.236926885056189
6	0.238619186083197	0.467913934572691
	0.661209386466265	0.360761573048139
	0.932469514203152	0.171324492379170

where the x_i are the $n+1$ zeros of q_{n+1} and w_i the corresponding weights, is exact for all polynomials of degree $2n+1$.

Proof: The argument is essentially the same as that for Theorem 6.3. ■

This gives us a variety of Gaussian rules for the evaluation of weighted integrals. Some of the more familiar ones involve the orthogonal polynomials introduced in Section 4.3.

Example 6.12 Integrals of the type

$$I = \int_{-1}^{1} \frac{1}{\sqrt{1-x^2}} f(x)dx$$

can be approximated by a Gaussian quadrature of the form

$$I_n = \sum_{i=0}^{n} w_i f(x_i)$$

where the quadrature points are the roots of the Chebyshev polynomials of the first kind. These are the *Gauss-Chebyshev* rules. The quadrature points that are easily found as the roots of the Chebyshev polynomials are explicitly given by (4.24). Rules of this type are a possible way to compute the integrals in Example 4.9. ■

Example 6.13 The integral

$$I = \int_0^\infty e^{-t} f(t) dt$$

can sometimes be evaluated efficiently by a Gaussian rule. For the interval $(0, \infty)$ and $w(t) = e^{-t}$, the orthogonal polynomials are the Laguerre polynomials. The resulting rules give the *Gauss-Laguerre* quadratures. Laguerre points and weights (to the accuracy of at least 10^{-12}) for $n = 2, 3, \ldots, 6, 7, 8$ are given in Table 6.5.

In the Gaussian rules above, all quadrature points are determined for maximum accuracy. In some instances, we need one or two points to be fixed, often at the end of the interval. In *Lobatto* quadrature both ends of the interval are used as quadrature points and the weights and $n-1$ interior points are selected to make the rule exact for polynomials of degree $2n-1$. The location of the interior quadrature points can be derived by an analysis similar to that of Theorem 6.3. When we do this for the interval $[-1, 1]$, we find that we want the integral of

$$r_{2n+1} = (x+1)(x-x_1)\ldots(x-x_{n-1})(x-1)u_{n-2}(x)$$

to vanish for all polynomials u_{n-2} of degree $n-2$. This will happen if the points $x_1, x_2, \ldots, x_{n-1}$ are the roots of a polynomial $P_{n-1}^{1,1}$ that is a member of the sequence orthogonal with respect to the weight $w(x) = (1+x)(1-x)$. These polynomials are a special case of the *Jacobi* polynomials $P_n^{\alpha,\beta}$ which are defined by the orthogonality condition

$$\int_{-1}^{1} (1+x)^\alpha (1-x)^\beta P_i^{\alpha,\beta}(x) P_j^{\alpha,\beta}(x) dx = 0, \tag{6.33}$$

for all $i \neq j$. Other cases of Jacobi polynomials that we have already encountered are the Legendre polynomials with $\alpha = \beta = 0$, the Chebyshev polynomials of the first kind with $\alpha = \beta = -1/2$, and the Chebyshev polynomials of the second kind with $\alpha = \beta = 1/2$. A great deal is known about Jacobi polynomials. In particular, we know how to find their roots, so we can get the points and weights for Lobatto quadrature. It turns out that the three-point Lobatto rule is just Simpson's method, while for $n = 3$, we get

$$x_0 = -1, \quad x_1 = -0.447214, \quad x_2 = 0.447214, \quad x_3 = 1$$

and

$$w_0 = w_3 = 0.166667, \quad w_1 = w_2 = 0.833333.$$

Table 6.5 Quadrature points and weights for Gauss-Laguerre rule.

n	Quadrature points (x_i)	Weights (w_i)
2	0.585786437627	0.853553390593
	3.414213562373	0.146446609407
3	0.415774556783	0.711093009929
	2.294280360279	0.278517733569
	6.289945082937	0.010389256502
4	0.322547689619	0.603154104342
	1.745761101158	0.357418692438
	4.536620296921	0.038887908515
	9.395070912301	0.000539294706
5	0.263560319718	0.521755610583
	1.413403059107	0.398666811083
	3.596425771041	0.075942449682
	7.085810005859	0.003611758680
	12.640800844276	0.000023369972
6	0.222846604179	0.458964673950
	1.188932101673	0.417000830772
	2.992736326059	0.113373382074
	5.775143569105	0.010399197453
	9.837467418383	0.000261017203
	15.982873980602	0.000000898548
7	0.19304367656	0.409318951701
	1.026664895339	0.421831277862
	2.567876744951	0.147126348658
	4.900353084526	0.020633514469
	8.182153444563	0.001074010143
	12.734180291798	0.000015865464
	19.395727862263	0.000000031703
8	0.170279632305	0.369188589342
	0.903701776799	0.418786780814
	2.251086629866	0.175794986637
	4.266700170288	0.033343492261
	7.045905402393	0.002794536235
	10.758516010181	0.000090765088
	15.740678641278	0.000000848575
	22.863131736889	0.000000001048

Table 6.6
Error in Gaussian quadrature for Example 6.14.

N	Error	Ratio
2	1.58×10^{-8}	
4	2.40×10^{-10}	65.8
8	3.73×10^{-12}	64.5
16	5.81×10^{-14}	64.2

Because Lobatto rules involve both endpoints of the interval in the quadrature, they are convenient for constructing efficient composite quadratures. Other rules, such as the *Radau* quadratures that use only one endpoint, are less convenient for this, but still have some special uses.

When Gaussian rules are used as composite formulas they give high orders of convergence. This follows directly from Theorem 6.2. An n-point Gauss-Legendre rule is exact for polynomials of degree $2n-1$, so its composite form has an order of $2n$ convergence.

Example 6.14 The three-point Gaussian quadrature in Example 6.11 can be used in composite form by re-scaling the points and weights. A simple linear transformation gives the rule

$$\int_{x_i}^{x_i+h} f(x)dx \cong \frac{h}{18}\left\{5f(x_i + \frac{1-\sqrt{3/5}}{2}h) + 8f(x_i + \frac{h}{2}) + 5f(x_i + \frac{1+\sqrt{3/5}}{2}h)\right\}. \tag{6.34}$$

The composite rule is then the sum of these, added over all panels.

As a test case, the function

$$f(x) = e^{-x}\cos(x)$$

was integrated over $[0, 1]$ using N panels. The correct answer is $\dfrac{e^{-1}\{\sin(1) - \cos(1)\} + 1}{2}$ and was compared with the result of the three point Gauss-Legendre rule. In Table 6.6 we show the errors in the results for a composite form of (6.34) with several values of N. The table also shows the ratios by which the error is reduced when we double the number of panels (or equivalently, halve h). As the order of convergence is six, the predicted ratio is 64. The observed results are close to the expected values. Note also the high accuracy of Gaussian quadratures for simple integrands. ■

The theory of Gaussian quadrature is extensive and a lot has been written on this subject. There are also tabulations of the weights and points for the different types of Gaussian quadratures, a variety of these can be found in Stroud and Secrest [24]. While the quadrature points can always be calculated by finding the roots of orthogonal polynomials, and the method of undetermined coefficients can be used to get the weights, the computation is tedious and involves some error. This makes the tabulations quite useful. Still, for relatively small n in situations where we do not need machine accuracy, the method of Example 6.11 works sufficiently well.

The success of Gaussian methods involves two important properties that we have implicitly assumed, but not proved. The first, used in Theorems 6.3 and 6.4, is that the roots of the polynomials involved are all real, distinct and lie in the interval of integration. We know from the theory of polynomials that this condition is satisfied and that all the roots of orthogonal polynomials are real. The second property is that the weights are all positive, which makes the methods stable. This again follows from what we know about orthogonal polynomials. Isaacson and Keller [12] is a good source for proofs and more details.

EXERCISES

1. If L_{n+1} denotes the Legendre polynomial of degree $n+1$, show that

$$\int_{-1}^{1} L_{n+1}(x)x^i\,dx = 0, \quad \text{for all } i = 0,\ 1,\ \ldots,\ n$$

is true for all n.

2. Verify that the quadrature method in Example 6.11 is exact for all polynomials of degree five or less.

3. Find the points and weights for the two-point Gauss-Legendre rule.

4. Find the points and weights for the three-point Gauss-Chebyshev rule.

5. Why is the positivity of the quadrature weights a desirable feature?

6. Make the discussion of the connection between Lobatto quadrature and the Jacobi polynomials $P_{n-1}^{1,1}$ more precise by modifying the proof of Theorem 6.3.

7. Derive the formula in Example 6.14 from the formula in Example 6.11.

8. (a) Use a single 5-point Gauss–Legendre rule to estimate

$$I = \int_0^2 \frac{\sin(x^2)}{1+x}\,dx.$$

(b) Use the same rule in composite on two panels.

(c) Use the results of (a) and (b) to estimate the error and to improve the results.

9. Repeat Exercise 8 with integrand

$$\int_0^2 \frac{\sin(\sqrt{x})}{1+x} dx.$$

Compare results with those in Exercise 8.

10. Use the Lobatto rule with $n = 3$ to approximate the integral in Exercise 8. Using a two-panel composite rule. Compare the accuracy achieved and the work involved with that in Exercise 8.

6.6 Dealing with Difficult Integrands*

Approximate integration is a simple matter as long as the integrand is well behaved. Any of the methods we discussed will work and adaptive strategies can deliver results automatically and accurately with a small amount of work. But the standard quadrature rules are developed on the assumptions that the domain of integration is finite, that the integrand is smooth, and that it has certain differentiability. When a problem does not satisfy these conditions, adjustments have to be made that often complicate matters. Here are some of the situations that merit special attention.

- The integrand has spikes or other regions of abrupt change. An example is e^{-100x^2}.
- The integrand fails to be differentiable at some points, for example $f(x) = \sqrt{|x|}$ in $(-1, 1)$.
- The integrand is not bounded but has an integrable singularity, such as $f(x) = 1/\sqrt{x}$ in $(0, 1)$.
- The domain of integration is not bounded.
- The integrand is highly oscillatory; for example, $f(x) = \cos(100x)$.

Each of these cases makes it necessary to modify the usual quadratures to take care of the difficulty. Sometimes it is not at all easy to see what the right approach is, but the first thing to try is some analysis to see if the situation can be improved. Many times a transformation of variables recasts the problem into a form for which standard quadrature is suitable.

Example 6.15 The integral

$$I = \int_0^1 \sqrt{x}\cos(x)dx$$

was computed by an adaptive Simpson's method with a specified accuracy of 10^{-6}. To achieve this, it needed 101 function evaluations. The relative inefficiency of the adaptive method has an obvious source, the lack of differentiability of the integrand at $x = 0$. By making the substitution

$$x = u^2,$$

we get the equivalent integral

$$I = 2\int_0^1 u^2 \cos(u^2)du,$$

whose integrand is better behaved. To evaluate this integral by the same method to the required accuracy took only 29 function evaluations.

Such simple change of variables can even convert problems for which many quadratures fail quite tractable. For example, in

$$I = \int_0^1 \frac{1}{\sqrt{x}} f(x)dx,$$

the substitution $x = u^2$ gives

$$I = 2\int_0^1 f(u^2)du.$$

If f is a well-behaved function, the new integral presents no numerical difficulties.

■

Ideally, adaptive methods work best when the integrand is smooth, but as the above example shows, they may do reasonably well when not all conditions are met. One method of dealing with poorly-behaved integrands is just to ignore the difficulty and hope that the adaptive strategy will take care of it. We usually pay a price, though. The number of quadrature points in the critical region can be so large as that it reduces the efficiency of the computation significantly.

In addition to efficiency, there may be other problems with adaptive methods. If the region of abrupt behavior is very narrow, the adaptive strategy may miss it altogether and consequently give incorrect answers. Alternatively, unless very carefully programmed, an adaptive method may get "hung up" in a difficult region and never stop. Because of such potential problems, it is sometimes better to remove the region of difficult behavior and treat it specially. One way to do this is to *subtract the singularity.*

Example 6.16 The integral

$$I = \int_0^1 \frac{\cos(x)}{\sqrt{x}} dx$$

can be rewritten as

$$I = \int_0^1 \frac{1}{\sqrt{x}} dx + \int_0^1 \frac{\cos(x) - 1}{\sqrt{x}} dx$$

$$= 2 + \int_0^1 \frac{\cos(x) - 1}{\sqrt{x}} dx.$$

The function $(\cos(x) - 1)/\sqrt{x}$ is bounded and differentiable near $x = 0$, so the new integral is more suitable for numerical treatment. ■

Another way of dealing with difficult regions is to split them out and use special quadrature rules for their evaluation.

Example 6.17 The integral

$$I = \int_0^1 \frac{1}{\sqrt{x}} f(x) dx$$

can be split it into two parts

$$I = \int_0^\varepsilon \frac{1}{\sqrt{x}} f(x) dx + \int_\varepsilon^1 \frac{1}{\sqrt{x}} f(x) dx.$$

If ε is sufficiently small, we can replace f by a constant and approximate

$$\int_0^\varepsilon \frac{1}{\sqrt{x}} f(x) dx \cong 2f(0)\sqrt{\varepsilon} \tag{6.35}$$

with an error less than

$$\frac{2}{3}\varepsilon^{3/2} \max_{0 \leq x \leq \varepsilon} |f'(x)|.$$

If we can put a bound on the first derivative of f near $x = 0$, we can determine how to choose ε to keep the contribution to the error of the first part suitably small. Since the integrand for the second part is now bounded, we can use the normal techniques, although the poor behavior near $x = \varepsilon$ will likely make adaptive integration expensive.

■

The approximation (6.34) involves a simple application of *product integration.* In product integration we approximate only the smooth part of the integrand and use analytical methods to deal with the singularity. We consider the integrand as composed of two parts, and write the integral as

$$I = \int_a^b p(x)f(x)dx, \tag{6.36}$$

where $p(x)$ is supposed to contain the troublesome part and $f(x)$ the smooth part of the entire integrand. We then replace f by a function $\hat{f}$and approximate I by

$$\hat{I} = \int_a^b p(x)\hat{f}(x)dx.$$

Normally $\hat{f}$ is a piecewise polynomial approximation, so that we can evaluate $\hat{I}$ exactly, provided we have a closed form expression for the moment integrals

$$m_i = \int_a^b p(x)x^i dx, \quad i = 0,\ 1,\ \ldots. \tag{6.37}$$

Example 6.18 To find a product trapezoidal rule for integrals of the type

$$I = \int_a^{a+h} \frac{f(x)}{\sqrt{x}}\, dx,$$

we first make the linear approximation

$$\hat{f}(x) = \frac{a+h-x}{h} f(a) + \frac{x-a}{h} f(a+h).$$

This gives an approximation $\hat{I}$ to I as

$$\hat{I} = w_0 f(a) + w_1 f(a+h),$$

with

$$w_0 = \int_a^{a+h} \frac{a+h-x}{h} \frac{1}{\sqrt{x}} dx$$
$$= \frac{4}{3h}(a+h)^{3/2} - \frac{4}{3h}a^{3/2} - 2a^{1/2}$$

and

$$w_1 = \int_a^{a+h} \frac{x-a}{h} \frac{1}{\sqrt{x}} dx$$
$$= -\frac{2a}{h}(a+h)^{1/2} + \frac{2}{3h}(a+h)^{3/2} + \frac{4}{3h}a^{3/2}.$$

This rule can be used in composite form to evaluate an integral over a large region, but product integration is most appropriate near the singularity $a = 0$.

■

Example 6.19 We have already encountered Fourier transformations in Chapter 4. This is one of many applications where we need to compute *Fourier* integrals, such as

$$I = \int_a^b \cos(\alpha x) f(x) dx.$$

When α is large, adaptive methods do poorly since the regions of rapid change are not confined to any part of the interval. Here product integration is superior. Since the moment integrals

$$m_i = \int_a^b \cos(\alpha x) x^i dx$$

have closed form solutions, the weights can be obtained easily.

The simplest weight pattern arises when the interval of integration is a full period of the cosine, particularly if $a = 0$ and $b = 2\pi/\alpha$. If we approximate f by quadratic interpolation on the points 0, π/α, $2\pi/\alpha$, we find the approximation

$$\int_0^{2\pi/\alpha} \cos(\alpha x) f(x) dx \cong \frac{2}{\pi\alpha} \left\{ f(0) - 2f(\pi/\alpha) + f(2\pi/\alpha) \right\}. \tag{6.38}$$

This is *Filon's* rule.

If α is very large, we may be able to use this in composite form with panels of the width $2\pi/\alpha$. If f can be approximated closely by a second degree polynomial we will get good accuracy. If a panel has to be made smaller, the weights are not quite as simple, but can be worked out without trouble.

■

Product integration can be very effective, but there are limitations. The first is that it is applicable only when the moment integrals in (6.37) can be found in closed form or can in some way be easily approximated. Since this can be done for only a few cases, the method is limited. Sometimes, by rearrangement we can extend the application of known formulas.

Example 6.20 To compute

$$I = \int_0^1 \frac{f(x)}{\sqrt{\sin(x)}}\, dx,$$

we can rewrite the integral as

$$I = \int_0^1 \frac{g(x)}{\sqrt{x}}\, dx,$$

where

$$g(x) = \frac{f(x)\sqrt{x}}{\sqrt{\sin(x)}}.$$

Since the function g is bounded near $x = 0$, we can use the product trapezoidal rule in Example 6.18.

■

Another difficulty with product integration is that the weights are complicated and have to be computed at each step, making product integration more expensive. Also, computing the weights by formulas like the ones in Example 6.18 may involve cancellation of nearly equal terms and therefore be susceptible to rounding. In general, a good bit of care is needed to apply product integration effectively.

Integration over an unbounded interval presents some special difficulty. If the integrand decays sufficiently rapidly; that is, when

$$f(x) = e^{-x} g(x),$$

where g is a bounded and smooth function, Gauss-Laguerre methods can be very effective, yielding accurate results with few computations.

Example 6.21 Applying the Gauss-Laguerre formula to

$$I = \int_0^\infty \frac{xe^{-x}}{1+x^3}\, dx,$$

using eight quadrature points and the corresponding weights given in Table 6.5, Section 6.5, we get an approximate value $I = 0.3141$. This is accurate to two significant digits. ■

Alternatively, we can try a change of variables to get an integral over a finite interval. Sometimes splitting the infinite interval makes finding a suitable transformation easier.

Example 6.22 Write

$$I = \int_0^\infty \frac{1}{1+x^4}\, dx$$

as

$$I = \int_0^1 \frac{1}{1+x^4}\, dx + \int_1^\infty \frac{1}{1+x^4}\, dx.$$

In the second integral on the right, make a change of variables by

$$x = \frac{1}{u},$$

which gives

$$\int_1^\infty \frac{1}{1+x^4}\, dx = \int_0^1 \frac{u^2}{1+u^4}\, du.$$

Both integrals can now be approximated by conventional methods. ■

Often the simplest way of dealing with infinite intervals is a truncation to a finite interval. An examination of the integrand should tell us where to truncate to get the needed accuracy.

Example 6.23 Compute

$$I = \int_0^\infty \frac{1}{1+x^4}\, dx$$

to five-digit accuracy. We first split the integral into two parts, an integration over a finite domain, and a tail, so that

$$I = I_F + I_T,$$

where

$$I_F = \int_0^A \frac{1}{1+x^4}\, dx,$$

and

$$I_T = \int_A^\infty \frac{1}{1+x^4}\, dx.$$

Since

$$\int_A^\infty \frac{1}{1+x^4}\, dx < \int_A^\infty \frac{1}{x^4}\, dx = \frac{A^{-3}}{3},$$

we can make I_T as small as desired by making A sufficiently large. In this case $A = 35$ is adequate to make the tail part negligible compared with the required accuracy. The finite part can evaluated by an adaptive Simpson's method. The result, I_F=1.11072, has the desired accuracy.

While numerical integration is easy and straightforward for smooth integrands, poorly behaved integrands or unbounded intervals of integration require careful handling. Most of the time some analysis has to be performed before deciding on a method, and sometimes several special techniques have to be combined for the most effective treatment.

EXERCISES

1. Improve the integration problem by subtracting the singularity in

$$I = \int_0^1 \frac{e^x}{x^{2/3}}\, dx.$$

2. Find the weights for the product trapezoidal rule with $p(x) = \log_e(x)$.
3. Find the weights of the product integration rule based on a piecewise quadratic approximation for integrals of the form

$$I = \int_{a}^{a+h} \frac{f(x)}{x^{3/2}}\, dx.$$

4. Suppose that in Example 6.17 we know both $f(0)$ and $f'(0)$. We can then use

$$\int_{0}^{\varepsilon} \frac{f(x)}{\sqrt{x}}\, dx \cong 2f(0)\sqrt{\varepsilon} + \frac{2}{3}f'(0)\,\varepsilon^{3/2}.$$

What is the error in this approximation?

5. For small h and $a >> h$, find a way of evaluating the weights in Example 6.18 that reduces the sensitivity to rounding.
6. Develop an error bound for the product trapezoidal rule in Example 6.18.
7. Show how (6.38) is derived.
8. Develop a Filon-like formula for

$$I = \int_{0}^{2\pi/\alpha} \sin(\alpha\, x) f(x) dx.$$

9. Find the weights of the Filon-like rule for

$$\int_{0}^{\pi/\alpha} \cos(\alpha\, x) f(x) dx.$$

10. Describe how you could approach the evaluation of

$$\int_{-\infty}^{\infty} \frac{e^{-x^2}}{1+x^4} dx$$

in an efficient way.

11. In Example 6.23, the transformation $x = u^2$ transforms the integral to

$$I = \int_{0}^{\infty} \frac{2u}{1+u^8}\, du$$

for which the integrand decays significantly faster than in the original form. Does this observation have any practical significance?

6.7 Multi-Dimensional Integration*

Numerical integration over regions in several dimensions is a more difficult matter than one-dimensional integration. It presents new problems, some of which are much different from one-dimensional quadratures. We will focus our discussion to the two-dimensional case, but the concepts extend to more dimensions. To evaluate

$$I = \int_a^b \int_c^d f(x,\ y)dxdy,$$

we can use tensor-type approximations for $f(x,\ y)$. For suitably chosen basis functions the approximation can then be integrated exactly. For example, if we use hat-function products, the resulting two-dimensional composite trapezoidal rule has a very simple form. Figure 6.4 shows the weight pattern when a uniform mesh size h is used in both dimensions.

The generalization of other Newton-Cotes or Gaussian quadratures is straightforward. However, when a region is not rectangular, the weights cannot be gotten simply as the product of one-dimensional quadratures. Here we can rely instead on the method of undermined coefficient, introduced in (6.15). We select a set of functions, such as $\{1,\ x,\ y,\ xy,\ x^2,\ y^2,\ \ldots\}$, for which we want to make the integration exact and choose a set of quadrature points. We then write down the equations corresponding to (6.15) and solve for the weights. The more points we choose, the better an approximation we expect to get.

Example 6.24 For the right-angled triangle in Figure 6.5, we take $(0,\ \frac{1}{2})$, $(\frac{1}{2},\ 0)$, and $(\frac{1}{2},\ \frac{1}{2})$ as sample points, and look for an approximation

$$I_n = w_0 f(\frac{1}{2},\ 0) + w_1 f(0,\ \frac{1}{2}) + w_2 f(\frac{1}{2},\ \frac{1}{2}). \tag{6.39}$$

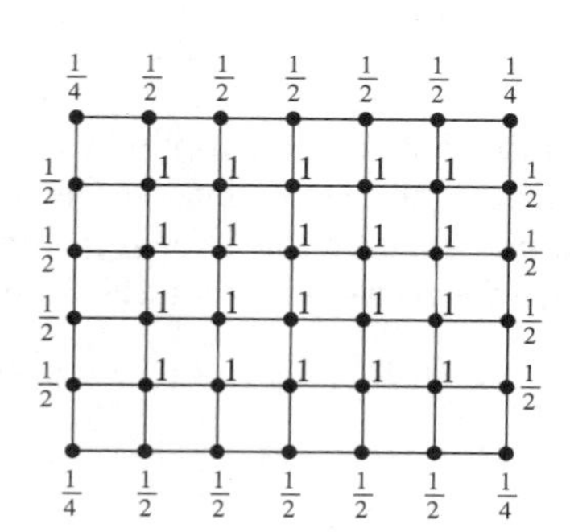

Figure 6.4 Weights for the two-dimensional trapezoidal rule. Each number is to be multiplied by h^2.

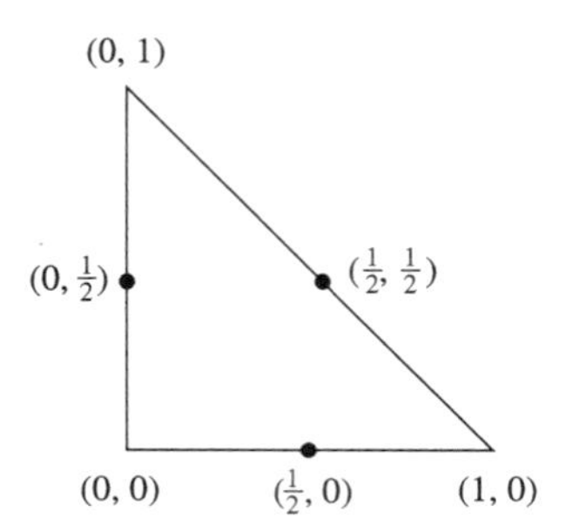

Figure 6.5 A three-point integration over a triangle.

We then have three weights to determine. We do this by making the rule exact for the functions 1, x, y over the triangle. The resulting system is

$$w_0 + w_1 + w_3 = \frac{1}{2},$$

$$\frac{1}{2}w_0 + \frac{1}{2}w_2 = \frac{1}{6}, \tag{6.40}$$

$$\frac{1}{2}w_1 + \frac{1}{2}w_2 = \frac{1}{6},$$

with the solution

$$w_0 = w_1 = w_2 = \frac{1}{6}.$$

By construction, rule (6.39) is exact for all planar functions, but it is actually more accurate than that. One can verify directly that the rule also integrates xy, x^2, and y^2 exactly.

The method of undetermined coefficients is very general and it is easy to develop rules for different types of regions. It is also not hard to scale the rules so that they can be applied over similar regions of different size and in composite form.

Example 6.25 The integration rule in Example 6.24 can be applied to the more complicated region in Figure 6.6, by splitting the latter into three similar pieces.

Subdividing the region into triangles as shown in the figure, scaling and applying (6.39) in composite form, we arrive at the formula

$$\int_\Omega f(x,\ y)dxdy \cong \tfrac{1}{6}\left\{f(0,\ \tfrac{1}{2}) + f(\tfrac{1}{2},\ 1) + 2f(\tfrac{1}{2},\ \tfrac{1}{2}) + f(\tfrac{1}{2},\ 0) + f(1,\ \tfrac{1}{2})\right\} + \tfrac{1}{12}\left\{f(1,\ \tfrac{1}{4}) + f(\tfrac{5}{4},\ 0) + f(\tfrac{5}{4},\ \tfrac{1}{4})\right\}. \tag{6.41}$$

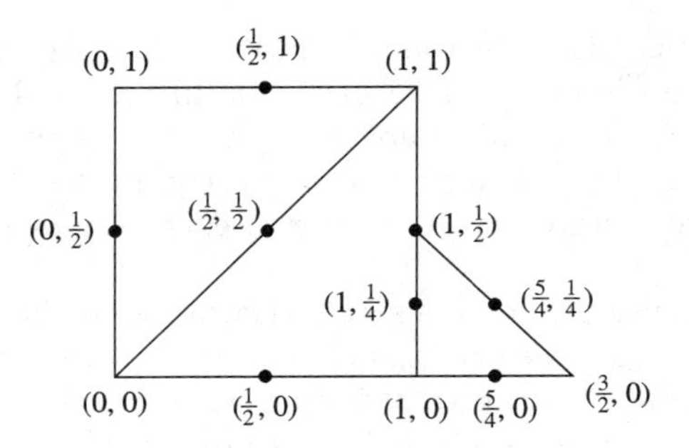

Figure 6.6
A region Ω that can be split into three right triangles.

One of the difficulties in multi-dimensional integration arises from the many varied shapes that can occur, so we need a large number of different rules. It gets worse when we try to integrate over regions with curved boundaries, since such regions cannot be subdivided into rectangles or triangles. In addition to needing different rules for the various rectangles and triangles, we cannot fit the boundaries exactly.

If we are given explicit expressions of the integrand and just want to find numerical values for a definite integral, we can do this more directly. The integral

$$I = \int_a^b \int_{g(y)}^{h(y)} f(x, y)dxdy \tag{6.42}$$

can be written as

$$I = \int_a^b u(y)dy, \tag{6.43}$$

where

$$u(y) = \int_{g(y)}^{h(y)} f(x, y)dx.$$

In this way, we have reduced the problem to a one-dimensional integration of a function which can be evaluated approximately by another one-dimensional integration. We can use adaptive techniques for this as long as we are a little careful. When we estimate the error in approximating (6.42), the function u is no longer exact, but is itself approximated by an adaptive quadrature and can only be computed with a certain error. This error is not necessarily smooth, so if we take the tolerance for the inner integral of the same order of magnitude as the tolerance for the outer integral, the error in the inner integration may make $u(y)$ look like a poorly behaved function. This in turn will make the adaptive quadrature of the outer integral very time-consuming. The usual practice is to require that the inner integral be evaluated more exactly than the outer integral, say by one or two orders of magnitude.

Three-dimensional integrals can also be done this way, but practical difficulties arise when we need to integrate over more than three dimensions. To use an n-point quadrature in k dimensions takes n^k quadrature points; even for $n = 10$, this gets prohibitive quite quickly. There is not much we can do about it.[4] If we need to evaluate a 12-dimensional integral to high accuracy, we have a problem that few people know how to solve.[5] If we are satisfied with very modest accuracy, there is some hope in the simple approach of the *Monte Carlo* method. This method is based on a random sampling of the integrand in the domain of integration Ω, using the statistical assumption that

$$\frac{\int_\Omega f(x_1,\ x_2,\ \ldots,\ x_n)dx_1\ \ldots\ dx_n}{\text{volume of } \Omega} \cong \text{average value of } f. \tag{6.44}$$

With randomly distributed points in Ω we compute the average value of f and use this to estimate the integral.

Example 6.26 The integral

$$\int_0^1\int_0^1 xydxdy = 0.25$$

was approximated with a Monte Carlo method with n sample points. The results are summarized in Table 6.7.

While the Monte Carlo method converges quite slowly, the amount of work needed to achieve a set level of accuracy does not strongly depend on the dimension of Ω. This makes it suitable for the low accuracy evaluation of high-dimensional integrals.

[4]The rapid growth of the work required for numerically solving many-dimensional problems is an example of the so-called *curse of dimensionality*. Any problem that involves more than three dimensions is likely to be difficult to handle.

[5]This is not entirely artificial. For example, the need for high-dimensional integrals arises in certain areas of quantum chemistry.

Table 6.7
Results computed by the Monte Carlo method.

n	Computed value
100	0.270
1000	0.255
10000	0.250

EXERCISES

1. Find the weights for integration over $[-1,\ 1] \times [-1,\ 1]$, using integration points $(z_1,\ z_1), (z_2,\ z_1), (z_1,\ z_2), (z_2,\ z_2)$, where z_1 and z_2 are the roots of the Chebyshev polynomial T_2.
2. Find the points and weights for a 3×3 Gauss-Legendre rule on the unit square.
3. Find the pattern of weights for the two-dimensional composite Simpson's rule on the unit square.
4. Show how the right side of (6.41) is arrived at.
5. Show that the rule in Example 6.24 also integrates the function $xy,\ x^2,\ y^2$ exactly.
6. In Example 6.24, use the corners of the triangle as sample points to find the integration weights. How would you expect the accuracy of this scheme to compare with that of Example 6.24?
7. Find weights for the integration on the triangle shown below using sample points $(0,0), (0,1), (1,0), (\frac{1}{2}, \frac{1}{2})$, such that the results are accurate for $\{1, x, y, xy\}$. Do you think this integration is more accurate than the one in Example 6.24?

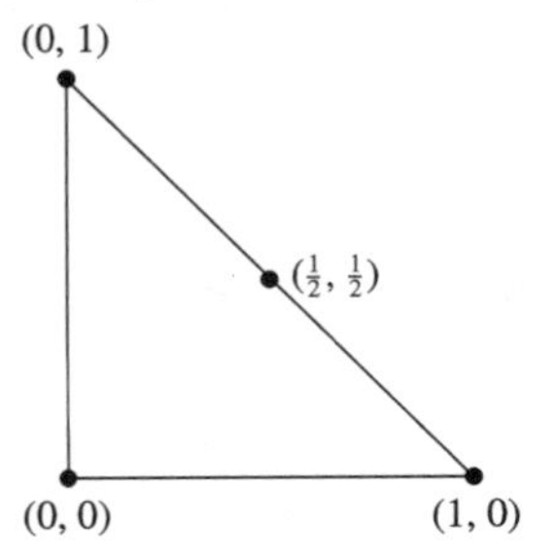

8. Establish a numerical integration rule for the unit square, using sample points at the corners and the center of the square. Select the weights so that the integration is exact for $1,\ x,\ y,\ xy,\ x^2 + y^2$.
9. Use the Monte Carlo method to approximate the value of

$$I = \int_0^2 \int_0^1 \int_0^1 \int_0^1 \frac{1}{1 + xyzw}\, dxdydzdw$$

to an accuracy of 1%.

Chapter 7

The Solution of Nonlinear Equations

In Chapter 3 we identified the solution of a system of linear equations as one of the fundamental problems in numerical analysis. It is also one of the easiest, in the sense that there are efficient algorithms for it, and that it is relatively straightforward to evaluate the results produced by them. Nonlinear equations, however, are more difficult, even when the number of unknowns is small.

Example 7.1 The following is an abstracted and greatly simplified version of a missile-intercept problem.

The movement of an object O_1 in the xy plane is described by the parameterized equations

$$\begin{aligned} x_1(t) &= t, \\ y_1(t) &= 1 - e^{-t}. \end{aligned} \tag{7.1}$$

A second object O_2 moves according to the equations

$$\begin{aligned} x_2(t) &= 1 - \cos(\alpha)t, \\ y_2(t) &= \sin(\alpha)t - 0.1t^2. \end{aligned} \tag{7.2}$$

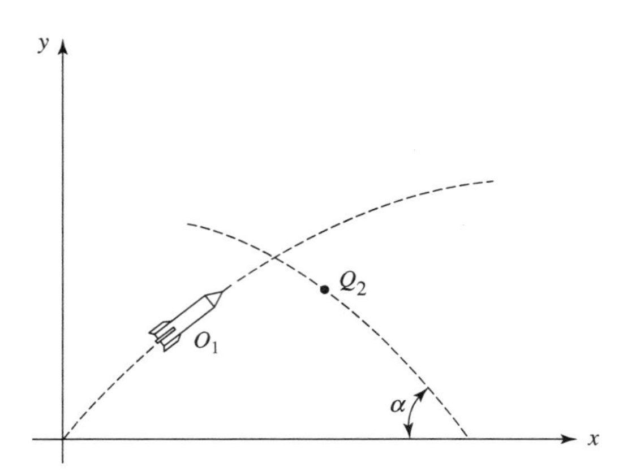

Figure 7.1 The missile-intercept problem.

Is it possible to choose a value for α so that both objects will be in the same place at some t? (See Figure 7.1.)

When we set the x and y coordinates equal, we get the system

$$\begin{aligned} t &= 1 - \cos(\alpha)t, \\ 1 - e^{-t} &= \sin(\alpha)t - 0.1t^2, \end{aligned} \tag{7.3}$$

that needs to be solved for the unknowns α and t. If real values exist for these unknowns that satisfy the two equations, both objects will be in the same place at some value of t. But even though the problem is a rather simple one that yields a small system, there is no obvious way to get the answers, or even to see if there is a solution.

The numerical solution of a system of nonlinear equations is one of the more challenging tasks in numerical analysis and, as we will see, no completely satisfactory method exists for it. To understand the difficulties, we start with what at first seems to be a rather easy problem, the solution of a single equation in one variable

$$f(x) = 0. \tag{7.4}$$

The values of x that satisfy this equation are called the *zeros* or *roots* of the function f. In what is to follow we will assume that f is continuous and sufficiently differentiable where needed.

7.1 Some Simple Root-Finding Methods

To find the roots of a function of one variable is straightforward enough—just plot the function and see where it crosses the x-axis. The simplest

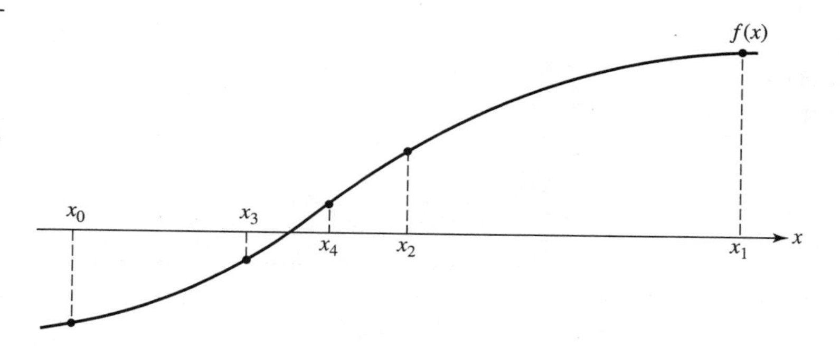

Figure 7.2
The bisection method. After three steps the root is known to lie in the interval $[x_3,\ x_4]$.

methods are in fact little more than that and only carry out this suggestion in a systematic and efficient way. Relying on the intuitive insight of the graph of the function f, we can discover many different and apparently viable methods for finding the roots of a function of one variable.

Suppose we have two values, x_0 and x_1, such that $f(x_0)$ and $f(x_1)$ have opposite signs. Then, because it is assumed that f is continuous, we know that there is a root somewhere in the interval $[x_0,\ x_1]$. To localize it, we take the midpoint x_2 of this interval and compute $f(x_2)$. Depending on the sign of $f(x_2)$, we can then place the root in one of the two intervals $[x_0,\ x_2]$ or $[x_2,\ x_1]$. We can repeat this procedure until the region in which the root is known to be located is sufficiently small (Figure 7.2). The algorithm is known as the *bisection method.*

The bisection method is very simple and intuitive, but has all the major characteristics of other root-finding methods. We start with an initial guess for the root and carry out some computations. Based on these computations we then choose a new and, we hope, better approximation of the solution. The term *iteration* is used for this repetition. In general, an iteration produces a sequence of approximate solutions; we will denote these iterates by $x^{[0]},\ x^{[1]},\ x^{[2]},\ \ldots$. The difference between the various root-finding methods lies in what is computed at each step and how the next iterate is chosen.

Suppose we have two iterates $x^{[0]}$ and $x^{[1]}$ that enclose the root. We can then approximate $f(x)$ by a straight line in the interval and find the place where this line cuts the x-axis (Figure 7.3). We take this as the new iterate

$$x^{[2]} = x^{[1]} - \frac{(x^{[1]} - x^{[0]})f(x^{[1]})}{f(x^{[1]}) - f(x^{[0]})}. \qquad (7.5)$$

When this process is repeated, we have to decide which of the three points $x^{[0]}$, $x^{[1]}$, or $x^{[2]}$, to select for starting the next iteration. There are two plausible choices. In the first, we retain the last iterate and one point from the previous ones so that the two new points enclose the solution (Figure 7.4). This is the *method of false position.*

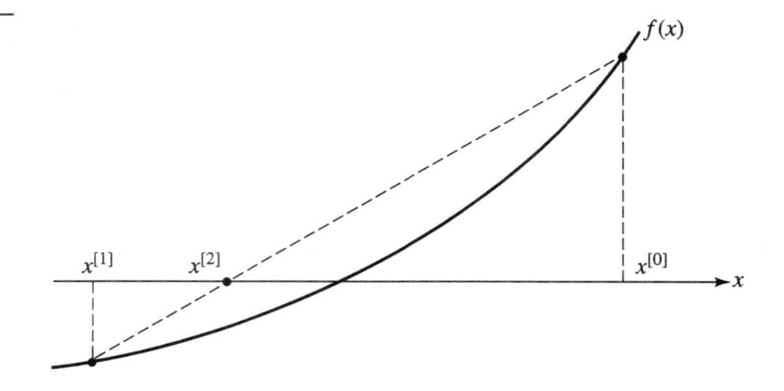

Figure 7.3
Approximating a root by linear interpolation.

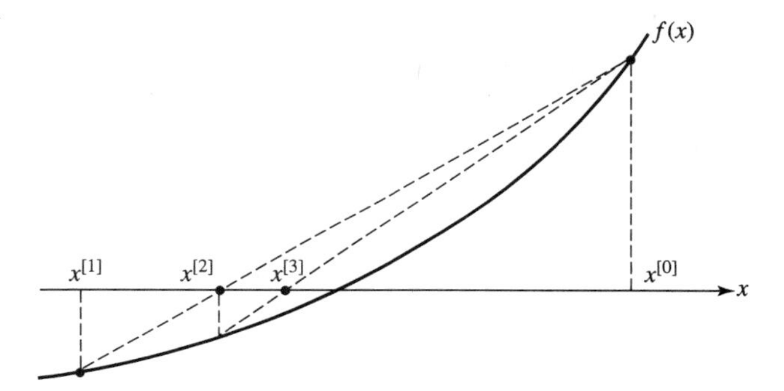

Figure 7.4
The method of false position. After the second iteration, the root is known to lie in the interval $(x^{[3]},\ x^{[0]})$.

The second choice is to retain the last two iterates, regardless of whether or not they enclose the solution. The successive iterates are then simply computed by

$$x^{[i+1]} = x^{[i]} - \frac{(x^{[i]} - x^{[i-1]})f(x^{[i]})}{f(x^{[i]}) - f(x^{[i-1]})}. \tag{7.6}$$

This is the *secant method.* Figure 7.5 illustrates how the secant method works and shows the difference between it and the method of false position. From this example we can see that now the successive iterates are no longer guaranteed to enclose the root.

Example 7.2 The function

$$f(x) = x^2 e^x - 1$$

has a root in the interval $[0,\ 1]$ since $f(0)f(1) < 0$. The results from the false position and secant methods, both started with $x^{[0]} = 0$ and $x^{[1]} = 1$,

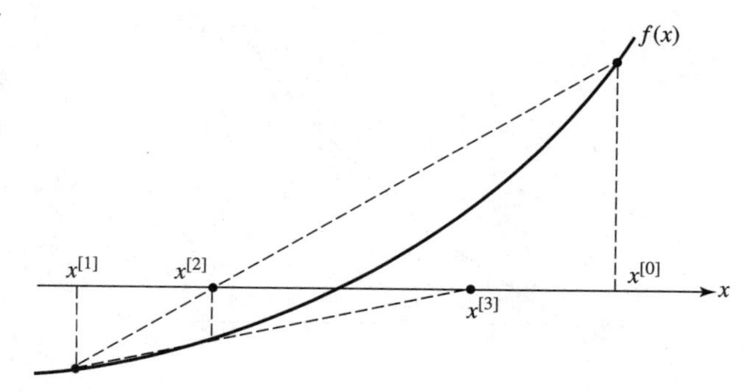

Figure 7.5
The secant method.

Table 7.1
Comparison of the false position and secant methods.

Iterates	False position	Secant
$x^{[2]}$	0.3679	0.3679
$x^{[3]}$	0.5695	0.5695
$x^{[4]}$	0.6551	0.7974
$x^{[5]}$	0.6868	0.6855
$x^{[6]}$	0.6978	0.7012
$x^{[7]}$	0.7016	0.7035

are shown in Table 7.1. It appears from these results that the secant method gives the correct result $x = 0.7035$ a little more quickly.

A popular iteration method can be motivated by Taylor's theorem. Suppose that x^* is a root of f. Then for any x near x^*,

$$\begin{aligned} f(x^*) &= f(x) + (x^* - x)f'(x) + \frac{(x^* - x)^2}{2} f''(x) + \ \ldots \\ &= 0. \end{aligned}$$

Neglecting the second order term on the right, we get

$$x^* \cong x - \frac{f(x)}{f'(x)}.$$

While this expression does not give the exact value of the root, it promises to give a good approximation to it. This suggests the iteration

$$x^{[i+1]} = x^{[i]} - \frac{f(x^{[i]})}{f'(x^{[i]})}, \tag{7.7}$$

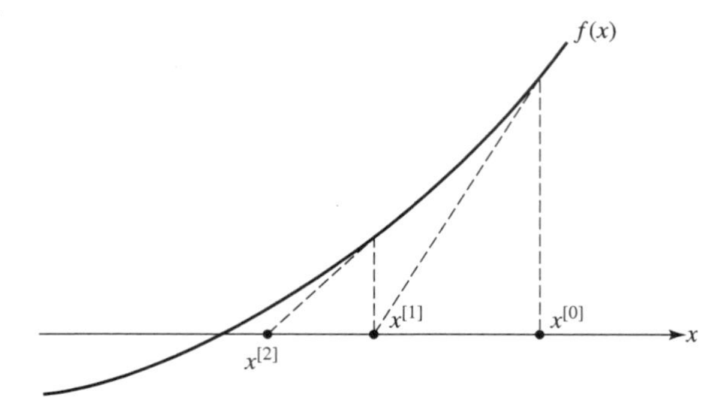

Figure 7.6
Two iterations of Newton's method.

Table 7.2
Results for Newton's method.

Iterates	Newton
$x^{[0]}$	0.3679
$x^{[1]}$	1.0071
$x^{[2]}$	0.7928
$x^{[3]}$	0.7133
$x^{[4]}$	0.7036
$x^{[6]}$	0.7035

starting with some initial guess $x^{[0]}$ (Figure 7.6). The process is called *Newton's method.*

Example 7.3

The equation in Example 7.2 was solved by Newton's method, with the starting guess $x^{[0]} = 0.3679$. Successive iterates are shown in Table 7.2. The results, though somewhat erratic in the beginning, give the root with four-digit accuracy quickly.

■

The false position and secant methods are both based on linear interpolation on two iterates. If we have three iterates, we can think of approximating the function by interpolating with a second degree polynomial and solving the approximating quadratic equation to get an approximation to the root (Figure 7.7).

If we have three points $x_0,\ x_1,\ x_2$ with the corresponding function values $f(x_0),\ f(x_1),\ f(x_2)$, we can use divided differences to find the second degree interpolating polynomial. From (4.12), with an interchange of x_0 and x_2, this is

$$p_2(x) = f(x_2) + (x - x_2)f[x_2,\ x_1] + (x - x_2)(x - x_1)f[x_2,\ x_1,\ x_0].$$

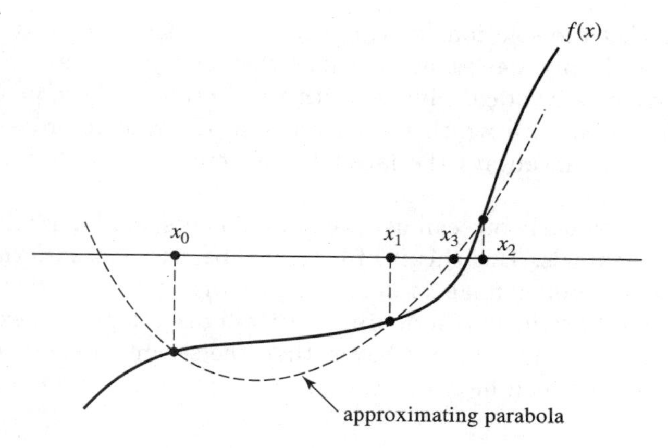

Figure 7.7 Approximating a root with quadratic interpolation.

If we write this in the form

$$p_2(x) = a(x - x_2)^2 + b(x - x_2) + c,$$

then

$$\begin{aligned} a &= f[x_2,\ x_1,\ x_0], \\ b &= f[x_2,\ x_1] + (x_2 - x_1)f[x_2,\ x_1,\ x_0], \\ c &= f(x_2). \end{aligned}$$

The equation

$$p_2(x) = 0$$

has two solutions

$$x = x_2 + \frac{-b \pm \sqrt{b^2 - 4ac}}{2a}.$$

Of these two we normally take the closest to x_2 (Figure 7.7), which we can write as

$$x_3 = x_2 + \frac{-b + sign(b)\sqrt{b^2 - 4ac}}{2a}.$$

Since the numerator may involve cancellation of two nearly equal terms, we prefer the more stable expression

$$x_3 = x_2 - \frac{2c}{b + sign(b)\sqrt{b^2 - 4ac}}. \tag{7.8}$$

As with the linear interpolation method, we have a choice in what points to retain when we go from one iterate to the next. If the initial three points bracket the solution, we can arrange it so that the next three points do so as well, and we get an iteration that always encloses the root. This has obvious advantages, but as with the method of false position, it can affect the speed with which the accuracy of the root improves. The alternative is to retain always the latest three iterates, an option known as *Muller's method.*

Obviously one can use even more points and higher degree interpolating polynomials. But this is of little use because closed form solutions for the roots of polynomials of degree higher than two are either very cumbersome or not known. Furthermore, as we will discuss in the next section, Muller's method is only slightly better than the secant method so little further improvement can be expected.

EXERCISES

1. How many iterations of bisection are needed in Example 7.2 to get 4-digit accuracy?
2. Use the bisection method to find a root of $f(x) = 1 - 2e^x$ to two significant digits.
3. Use Newton's method to find the root in Exercise 2 to six significant digits.
4. Give a graphical explanation for the irregularity of the early iterates in the secant and Newton's methods observed in Examples 7.2 and 7.3.
5. Consider the following suggestion: We can modify the bisection method to get a *trisection* method by computing the value of f at the one-third and two-thirds points of the interval, then taking the smallest interval over which there is a sign change. This will reduce the interval in which the root is known to be located by a factor three in each step and so give the root more quickly. Is there any merit to this suggestion?
6. Suppose that a computer program claims that the root it produced has an accuracy of 10^{-6}. How do you verify this claim?
7. Use Muller's method to get a rough location of the root of a function f whose values are tabulated as follows.

x	$f(x)$
0	1.20
0.5	0.65
1.0	−0.50

8. Find the three smallest positive roots of

$$x - \cot(x) = 0$$

to an accuracy of 10^{-4}.

9. In the system (7.3), use the first equation to solve for $\cos(\alpha)$ in terms of t. Substitute this into the second equation to get a single equation in the unknown t. Use the secant method to solve for t and from this get a value for α.

7.2 Convergence Rates of Root-Finding Methods

To put these preliminary results into perspective, we need to develop a way of comparing the different methods by how well they work and how quickly they will give a desired level of accuracy.

Definition 7.1

Let $x^{[0]}$, $x^{[1]}$, ... be a sequence of iterates produced by some root-finding method for the equation $f(x) = 0$. Then we say that the method produces *convergent* iterates (or just simply that it is convergent) if there exists an x^* such that

$$\lim_{i \to \infty} x^{[i]} = x^*, \tag{7.9}$$

with $f(x^*) = 0$. If a method is convergent and there exists a constant c such that

$$|x^{[i+1]} - x^*| \leq c|x^{[i]} - x^*|^k \tag{7.10}$$

for all i, then the method is said to have *iterative order of convergence k*.

For a first-order method, convergence can be guaranteed only if $c < 1$; for methods of higher order the error in the iterates will decrease as long as the starting guess $x^{[0]}$ is sufficiently close to the root.

An analysis of the bisection method is elementary. At each step, the interval in which the root and the iterate are located is halved, so we know that the method converges and the error is reduced by about one-half at each step. The bisection method is therefore a first-order method. To reduce the interval to a size ε, and thus guarantee the root to this accuracy, we must repeat the bisection process k times such that

$$2^{-k}|x^{[0]} - x^{[1]}| \leq \varepsilon,$$

or

$$k \geq \log_2 \frac{|x^{[0]} - x^{[1]}|}{\varepsilon}.$$

If the original interval is of order unity, it will take about 50 bisections to reduce the error to 10^{-15}.

To see how Newton's method works, let us examine the error in successive iterations,

$$\varepsilon_i = x^* - x^{[i]}.$$

From (7.6), we get that

$$x^* - x^{[i+1]} = x^* - x^{[i]} + \frac{f(x^{[i]}) - f(x^*)}{f'(x^{[i]})},$$

and expanding the last term on the right by Taylor's theorem,

$$x^* - x^{[i+1]} = x^* - x^{[i]} + \frac{(x^{[i]} - x^*)f'(x^{[i]})}{f'(x^{[i]})} + \frac{(x^{[i]} - x^*)^2 f''(x^{[i]})}{2f'(x^{[i]})} + \ \ldots$$

After canceling terms, this gives that, approximately,

$$\varepsilon_{i+1} \cong c\varepsilon_i^2, \tag{7.11}$$

where $c = f''(x^{[i]})/2f'(x^{[i]})$. If c is of order unity, then the error is roughly squared on each iteration; to reduce it from 0.5 to 10^{-15} takes about 6 or 7 iterations. This is potentially much faster than the bisection method.

This discussion suggests that Newton's method has second-order convergence, but the informality of the arguments does not quite prove this. To produce a rigorous proof is a little involved and is of no importance here. All we need to remember is the somewhat vague, but nevertheless informative statement that Newton's method has iterative order of convergence two, provided the starting value is sufficiently close to a zero. Arguments can also be made to show that the secant method has an order of convergence of about 1.62 and Muller's method an order approximately 1.84. While this makes Muller's method faster in principle, the improvement is not very great. This often makes the simplicity of the secant method preferable. The method of false position, on the other hand, has order of convergence one and can be quite slow.

The arguments for establishing the convergence rates for the secant method and for Muller's method are quite technical and we need not pursue them here. A simple example will demonstrate the rate quite nicely. If a method has order of convergence k, then

$$\varepsilon_{i+1} \cong c\varepsilon_i^k.$$

A little bit of algebra shows that

$$k \cong \frac{\log \varepsilon_{i+1} - \log \varepsilon_i}{\log \varepsilon_i - \log \varepsilon_{i-1}}. \tag{7.12}$$

Table 7.3 Estimation of the order of convergence for Newton's method using (7.12).

Iterates	Errors	k
$x^{[0]}$	1.315×10^{0}	--
$x^{[1]}$	8.282×10^{-1}	--
$x^{[2]}$	3.836×10^{-1}	1.66
$x^{[3]}$	7.532×10^{-2}	2.12
$x^{[4]}$	1.140×10^{-3}	2.57
$x^{[5]}$	1.001×10^{-7}	2.23
$x^{[6]}$	7.772×10^{-16}	2.00

For examples with known roots, the quantity on the right can be computed to estimate the convergence rate.

Example 7.4 The function

$$f(x) = e^{x^2} - \frac{5}{e^{2x}}$$

has a known positive root $x^* = \sqrt{1 + \log_e(5)} - 1$. Using Newton's method, with the starting guess $x^{[0]} = -0.7$, errors in successive iterates and the estimates for the order of convergence k are shown in Table 7.3. ∎

It is possible to construct methods that have iterative order larger than two, but the derivations get quite complicated. In any case, second-order methods converge so quickly that higher order methods are rarely needed.

EXERCISES

1. Suppose that f has a root x^* in some interval $[a, b]$ and that $f'(x) > 0$ and $f''(x) > 0$ in this interval. If $x^{[0]} > x^*$, show that convergence of Newton's method is monotone toward the root; that is

$$x^{[0]} > x^{[1]} > x^{[2]} > \ \ldots .$$

2. Describe what happens with the method of false position under the conditions of Exercise 1.

3. Give a rough estimate of how many iterations of the secant method will be needed to reduce the error from 0.5 to about 10^{-15}.

4. Give an estimate of how many iterations with Muller's method are needed under the conditions of the previous exercise.

5. Roughly how many iterations would you expect a third-order method to take to reduce the error from 0.5 to 10^{-15}? Compare this with the number of iterations from Newton's method.

6. Draw a graph to argue that the false position method will not work very well for finding the root of

$$f(x) = e^{-20|x|} - 10^{-4}.$$

7. A common use of Newton's method is in algorithms for computing the square root of a positive number a. Applying the method to the equation

$$x^2 - a = 0,$$

show that the iterates are given by

$$x^{[i+1]} = \frac{1}{2}\left(x^{[i]} + \frac{a}{x^{[i]}}\right).$$

Prove that, for any positive $x^{[0]}$, this sequence converges to $\sqrt{a}$.

8. Use Newton's method to design an algorithm to compute $\sqrt[3]{a}$ for positive a. Verify the second-order convergence of the algorithm.

9. Use the equation in Example 7.4 to investigate the convergence rate for Muller's method.

7.3 Difficulties with Root-Finding

Most root-finding methods for a single equation are intuitive and easy to apply, but they are all based on the assumption that the function whose roots are to be found is well behaved in some sense. When this is not the case, the computations can become complicated. For example, both the bisection method and the method of false position require that we start with two guesses that bracket a root; that is, that

$$f(x^{[0]})f(x^{[1]}) \leq 0,$$

while for Newton's method we need a good initial guess for the root. We can try to find suitable starting points by a sampling of the function at various points, but there is a price to pay. If we sample at widely spaced points, we may miss a root; if our spacing is very close we will have to expend a great deal of work. But even with very close sampling we may miss some roots. A situation like that shown in Figure 7.8 defeats many algorithms.

This situation illustrates how difficult it is to construct an automatic root-finder. Even the most sophisticated algorithm will occasionally miss if it is used without an analysis that tells something about the structure of the function. This situation is not unique to root-finding, but applies to many other numerical algorithms. Whatever method we choose, and no

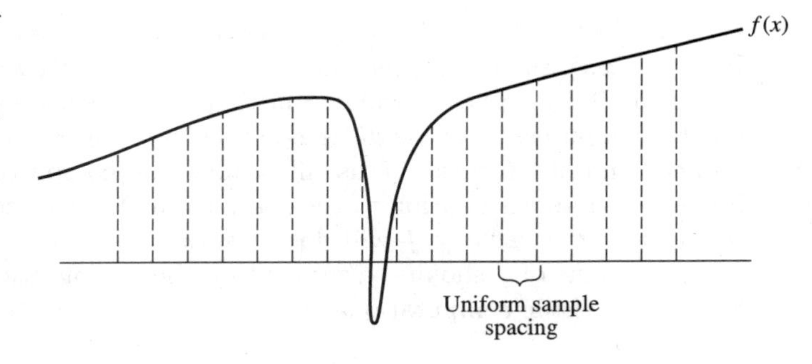

Figure 7.8
A case when two roots can be missed entirely by sampling.

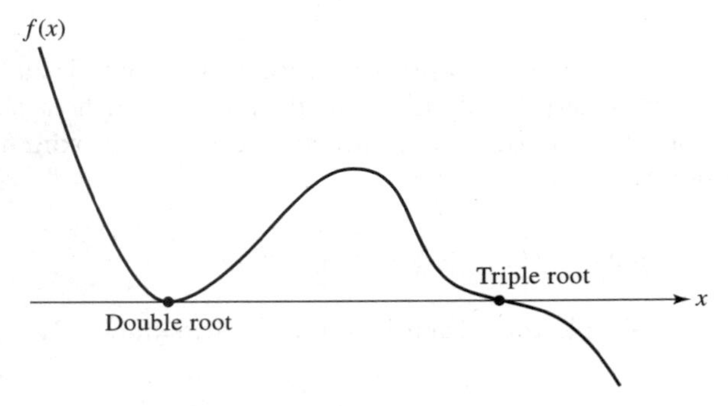

Figure 7.9
Two roots of higher multiplicity.

matter how carefully we implement it, there will always be some problems for which the method fails.

Newton's method, in addition to requiring a good initial approximation, also requires that $f'(x^{[i]})$ does not vanish. This creates a problem for the situation shown in Figure 7.9, where $f'(x^*) = 0$. Such roots require special attention.

Definition 7.2

A root x^* of f is said to have a multiplicity p if

$$f(x^*) = f'(x^*) = \ \dots \ = f^{(p-1)}(x^*) = 0,$$
$$f^{(p)}(x^*) \neq 0.$$

A root of multiplicity one is called a *simple* root, a root of multiplicity two is a *double* root, and so on.

It can be shown that Newton's method still converges to roots of higher multiplicity, but the order of convergence is reduced. Methods like bisection are not directly affected, but cannot get started for roots of even multiplicity. In either case, there is a problem of the accuracy with which a root can be obtained. To see this, consider what happens when we try to find a root on the computer. Because x^* is not necessarily a computer-representable number and the function f usually cannot be computed exactly, we can find only an approximation to the root. Even at the closest approximation to the root the value of f will not be zero. All we can say is that at the root, the value of f should be zero within the computational accuracy. If ε represents all the computational error, we can only agree that as long as

$$|f(\hat{x})| \leq \varepsilon \tag{7.13}$$

then $\hat{x}$ is an acceptable approximation to the root. Usually, there is a region around x^* where (7.13) is satisfied and the width of this region defines a limit on the accuracy with which one can approximate the root. From Taylor's theorem

$$f(\hat{x}) = f(x^*) + (\hat{x} - x^*)f'(x^*) + \frac{(\hat{x} - x^*)^2}{2} f''(x^*) + \ldots .$$

If x^* is a simple root, then it follows that, sufficiently close to the root,

$$|\hat{x} - x^*| \leq \frac{\varepsilon}{|f'(x^*)|},$$

so that the accuracy with which a simple root can be located is $O(\varepsilon)$. This makes finding a simple root a well-conditioned problem, unless $f'(x^*)$ is very close to zero.

When x^* is a zero of multiplicity two, then $f'(x^*) = 0$ and

$$|\hat{x} - x^*| \leq \sqrt{\frac{2\varepsilon}{|f''(x^*)|}},$$

so we can guarantee only an $O(\sqrt{\varepsilon})$ accuracy. In a similar way, we can show that if a root is of multiplicity p, then we can get its location only to an accuracy $O(\varepsilon^{1/p})$. Finding roots of high multiplicity is an ill-conditioned problem.

EXERCISES

1. Show that the accuracy with which a root of multiplicity p can be found is $O(\varepsilon^{1/p})$.
2. Show that if f has a simple root at x^* then f^2 has a double root at that point.
3. Draw a graph that illustrates why Newton's method converges slowly to a double root.

4. Repeat the analysis leading to (7.11) to lend substance to the conjecture that Newton's method converges linearly to a double root.

5. Do you think that Newton's method tends to converge slower near a triple root than near a double root?

6. How do you think the false position method will behave near a triple root?

7. If we know that f has a root x^* of multiplicity p, then the following modification of Newton's method (7.6) will still have order two convergence:

$$x^{[i+1]} = x^{[i]} - p\frac{f(x^{[i]})}{f'(x^{[i]})}.$$

Experimentally examine the claim that this modification of Newton's method has order two convergence by applying the function

$$f(x) = (x-1)^3 \sin(x)$$

with initial iterate $x^{[0]} = 1.5$.

7.4 The Solution of Simultaneous Equations by Newton's Method

Not all methods for a single equation carry over to systems of equations. For example, it is hard to see how one can adapt the bisection method to more than one equation. The one-dimensional method that most easily extends to several dimensions is Newton's method.

We consider the solution of a system of nonlinear equations

$$\begin{aligned} f_1(x_1,\ x_2,\ \ldots,\ x_n) &= 0,\\ f_2(x_1,\ x_2,\ \ldots,\ x_n) &= 0,\\ &\vdots\\ f_n(x_1,\ x_2,\ \ldots,\ x_n) &= 0, \end{aligned} \tag{7.14}$$

which we write in vector notation as

$$\mathbf{F}(\mathbf{x}) = 0. \tag{7.15}$$

We will assume that the number of unknowns and the number of equations are the same.

Using a multi-dimensional Taylor expansion, one can show that in n dimensions (7.7) becomes

$$\mathbf{x}^{[i+1]} = \mathbf{x}^{[i]} - \mathbf{J}^{-1}(\mathbf{x}^{[i]})\mathbf{F}(\mathbf{x}^{[i]}), \tag{7.16}$$

where $\mathbf{J}$ is the *Jacobian*

$$\mathbf{J}(\mathbf{z}) = \begin{bmatrix} \frac{\partial f_1}{\partial x_1} & \frac{\partial f_1}{\partial x_2} & \cdots & \frac{\partial f_1}{\partial x_n} \\ \frac{\partial f_2}{\partial x_1} & \frac{\partial f_2}{\partial x_2} & \cdots & \frac{\partial f_2}{\partial x_n} \\ \vdots & \vdots & & \vdots \\ \frac{\partial f_n}{\partial x_1} & \frac{\partial f_n}{\partial x_2} & \cdots & \frac{\partial f_n}{\partial x_n} \end{bmatrix}_{\mathbf{x}=\mathbf{z}}.$$

For practical computations, we do not usually use (7.16) but prefer the form

$$\mathbf{x}^{[i+1]} = \mathbf{x}^{[i]} + \Delta^{[i]}, \tag{7.17}$$

where $\Delta^{[i]}$ is the solution of

$$\mathbf{J}(\mathbf{x}^{[i]})\Delta^{[i]} = -\mathbf{F}(\mathbf{x}^{[i]}). \tag{7.18}$$

Each step in the multi-dimensional Newton's method involves the solution of an n-dimensional linear system. In addition, in each step we have to evaluate the n^2 elements of the Jacobian. It can be shown that, as in the one-dimensional case, the order of convergence of the multi-dimensional Newton's method is two, provided we have a good starting guess.

Example 7.5

Consider the solution of Example 7.1. Setting the x and y coordinates equal to each other, we get a system in two unknowns

$$\begin{aligned} 1 - \cos(\alpha)t - t &= 0, \\ \sin(\alpha)t - 0.1t^2 - 1 + e^{-t} &= 0. \end{aligned}$$

Using (7.16) with initial guess $(t,\ \alpha) = (1,\ 1)$, Newton's method converges to (0.6278603030418165, 0.9363756944918756) in four iterations. This can be shown to agree with the true solution to thirteen significant digits. The 2-norm of the error in successive iterates and the estimates for the order of convergence k are shown in Table 7.4.

■

But things do not always work as smoothly as this. Even for some small, innocent-looking systems we may get into some difficulties.

Example 7.6

For the system

$$\begin{aligned} x_1^2 + 2\sin(x_2) + x_3 &= 0, \\ \cos(x_2) - x_3 &= 2, \\ x_1^2 + x_2^2 + x_3^2 &= 2, \end{aligned}$$

Table 7.4
Estimate of order of convergence for Newton's method in calculating the solution of Example 7.1 using (7.12).

Iterates	Errors	k
0	3.78×10^{-1}	--
1	4.55×10^{-2}	--
2	6.77×10^{-4}	1.99
3	3.90×10^{-7}	1.77
4	8.98×10^{-14}	2.05

the Jacobian is

$$\mathbf{J} = \begin{bmatrix} 2x_1 & 2\cos(x_2) & 1 \\ 0 & -\sin(x_2) & -1 \\ & 2x_2 & 2x_3 \end{bmatrix}.$$

... produced by Newton's

... 992723588),
... .999751903),
... 0.999999795),

... , 0, −1) to more than ten

... 1, 0) we obtained the fol-

... .0443),
... −1.4936),
... −1.2896).

... approximations and no conver-

... method and illustrates its main ... rks exceedingly well. The second ... y accurate results with just a few ... we have a good starting value, the ... d we may do a lot of work without

■

... ed extensively because theoretically it ... der convergence and one can establish theorems that tell us exactly how close the first guess has to be to assure convergence. From a practical point of view, there are some immediate difficulties. First, we have to get the Jacobian which requires explicit expressions for n^2 partial derivatives. If this proves too cumbersome, we can use numerical differentiation but, as suggested in Chapter 6, this will lead to some loss of accuracy. Also, each step requires the solution of a linear system and this can be expensive. There are ways in which some work can be saved, say, by using the same Jacobian for a few steps before recomputing it. This slows the convergence but can improve the overall efficiency. The main difficulty, though, with Newton's method is that it requires a good starting guess. When we do not have a good starting vector, the iterates can behave quite erratically and wander around in n-space for a long time. This problem can be alleviated by monitoring the iterates and restarting the computations when it looks like convergence has failed, but on the whole, there is no entirely satisfactory solution to the starting problem. In some applications, the physical origin of the system might suggest a good initial value. In other cases, we may need to solve a sequence of nonlinear problems that are closely related, so one problem could suggest starting values for the next. In these situations, Newton's method can be very effective. In general, though, locating the solution roughly is much harder than to refine its accuracy. This is the main stumbling block to the effective solution of (7.15) by Newton's method.

EXERCISES

1. Use the Taylor expansion in two dimensions to derive (7.16) for $n = 2$.
2. Why is $\mathbf{x}^{[0]} = (0,\ 0,\ 0)$ not a good starting vector for Example 7.6?
3. Investigate whether Example 7.5 has any other solutions.
4. Find all the solutions of the system

$$x_1 + 2x_2^2 = 1,$$
$$|x_1| - x_2^2 = 0.$$

5. What can you expect from Newton's method near a multiple root?
6. An interesting biological experiment concerns the determination of the maximum water temperature at which various species of hydra can survive without shortened life expectancy. The problem can be solved by finding the weighted least-squares approximation for the model $y = a/(x-b)^c$ with the unknown parameters a, b, and c, using a collection of experimental data. The x refers to water temperature, and y refers to the average life expectancy at that temperature. More precisely, a, b, and c are chosen to minimize

$$\sum_{i=1}^{n} \left[y_i w_i - \frac{a}{(x_i - b)^c} \right]^2.$$

(a) Show that $a,\ b,\ c$ must be a solution of the following nonlinear system:

$$a = \Big(\sum_{i=1}^{n} \frac{y_i w_i}{(x_i - b)^c}\Big) \Big/ \Big(\sum_{i=1}^{n} \frac{1}{(x_i - b)^{2c}}\Big),$$

$$0 = \sum_{i=1}^{n} \frac{y_i w_i}{(x_i - b)^c} \cdot \sum_{i=1}^{n} \frac{1}{(x_i - b)^{2c+1}} - \sum_{i=1}^{n} \frac{y_i w_i}{(x_i - b)^{c+1}} \cdot \sum_{i=1}^{n} \frac{1}{(x_i - b)^{2c}},$$

$$0 = \sum_{i=1}^{n} \frac{y_i w_i}{(x_i - b)^c} \cdot \sum_{i=1}^{n} \frac{\log_e(x_i - b)}{(x_i - b)^{2c}} - \sum_{i=1}^{n} \frac{y_i w_i \log_e(x_i - b)}{(x_i - b)^{c+1}} \cdot \sum_{i=1}^{n} \frac{1}{(x_i - b)^{2c}}.$$

(b) Solve the nonlinear system for the species with the following data. Use the weights $w_i = \log_e(y_i)$.

i	x_i	y_i
1	30.2	21.6
2	31.2	4.75
3	31.5	3.80
4	31.8	2.40

7. Find a nonzero solution for the following system to an accuracy of at least 10^{-12}, using Newton's method.

$$\begin{aligned} x_1 + 10x_2 &= 0, \\ x_3 - \sqrt{|x_4|} &= 0, \\ (x_2 - 2x_3)^2 - 1 &= 0, \\ x_1 - x_4 &= 0. \end{aligned}$$

7.5 The Method of Successive Substitution*

Note one difference between Newton's method and the secant method: Newton's method obtains the next iterate $x^{[i+1]}$ from $x^{[i]}$ only, while the secant method needs both $x^{[i]}$ and $x^{[i-1]}$. We say that Newton's method is a *one-point* iteration, while the secant method is a *two-point* scheme. The analysis for one-point methods is straightforward and considerably simpler than the analysis for multipoint methods.

We begin by rewriting (7.4) as

$$x = G(x) \tag{7.19}$$

in such a way that x is a solution of (7.4) if and only if it satisfies (7.19). There are many ways such a rearrangement can be done; as we will see, some are more suitable than others.

Equation (7.19) can be made into a one-point iterative scheme by substituting $x^{[i]}$ into the right side to compute $x^{[i+1]}$; that is

$$x^{[i+1]} = G(x^{[i]}). \tag{7.20}$$

We start with some $x^{[0]}$ and use (7.20) repeatedly with $i = 0, 1, 2, \ldots$. This approach is sometimes called the *method of successive substitution*, but is actually the general form of any one-point iteration. The obvious question is what happens as $i \to \infty$.

Let x^* satisfy equation (7.19). Such a point is called a *fixed point* of the equation. Then

$$\begin{aligned} x^{[i+1]} - x^* &= G(x^{[i]}) - x^* \\ &= G(x^{[i]}) - G(x^*). \end{aligned}$$

Using a Taylor expansion, we get

$$x^{[i+1]} - x^* = G'(x^*)(x^{[i]} - x^*) + \frac{1}{2}G''(x^*)(x^{[i]} - x^*)^2 + \ldots. \tag{7.21}$$

If $G'(x^*) \neq 0$, each iteration multiplies the magnitude of the error roughly by a factor $|G'(x^*)|$. The method is therefore of order one and converges if $|G'(x^*)| < 1$. If $|G'(x^*)| > 1$ the iteration may diverge.

Example 7.7

Both equations $x = 1 - \frac{1}{4}\sin \pi x$ and $x = 1 - \sin \pi x$ have a solution $x^* = 1$. In Table 7.5, the results of the first ten iterates for both these equations are shown, using (7.20) with the initial guess $x^{[0]} = 0.9$. Clearly the iterations converge in the second column of Table 7.5, although not very quickly. It is easy to verify that $|G'(x^*)| = \frac{\pi}{4} < 1$. On the other hand, the third column in the table reveals a divergent sequence. This is because $|G'(x^*)| = \pi > 1$. ∎

Since Newton's method is a single-point iteration, (7.21) is applicable with

$$G(x) = x - \frac{f(x)}{f'(x)}$$

Table 7.5 Iterates for equations in Example 7.7.

Iterates	$x = 1 - 0.25 \sin \pi x$	$x = 1 - \sin \pi x$
$x^{[0]}$	0.9000	0.9000
$x^{[1]}$	0.9227	0.6910
$x^{[2]}$	0.9399	0.1747
$x^{[3]}$	0.9531	0.4784
$x^{[4]}$	0.9633	0.0023
$x^{[5]}$	0.9712	0.9928
$x^{[6]}$	0.9774	0.9773
$x^{[7]}$	0.9823	0.9288
$x^{[8]}$	0.9861	0.7782
$x^{[9]}$	0.9891	0.3582
$x^{[10]}$	0.9914	0.0976

and

$$G'(x) = \frac{f(x)f''(x)}{(f'(x))^2}.$$

Now $G'(x^*) = 0$ and the situation is changed. Provided $f'(x^*) \neq 0$,

$$x^{[i+1]} - x^* = \frac{1}{2}G''(x^*)(x^{[i]} - x^*)^2 + \ \ldots$$

and the magnitude of the error is roughly squared in each iteration; that is, the method has order of convergence two.

The condition $f'(x^*) \neq 0$ means that x^* is a simple root, so second order convergence holds only for that case. For roots of higher multiplicity, we reconsider the analysis. Suppose that f has a root of multiplicity two at x^*. From Taylor's theorem we know that

$$f(x) = \frac{1}{2}(x - x^*)^2 f''(x^*) + \ \ldots$$
$$f'(x) = (x - x^*)f''(x^*) + \ \ldots$$

so that

$$\lim_{x \to x^*} \frac{f(x)f''(x)}{(f'(x))^2} \cong \frac{1}{2}.$$

We expect then that near a double root, Newton's method will converge with order one, and that the error will be reduced by a factor of one half on each iteration.

It can also be shown (in Exercise 5 at the end of this section) that for roots of higher multiplicity we still get convergence, but that the rate gets slower as the multiplicity goes up. For roots of high multiplicity, Newton's method becomes quite inefficient.

These arguments can be extended formally to nonlinear systems. Using vector notation, we write the problem in the form

$$\mathbf{x} = \mathbf{G}(\mathbf{x}).$$

Then, starting with some initial guess $\mathbf{x}^{[0]}$, we compute successive values by

$$\mathbf{x}^{[i+1]} = \mathbf{G}(\mathbf{x}^{[i]}). \tag{7.22}$$

From (7.22), we get

$$\mathbf{x}^{[i+1]} - \mathbf{x}^{[i]} = \mathbf{G}(\mathbf{x}^{[i]}) - \mathbf{G}(\mathbf{x}^{[i-1]}),$$

and from the multi-dimensional Taylor's theorem,

$$\mathbf{x}^{[i+1]} - \mathbf{x}^{[i]} = \mathbf{G}'(\mathbf{x}^{[i]})(\mathbf{x}^{[i]} - \mathbf{x}^{[i-1]}) + O(||\mathbf{x}^{[i]} - \mathbf{x}^{[i-1]}||^2),$$

where $\mathbf{G}'$ is the Jacobian associated with $\mathbf{G}$. If $\mathbf{x}^{[i+1]}$ and $\mathbf{x}^{[i]}$ are so close that we can neglect higher order terms,

$$||\mathbf{x}^{[i+1]} - \mathbf{x}^{[i]}|| \;\cong\; ||\mathbf{G}'(\mathbf{x}^{[i]})||\;||\mathbf{x}^{[i]} - \mathbf{x}^{[i-1]}||. \tag{7.23}$$

If $||\mathbf{G}'(\mathbf{x}^{[i]})|| < 1$, then the difference between successive iterates will diminish, suggesting convergence.

It takes a good bit of work to make this into a precise and provable result and we will not do this here. The intuitive rule of thumb, which can be justified by precise arguments, is that if $||\mathbf{G}'(\mathbf{x}^*)|| < 1$, and if the starting value is close to a root x^*, then the method of successive substitutions will converge to this root. The order of convergence is one. The inequality (7.23) also suggests that the error in each iteration is reduced by a factor of $||\mathbf{G}'(\mathbf{x}^{[i]})||$, so that the smaller this term, the faster the convergence.

The simple form of the method of successive substitution, as illustrated in Example 7.7, is rarely used for single equations. However, for systems there are several instances where the approach is useful. For one of them we need a generalization of (7.22). Suppose that $\mathbf{A}$ is a matrix and we want to solve

$$\mathbf{A}\mathbf{x} = \mathbf{G}(\mathbf{x}). \tag{7.24}$$

Then, by essentially the same arguments, we can show that the iterative process

$$\mathbf{A}\mathbf{x}^{[i+1]} = \mathbf{G}(\mathbf{x}^{[i]}) \tag{7.25}$$

converges to a solution of (7.24), provided that $||\mathbf{A}^{-1}||\ ||\mathbf{G}'(\mathbf{x}^*)|| < 1$ and $\mathbf{x}^{[0]}$ is sufficiently close to x^*.

Example 7.8 Find a solution of the system

$$\begin{aligned} x_1 + 3x_2 + 0.1x_3^2 &= 1, \\ 3x_1 + x_2 + 0.1\sin(x_3) &= 0, \\ -0.25x_1^2 + 4x_2 - x_3 &= 0. \end{aligned}$$

We first rewrite the system as

$$\begin{bmatrix} 1 & 3 & 0 \\ 3 & 1 & 0 \\ 0 & 4 & -1 \end{bmatrix} \begin{bmatrix} x_1 \\ x_2 \\ x_3 \end{bmatrix} = \begin{bmatrix} 1 - 0.1x_3^2 \\ -0.1\sin(x_3) \\ 0.25x_1^2 \end{bmatrix}$$

and carry out the suggested iteration. As a starting guess, we can take the solution of the linear system that arises from neglecting all the small nonlinear terms. The solution of

$$\begin{bmatrix} 1 & 3 & 0 \\ 3 & 1 & 0 \\ 0 & 4 & -1 \end{bmatrix} \begin{bmatrix} x_1 \\ x_2 \\ x_3 \end{bmatrix} = \begin{bmatrix} 1 \\ 0 \\ 0 \end{bmatrix}$$

is $(-0.1250,\ 0.3750,\ 1.5000)$ which we take as $\mathbf{x}^{[0]}$. This gives the iterates

$$\begin{aligned} \mathbf{x}^{[1]} &= (-0.1343,\ 0.3031,\ 1.2085), \\ \mathbf{x}^{[2]} &= (-0.1418,\ 0.3319,\ 1.3232), \\ \mathbf{x}^{[3]} &= (-0.1395,\ 0.3215,\ 1.2808). \end{aligned}$$

Clearly, the iterations converge, although quite slowly. ■

The above example works, because the nonlinear part is a small effect compared to the linear part. This is not uncommon in practice where a linear model can sometimes be improved by incorporating small nonlinear perturbations. There are other special instances from partial differential equations where the form of the equations assures convergence. We will see some of this in Chapter 13.

EXERCISES

1. Use the method of successive substitution to find the positive root of

$$x^2 - e^{0.1x} = 0$$

to three-digit accuracy. Are there any negative roots?

2. Estimate the number of iterations required to compute the solution in Example 7.8 to four significant digits.

3. Rewrite the equations in the form (7.22) and iterate to find a solution for

$$10x_1^2 + \cos(x_2) = 12,$$

$$x_1^4 + 6x_2 = 2,$$

to three significant digits.

4. Use the method suggested in Example 7.8 to find a solution, accurate to 10^{-2}, for the system of equations

$$10x_1 + 2x_2 + \tfrac{1}{10}x_3^2 = 1,$$

$$2x_1 + 3x_2 + x_3 = 0,$$

$$\tfrac{1}{10}x_1 + x_2 - x_3 = 0.$$

Estimate how many iterations would be required to get an accuracy of 10^{-6}.

5. Show that for $f(x) = x^n$, $n > 2$, Newton's method converges for x near zero with order one. Show that the error is reduced roughly by a factor of $\frac{n-1}{n}$ in each iteration.

6. Consider the nonlinear system

$$3x_1 - \cos(x_2x_3) - \frac{1}{2} = 0,$$

$$x_1^2 - 81(x_2 + 0.1)^2 + \sin x_3 + 1.06 = 0,$$

$$e^{-x_1x_2} + 20x_3 + \frac{10\pi - 3}{3} = 0.$$

(a) Show that the system can be changed into the following equivalent problem

$$x_1 = \frac{1}{3}\cos(x_2x_3) + \frac{1}{6},$$

$$x_2 = \frac{1}{9}\sqrt{x_1^2 + \sin x_3 + 1.06} - 0.1,$$

$$x_3 = -\frac{1}{20}e^{-x_1x_2} - \frac{10\pi - 3}{60}.$$

(b) Use the method given by (7.25) to find a solution for the system in (a), accurate to 10^{-5}.

7.6 Minimization Methods*

A conceptually simple way of solving the multi-dimensional root-finding problem is by *minimization*. We construct

$$\Phi(x_1, x_2, \dots, x_n) = \sum_{i=1}^{n} f_i^2(x_1, x_2, \dots, x_n) \tag{7.26}$$

and look for a minimum of Φ. Clearly, any solution of (7.15) will be a minimum of Φ.

Optimization, with the special case of minimization, is an extensive topic that we consider here only in connection with root-finding. As with all nonlinear problems, it involves many practical difficulties. The main obstacle to minimization is the distinction between a *local* minimum and a *global* minimum. A global minimum is the place where the function Φ takes on its smallest value, while a local minimum is a minimum only in some neighborhood. In Figure 7.10, we have a function with one global and several local minima. In minimization it is usually much easier to find a local minimum than a global one. In root-finding, we unfortunately need the global minimum and any local minimum, at which Φ is not zero is of no interest.

To find the minimum of a function, we can take its derivative and then use root-finding algorithms. But this is of limited use and normally we approach the problem more directly. For minimization in one variable, we can use search methods reminiscent of bisection, but now we need three points at each step. The process is illustrated in Figure 7.11. We take three points, perhaps equally spaced, and compute the value of Φ. If the minimum value is at an endpoint, we proceed in the direction of the smallest value until

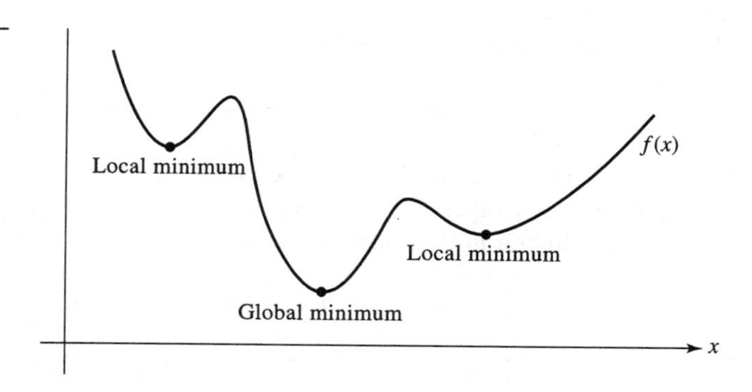

Figure 7.10 Global and local minima of a function.

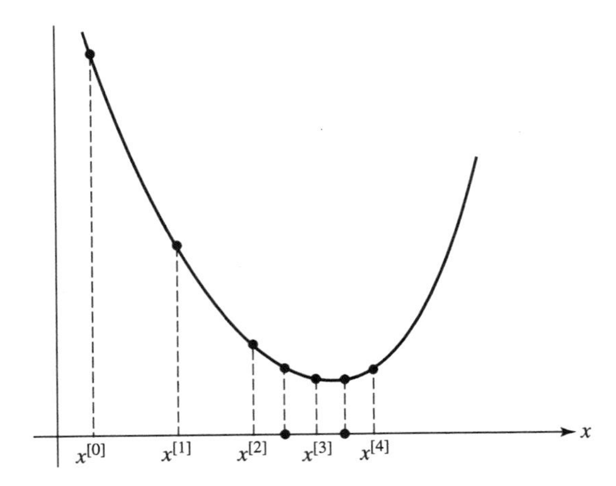

Figure 7.11 One-dimensional minimization.

the minimum occurs at the interior point. When this happens, we reduce the spacing and search near the center, continuing until the minimum is located with sufficient accuracy. Alternatively, if we have three trial points we can fit a parabola to these points and find its minimum. This will give us an approximation which can be improved by further iterations.

The main use of one-dimensional minimization is for the solution of problems in many dimensions. To find a minimum in several dimensions, we use an iterative approach in which we choose a direction, perform a one-dimensional minimization in that direction, then change direction, continuing until the minimum is found. The point we find is a local minimum, but because of computational error this minimum may be attainable only within a certain accuracy. The main difference between minimization algorithms is the choice of the direction in which we look. The process is easily visualized in two dimensions through contour plots.

In one approach we simply cycle through all the coordinate directions $x_1, x_2, \ldots, x_n$ in some order. This gives the iterates shown in Figure 7.12.

Example 7.9 Find a minimum point of the function

$$f(x, y) = x^2 + y^2 - 8x - 10y + e^{-xy/10} + 41,$$

using the strategy of search in the coordinate axes directions.

For the lack of any better guess, we start at (0, 0) and take steps of unit length in the y-direction. After a few steps, we find

$$f(0, 4) = 18, \; f(0, 5) = 17, \; f(0, 6) = 18,$$

indicating a directional minimum in the interval $4 \le y \le 6$. We fit the three points with a parabola and find its vertex, which in this case is at $y = 5$.

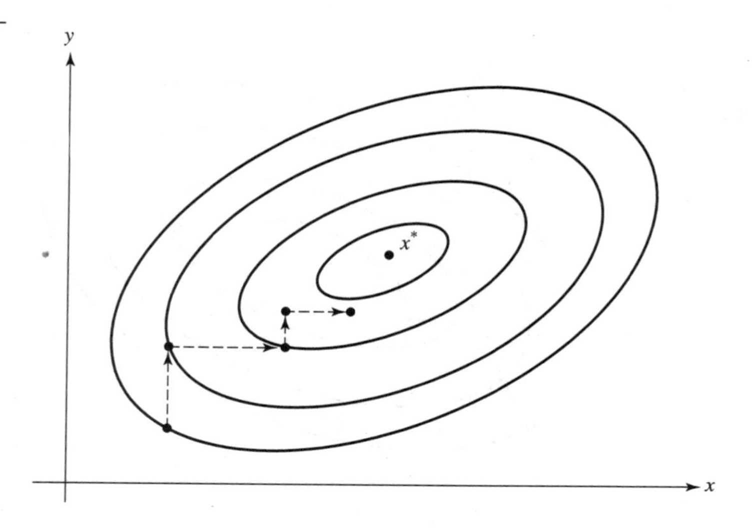

Figure 7.12 Minimization in direction of the coordinate axes.

We now change direction and continue the search. After a few more computations, we find that

$$f(3,\ 5) = 1.2231,\ f(4,\ 5) = 0.1353,\ f(5,5) = 1.0821,$$

with a resulting directional minimum at $x = 4.0347$.

Returning once more to the y-direction, we conduct another search, now with a smaller step size. We get

$$f(4.0347,\ 4.5) = 0.4139,\ f(4.0347,\ 5) = 0.1342,\ f(4.0347,\ 5.5) = 0.3599.$$

The computed directional minimum is now at $y = 5.0267$. Obviously, the process can be continued to get increasingly better accuracy.

If we stop at this point, the approximation for the minimal point is (4.0347, 5.0267). This compares well with the more accurate result[1] (4.0331, 5.0266).

In another way we proceed along the gradient

$$\Delta\Phi = \begin{bmatrix} \dfrac{\partial\,\Phi}{\partial\,x_1} \\ \dfrac{\partial\,\Phi}{\partial\,x_2} \\ \vdots \\ \dfrac{\partial\,\Phi}{\partial\,x_n} \end{bmatrix}.$$

[1]This result was produced by the MATLAB function fmins.

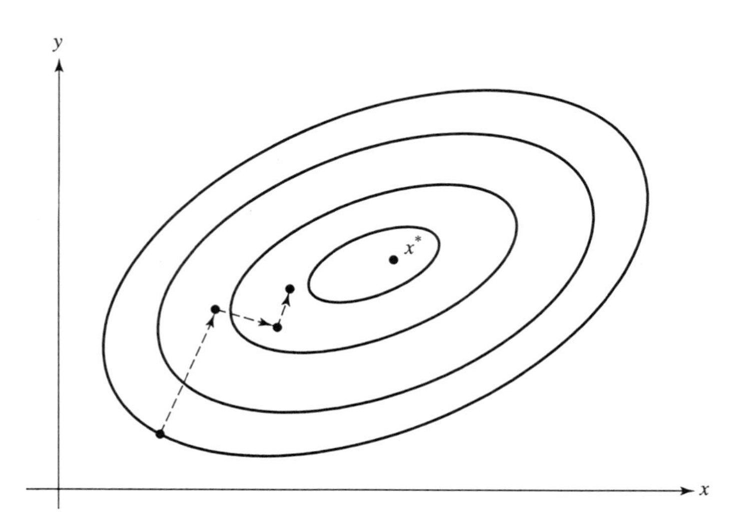

Figure 7.13 Minimization by steepest descent.

This is called the *steepest descent method* because moving along the gradient reduces Φ locally as rapidly as possible (Figure 7.13).

In a more brute force search we systematically choose a set of points and use the one where Φ is smallest for the next iteration. If triangles (or simplexes in more than two dimensions) are used, we can compute the value of Φ at the corners and at the center. If the minimum occurs at a vertex, we take this as the new center. We proceed until the center has a smaller value than all the vertices. When this happens, we start to shrink the triangle and repeat until the minimum is attained. (Figure 7.14).

There are many other more sophisticated minimization methods that one can think of and that are sometimes used. But, as in root-finding, there are always some cases that defeat even the best algorithm.

Minimization looks like a safe road toward solving nonlinear systems, and this is true to some extent. We do not see the kind of aimless wandering about that we see in Newton's method, but a steady progress toward a minimum. Unfortunately, minimization is not a sure way of getting the roots either. The most obvious reason is that we may find a local minimum instead of a global minimum. When this happens, we need to restart to find another minimum. Whether or not this eventually gives a solution of (7.15) is not clear; in any case, a lot of searching might have to be done. A second problem is that at a minimum all derivatives are zero, so it acts somewhat like a double root. When the minimum is very flat, it can be hard to locate it with any accuracy.

The simplex search illustrated in Figure 7.14 is easy to visualize, but not very efficient because each iteration requires $n + 1$ new values of Φ. In practice we use more sophisticated strategies that reduce the work considerably. For a simple discussion of this, see the elementary treatment of Murray [19]. For a more up-to-date and thorough analysis of the methods and difficulties in optimization, see Dennis and Schnabel [7].

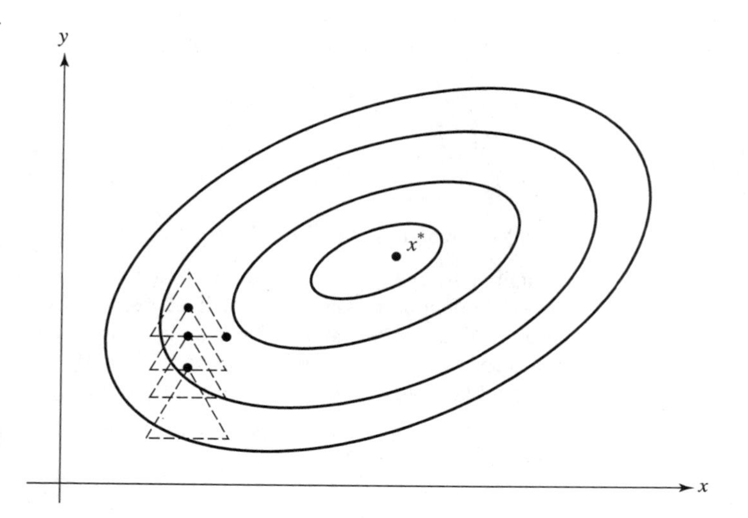

Figure 7.14
Minimization using a simplex.

EXERCISES

1. Set up the system in Exercise 3, Section 7.5 as a minimization problem of the form given by (7.26). Plot the contour graph for the function to be minimized.
2. Starting at (0, 0), visually carry out three iterations with the method of steepest descent for the problem in the previous exercise.
3. Starting at (0, 0), visually carry out three iterations with the descent along the coordinate axes for the problem in Exercise 1.
4. Elaborate a strategy for using the simplex-searching method for the minimization problem.
5. Assume that for all i, $f_i'(x^*) \neq 0$ at the solution x^* of (7.15). Suppose that the inherent computational errors are of order $O(\varepsilon)$. What is the order of accuracy with which we can expect to locate the minimum of Φ in (7.26)?
6. Instead of minimizing the Φ in (7.26), we could minimize other functions such as

$$\Phi(x_1, x_2, \ldots, x_n) = \sum_{i=1}^{n} |f_i(x_1, x_2, \ldots, x_n)|.$$

   Do you think that this is a good idea?
7. Set up the system in Example 7.1 as a minimization problem. Then, starting from (0, 0), carry out four steps of minimization in the direction of the coordinate axes.
8. Repeat Exercise 7, using four steps with the steepest descent method.

9. Locate an approximate root of the system

$$\begin{aligned} 10x + y^2 + z^2 &= 1, \\ x - 20y + z^2 &= 2, \\ x + y + 10z &= 3. \end{aligned}$$

Find a good first guess for the solution, then use minimization to improve it.

10. Implement a search procedure for finding the minimum of a function of n variables, analogous to the method illustrated in Figure 7.14, but using the 2^n corners of a n-dimensional cube as sample points. Select a set of test examples to investigate the effectiveness of this algorithm. Does this method have any advantages or disadvantages compared to fmins?

11. Use the program created in the previous exploration to implement an n-dimensional root-finder.

7.7 Roots of Polynomials*

In our discussion of Gaussian quadrature, we encountered the need for finding roots of orthogonal polynomials. This is just one of many instances in numerical analysis where we want algorithms for approximating the roots of polynomials. These roots can of course be found by the standard methods, but there are several reasons why we need to pay special attention to polynomials. On one hand, because of the simple nature of polynomials, some of the difficulties mentioned in Section 7.3 can be overcome and we can find generally useful and efficient algorithms. A little bit of analysis often gives us information that is not always available for general functions. For example, it is possible to locate roots roughly. If we write the polynomial equation in the form

$$x^n = -\frac{a_{n-1}}{a_n}x^{n-1} - \frac{a_{n-2}}{a_n}x^{n-2} - \ \ldots \ - \frac{a_0}{a_n}, \tag{7.27}$$

then it follows easily that, provided $|x| \geq 1$,

$$|x| \leq \sum_{i=0}^{n-1} \left| \frac{a_i}{a_n} \right|. \tag{7.28}$$

This shows that the roots must be confined to some finite part of the complex plane, eliminating much of the need for extensive searching.

Also, once a root, say z, has been found, it can be used to simplify the finding of the other roots. The polynomial $p_n(x)/(x - z)$ has all the roots of $p_n(x)$ except z. This lets us reduce the degree of the polynomial every time a root is found. Reduction of the degree, or *deflation*, can be done by *synthetic division*, which can be implemented recursively by

$$\frac{p_n(x)}{x-z} = b_{n-1}x^{n-1} + b_{n-2}x^{n-2} + \ \dots \ + b_0,$$

where the b_i are computed by

$$\begin{aligned} b_{n-1} &= a_n, \\ b_{n-i} &= a_{n-i+1} + z b_{n-i+1}, \quad i = 2, 3, \ \dots \ , n. \end{aligned} \tag{7.29}$$

On the other hand, polynomial root-finding has special needs. In many applications where polynomials occur we are interested not only in the real roots, but all the roots in the complex plane. This makes it necessary to extend root-finding methods to the complex plane.

While some of the methods we have studied can be extended to find complex roots, it is not always obvious how to do this. An exception is Muller's method if we interpret all operations as complex.

Example 7.10 Find all five roots $z_1, z_2, \ \dots \ , z_5$ of the polynomial

$$p_5(x) = x^5 - 2x^4 - \frac{15}{16}x^3 + \frac{45}{32}x^2 + x + \frac{3}{16}.$$

We use equation (7.8) with a stopping criteria $|x_3 - x_2| < 10^{-8}$ and initial points $x^{[0]} = -s$, $x^{[1]} = 0$, and $x^{[2]} = s$. Here $s = 5.53125$ is the right side of equation (7.28). Muller's method then produces the following sequence that converges to a complex root:

$$\begin{aligned} x^{[3]} &= -0.00020641435017, \\ x^{[4]} &= -0.00489777122243 + 0.04212476080079\, i, \\ x^{[5]} &= -0.01035051976510 + 0.06760459238812\, i, \\ x^{[6]} &= -0.31296331041835 + 0.08732823246456\, i, \\ x^{[7]} &= -0.34558627541781 + 0.13002738461112\, i, \\ x^{[8]} &= -0.36010841180409 + 0.15590007220620\, i, \\ x^{[9]} &= -0.35624447399641 + 0.16301914734518\, i, \\ x^{[10]} &= -0.35606105305313 + 0.16275737859175\, i, \\ x^{[11]} &= -0.35606176171231 + 0.16275838282328\, i, \\ x^{[12]} &= -0.35606176174733 + 0.16275838285138\, i. \end{aligned}$$

We take the last iteration as the first root.[2] Applying the synthetic division procedure given in (7.29) with $z_1 = x^{[12]}$, the resulting deflated polynomial is:

$$\begin{aligned} p_4(x) = x^4 &+ (-2.35606176174733 + 0.16275838285138i)x^3 \\ &+ (-0.12508678951512 - 0.44142083877717i)x^2 \\ &+ (1.52263356452234 + 0.13681415794943i)x \\ &+ 0.43558075942153 + 0.19910708652543i. \end{aligned}$$

Repeating the application of Muller's method, now using initial points $x_0 = -s$, $x_1 = 0$, and $x_2 = s$, with $s = 4.82817669929469$, we get the second root after nine steps at

$$z_2 = 1.97044607872988.$$

Again the deflated polynomial for z_2 is

$$\begin{aligned} p_3(x) = x^3 &+ (-0.38561568301745 + 0.16275838285138i)x^2 \\ &+ (-0.88492170001360 - 0.12071422150726i)x \\ &- 0.22105692925244 - 0.10104670646647i. \end{aligned}$$

Continuing in the same fashion, we find

$$z_3 = -0.35606176174735 - 0.16275838285137i.$$

After deflation, we obtain the quadratic equation:

$$p_2(x) = x^2 - 0.74167744476478x - 0.62083872238239.$$

The two roots for $p_2(x)$ can then be obtained by applying the quadratic formula. They are

$$z_4 = 1.24167744476478$$

and

$$z_5 = -0.5.$$

■

Many special methods have been developed for the effective computation of the roots of a polynomial, but this is a topic that is technically difficult and we will not pursue it. Polynomial root finding was a popular topic in early numerical analysis texts. We refer the interested reader to

[2] We could use a shortcut here. The theory of polynomials tells us that for polynomials with real coefficients the roots occur in complex conjugate pairs. Thus, we can claim that another root is $z_3 = -0.35606176174735 - 0.16275838285137i$.

Blum[5], Isaacson and Keller[12], or similar books written in the 1960s. We describe only the *companion matrix* method, which is easy to understand and can be used if we need to deal with this special problem.

Consider the polynomial

$$p_n(\lambda) = \lambda^n + a_{n-1}\lambda^{n-1} + \ \ldots\ + a_1\lambda + a_0. \tag{7.30}$$

Then the companion matrix of this polynomial is

$$\mathbf{C} = \begin{bmatrix} 0 & 1 & 0 & & \cdots & 0 \\ 0 & 0 & 1 & 0 & \cdots & \\ 0 & 0 & 0 & 1 & & \vdots \\ \vdots & \ddots & \ddots & \ddots & \ddots & 0 \\ 0 & \cdots & 0 & 0 & 0 & 1 \\ -a_0 & -a_1 & \cdots & & -a_{n-2} & -a_{n-1} \end{bmatrix}. \tag{7.31}$$

One can then show that the determinant

$$\det(\mathbf{C} - \lambda\mathbf{I}) = (-1)^n p_n(\lambda), \tag{7.32}$$

so that the roots of the polynomial in (7.30) are the eigenvalues of its companion matrix. Anticipating results in the next chapter, we know that there are effective methods to solve this eigenvalue problem. The companion matrix method therefore gives us a convenient way to get the complex roots of any polynomial.

Now, as we will also see in the next chapter, finding the eigenvalues of a matrix often involves computing the roots of a polynomial. So it seems we have cheated, explaining how to find the roots of a polynomial by some other method that does the same thing! But from a practical viewpoint this objection has little force. Programs for the solution of the eigenvalue problem are readily available; these normally use polynomial root-finders (using methods we have not described here), but this is a matter for the expert. If, for some reason, we need to find the roots of a polynomial, the companion matrix is the most convenient way to do so.

Example 7.11 The Legendre polynomial of degree four is

$$P_4(x) = \frac{35x^4 - 30x^2 + 3}{8}.$$

Its companion matrix is

$$\mathbf{C} = \begin{bmatrix} 0 & 1 & 0 & 0 \\ 0 & 0 & 1 & 0 \\ 0 & 0 & 0 & 1 \\ -3/35 & 0 & 30/35 & 0 \end{bmatrix}.$$

Using the techniques for matrix eigenvalue problems that we will discuss in the next chapter, we find that the eigenvalues of $\mathbf{C}$ are ± 0.8611 and ± 0.3400. These are then the roots of P_4, which, as expected, are all real. They are also the quadrature points for a four-point Gauss-Legendre quadrature over $[-1,\ 1]$.

EXERCISES

1. By finding bounds on the derivative, show that the polynomial

$$p_3(x) = x^3 - 3x^2 + 2x - 1$$

has no real root in $[-0.1,\ 0.1]$. Can you extend this result to a larger interval around $x = 0$?

2. Prove that the polynomial in Exercise 1 has at least one real root.

3. Let z be a root of the polynomial

$$p_n(x) = a_n x^n + a_{n-1} x^{n-1} + \ \ldots \ + a_0.$$

Show that the recursive algorithm suggested in (7.29) produces the polynomial

$$p_{n-1}(x) = \frac{p_n(x)}{x - z}.$$

4. Suppose that in Exercise 3, z is complex; the process then creates a polynomial p_{n-1} with complex coefficients. How does this affect the usefulness of deflation in root-finding?

5. Use Newton's method to find a real root of the polynomial in Exercise 1. Then use deflation to find the other two roots.

6. Prove (7.32).

7. Use the companion matrix method to check the results of Example 7.10.

8. Consider the ellipse

$$x^2 + 2(y - 0.1)^2 = 1$$

and the hyperbola

$$xy = \frac{1}{a}.$$

(a) For $a = 10$, find all the points at which the ellipse and the hyperbola intersect. How can you be sure that you have found all the intersections?

(b) For what value of a is the positive branch of the hyperbola tangent to the ellipse?

7.8 Selecting a Root-Finding Method*

Even though there are many methods for solving nonlinear equations, selecting the most suitable algorithm is not a simple matter. To be successful one has to take into account the specific problem to be solved and choose or modify the algorithm accordingly. This is especially true for systems of equations.

For a single equation, one can write fairly general software that deals effectively with simple roots and functions that do not have the pathological behavior exhibited in Figure 7.8. Here is some of the reasoning we might use in designing such software: We assume that to use this software we need only to supply the name of the function, a rough location of the root of interest, and an error tolerance, and that the program returns the root to an accuracy within this tolerance. The first step is to decide on the basic root-finding method. Most likely, our choice would be between Newton's method and the secant method. The method of false position often converges very slowly, therefore we rule it out for general use. Newton's method converges faster than the secant method, but requires the computation of the derivative, so each step of the secant method takes less work. If the derivative is easy to evaluate we might use Newton's method, but in a general program we cannot assume this. Also, because the error in two applications of the secant method is reduced roughly by

$$\begin{aligned} e_i &= e_{i-1}^{1.62} \\ &= (e_{i-2}^{1.62})^{1.62} \\ &= e_{i-2}^{2.6}, \end{aligned}$$

we see that two iterations of the secant method are better than one application of Newton's method. Since the secant method additionally relieves the user of having to supply the derivative, we might go with it. To ensure convergence of the method, we will design the program so that the iterates never get outside a region known to contain a root. We can ask that the user supply two guesses that enclose a root or just one guess of the approximate solution. In the latter case, we need to include in our algorithm a part that searches for a sign change in the solution. Once we have a starting point and an enclosure interval, we apply the secant method repeatedly to get an acceptable answer. The first problem is that the method may not converge and give iterates that are outside the enclosure interval. To solve this problem, we test the prospective iterate and, if it is outside the interval, apply one bisection. After that, we revert to the secant method. In this way we always have two points that bracket the solution, and the process should converge. The next issue is when to terminate the iteration. A suggestion is to compare two successive iterates and stop if they are within the error tolerance. This, however, does not guarantee the accuracy unless the two last iterates bracket the solution; what we really want is a bracketing interval within the tolerance level. To achieve this we might decide to apply the secant method until two iterates are within half the tolerance, and if the two last iterates do not bracket the solution, take a small step (say the difference between the last two iterates) in the direction toward the root (Figure 7.15). This often resolves the situation, although we still have to decide what to do if it does not.

A number of widely available programs, including MATLAB's `fzero` have been designed in such a manner. Most of them combine several simple ideas in a more sophisticated way than what we have described here.

For systems of equations, writing general purpose software is much more difficult and few attempts have been made. Most software libraries just give the basic algorithms, Newton's method, minimization, or other algorithms, leaving it to the user to find good starting points and to evaluate the results. If we want to write a more automatic program, we need safeguards against

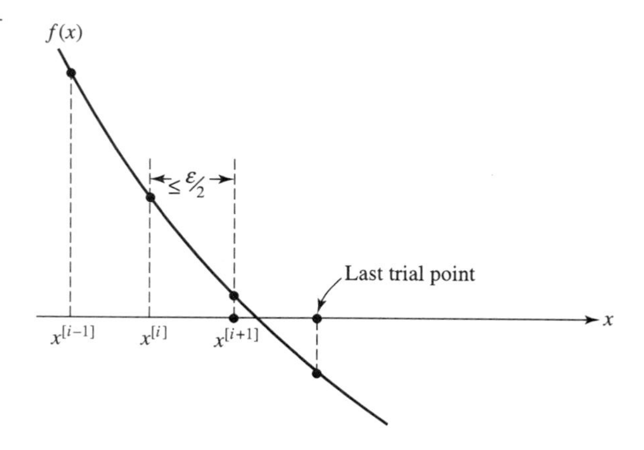

Figure 7.15 Using the secant method with bracketing of the root.

iterating without any progress. Typically, we can do two things. The first is to check that we are getting closer to a solution. We can do this by accepting the next iteration only if

$$||\mathbf{F}(\mathbf{x}^{[i+1]})|| < ||\mathbf{F}(\mathbf{x}^{[i]})||.$$

If we are using Newton's method, the second thing we can do is put a limit on the number of iterations. If the method converges, it does so quite quickly; a large number of iterations is usually a sign of trouble. For minimization on the other hand, a fair number of iterations is not unusual. To be effective for a large number of cases, multi-dimensional root-finders have to have very sophisticated search strategies. Combining a good strategy with knowledge of the specific case often gives some solutions of nonlinear systems. But unless the problem has some illuminating origin, we generally will not know that we have found all solutions. This is the harder problem for which we have no good answer.

Actually, the last statement has to be qualified. There are indeed methods that, in principle, can give us a complete solution to the nonlinear problem. These methods are based on an interval arithmetic approach, constructing functions whose domains and ranges are intervals. Interval programs have been produced that work quite well in three or four dimensions but, unfortunately, they become impractical for larger n. This is still a matter of research, but for the moment the problem is still not solved.

EXERCISES

1. Under what conditions are two iterates of the secant method on the same side of the root?
2. Can you think of a situation where the algorithm for solving a single equation sketched in this section might fail?
3. Is it possible for Newton's method to be on a convergent track, yet require a large number of iterations to get high accuracy?

Chapter 8

Advanced Matrix Algorithms

While the Gauss Elimination method is suitable for solving most systems of linear equations, there are situations where the special nature of a problem makes it possible, or even necessary, to find other methods. We now study some of the more advanced, but practically important, matrix algorithms.

One of the key elements in understanding these methods is the idea of a *factorization* or *decomposition* of a matrix. By this we mean that a matrix can be written as the product of two or more matrices of a special form, in such a way that the solution of the problem at hand is simplified.

8.1 The LU and Cholesky Factorizations

In the discussion of the GEM we saw the importance of triangular forms. By a sequence of operations we first reduce a matrix to upper triangular form; after that, the solution can be obtained easily. Triangular factorization is at the root of the closely related LU decomposition.

Under certain circumstances we can factor the matrix

$$\mathbf{A} = \begin{bmatrix} a_{11} & a_{12} & \cdots & a_{1n} \\ a_{21} & & & \vdots \\ \vdots & & \ddots & \\ a_{n1} & & \cdots & a_{nn} \end{bmatrix}$$

as

$$\mathbf{A} = \mathbf{L}\mathbf{U},$$

where $\mathbf{L}$ is a lower triangular matrix of the form

$$\mathbf{L} = \begin{bmatrix} 1 & 0 & \cdots & 0 \\ l_{21} & 1 & \ddots & \vdots \\ \vdots & \ddots & \ddots & 0 \\ l_{n1} & \dots & l_{n,n-1} & 1 \end{bmatrix} \tag{8.1}$$

and $\mathbf{U}$ is an upper triangular matrix

$$\mathbf{U} = \begin{bmatrix} u_{11} & u_{12} & \cdots & u_{1n} \\ 0 & u_{22} & \ddots & u_{2n} \\ \vdots & \ddots & \ddots & \vdots \\ 0 & \cdots & 0 & u_{nn} \end{bmatrix}. \tag{8.2}$$

To find formulas for the l_{ij} and u_{ij}, we multiply the elements of $\mathbf{L}$ and $\mathbf{U}$ and equate the results to the elements of the matrix $\mathbf{A}$. Multiplying the first row of $\mathbf{L}$ and the first column of $\mathbf{U}$ shows that

$$u_{11} = a_{11}.$$

Then, from the second row of $\mathbf{L}$ and the first column of $\mathbf{U}$,

$$l_{21}u_{11} = a_{21},$$

so that

$$l_{21} = \frac{a_{21}}{a_{11}}.$$

We can see without too much difficulty how to proceed:

$$\begin{aligned} u_{12} &= a_{12}, \\ l_{21}u_{12} + u_{22} &= a_{22}, \end{aligned}$$

giving

$$u_{22} = a_{22} - \frac{a_{21}}{a_{11}} a_{12},$$

and so on. The general scheme is

$$l_{ij} = \frac{1}{u_{jj}} \left\{ a_{ij} - \sum_{k=1}^{j-1} l_{ik} u_{kj} \right\}, \quad i = 2,\ 3,\ \ldots,\ n, \quad j = 1,\ 2,\ \ldots,\ i-1, \tag{8.3}$$

$$u_{ij} = a_{ij} - \sum_{k=1}^{j-1} l_{ik} u_{kj}, \quad i = 1,\ 2,\ \ldots,\ n,\quad j = i,\ i+1,\ \ldots,\ n. \tag{8.4}$$

If none of the denominators u_{jj} are zero, this determines the matrices **L** and **U** uniquely.

Although not entirely obvious, LU factorization is essentially equivalent to the GEM. To see this, assume for the moment that no row interchanges are needed and look at the element in row i and column j, as computed by equations (3.7) to (3.9). Since the elements in row i change only during the first $i-1$ stages, we know that

$$[\mathbf{A}^{(k)}]_{i,j} = [\mathbf{A}^{(n)}]_{i,j}$$

whenever $k \geq i$. Also from the way the elements are computed, we can see that

$$\begin{aligned} [\mathbf{A}^{(n)}]_{i,j} &= a_{ij} - \sum_{k=1}^{i-1} m_{i,k+1} [\mathbf{A}^{(k)}]_{k,j} \\ &= a_{ij} - \sum_{k=1}^{i-1} m_{i,k+1} [\mathbf{A}^{(n)}]_{k,j} \end{aligned} \tag{8.5}$$

and

$$\begin{aligned} m_{i,j+1} &= \frac{[\mathbf{A}^{(n)}]_{i,j}}{[\mathbf{A}^{(n)}]_{j,j}} \\ &= \frac{1}{[\mathbf{A}^{(n)}]_{j,j}} \left\{ a_{ij} - \sum_{k=1}^{i-1} m_{i,k+1} [\mathbf{A}^{(n)}]_{k,j} \right\}. \end{aligned} \tag{8.6}$$

If we compare (8.5) and (8.6) with (8.3) and (8.4), we find that they are the same if we identify

$$u_{ij} = [\mathbf{A}^{(n)}]_{i,j} \tag{8.7}$$

and

$$l_{ij} = m_{i,j+1}. \tag{8.8}$$

The upper triangular matrix in the LU decomposition is the reduced matrix of the GEM, while the lower triangular matrix $\mathbf{L}$ contains all the multipliers used in the GEM.

As described, the LU factorization breaks down if one of the elements u_{ii} is zero, but as in the GEM, this can be fixed by a row interchange. To express row interchanges formally, we introduce the concept of a *permutation matrix.* The square matrix $\mathbf{P}$ is said to be a permutation matrix if it consists entirely of zeros and ones, and each row and column has exactly one nonzero element.

Example 8.1 The matrix

$$\mathbf{P} = \begin{bmatrix} 0 & 1 & 0 \\ 0 & 0 & 1 \\ 1 & 0 & 0 \end{bmatrix}$$

is a permutation matrix. If we pre-multiply a 3×3 matrix $\mathbf{A}$ by $\mathbf{P}$, the second row of $\mathbf{A}$ becomes the first row of $\mathbf{PA}$, the third row of $\mathbf{A}$ becomes the second row of $\mathbf{PA}$, and the first row of $\mathbf{A}$ becomes the third row of $\mathbf{PA}$.

Post-multiplying by $\mathbf{P}$ rearranges columns, putting the third column into the first, the first into the second, and the second into the third. ■

Theorem 8.1 Let $\mathbf{A}$ be an $n \times n$ nonsingular matrix. Then there exists some permutation matrix $\mathbf{P}$ such that

$$\mathbf{PA} = \mathbf{LU},$$

where $\mathbf{L}$ is a lower triangular matrix of the form (8.1) and $\mathbf{U}$ is an upper triangular matrix of the form (8.2).

Proof: The proof involves working out the details leading to (8.3) and (8.4) to show that, with the proper permutation of rows, none of the u_{ii} vanish. The permutation matrix involved is the matrix that corresponds to the row interchanges in the GEM. We will leave this as an exercise. ■

Because the GEM and LU method are effectively identical, the total number of basic additions and multiplications to get the decomposition is approximately $n^3/3$. Suppose we have found the decomposition

$$\mathbf{PA} = \mathbf{LU}.$$

Then the linear system

$$\mathbf{Ax} = \mathbf{b}$$

leads to

$$\mathbf{LUx} = \mathbf{Pb}.$$

Now we set $\mathbf{Ux} = \mathbf{z}$ and solve

$$\mathbf{Lz} = \mathbf{Pb}$$

for $\mathbf{z}$. Since $\mathbf{L}$ is triangular, solving for $\mathbf{z}$ can be done in $O(n^2)$ operations. Next, we solve

$$\mathbf{Ux} = \mathbf{z}$$

for $\mathbf{x}$, which can also be done in $O(n^2)$ operations. Thus, once the LU decomposition has been found, solving a linear system is a fast process. The implication of this is that if we have to solve a linear system many times with different right sides, we should probably do an LU decomposition first. It is also possible to use the GEM in some such situations by putting several right sides in the augmented matrix, but this is possible only if all of the right sides are available at the same time. The LU method does not require this and is therefore more flexible.

Example 8.2 In Example 7.8 we outlined a method that is sometimes useful for nonlinear systems. The method is an iteration in which each step requires the solution of a linear system. Since the matrix for the linear system is the same for each iteration, an LU decomposition can save some work. ■

While the GEM and LU methods are the most suitable for solving general linear equations, in practice one often deals with systems that have a special structure. In such cases, we can use variants of the LU method to increase the efficiency.

Definition 8.1

If for a symmetric matrix $\mathbf{A}$ there exists a lower triangular matrix $\mathbf{L}$ such that

$$\mathbf{A} = \mathbf{L}\mathbf{L}^T,$$

then $\mathbf{A}$ is said to have a *Cholesky factorization.*

If a matrix has a Cholesky factorization, it can be found in a straightforward manner. From

$$\begin{bmatrix} a_{11} & a_{12} & \cdots & a_{1n} \\ a_{21} & a_{22} & & \vdots \\ \vdots & & \ddots & \\ a_{n1} & \cdots & & a_{nn} \end{bmatrix} = \begin{bmatrix} l_{11} & 0 & \cdots & 0 \\ l_{21} & l_{22} & \ddots & \vdots \\ \vdots & & \ddots & 0 \\ l_{n1} & \cdots & & l_{nn} \end{bmatrix} \begin{bmatrix} l_{11} & l_{21} & \cdots & l_{n1} \\ 0 & l_{22} & & \vdots \\ \vdots & \ddots & \ddots & \\ 0 & \cdots & 0 & l_{nn} \end{bmatrix},$$

we see that we must have

$$l_{11} = \sqrt{a_{11}},$$
$$l_{21}l_{11} = a_{21},$$

and so on, up to

$$l_{n1}l_{11} = a_{n1}.$$

From these, for consecutive values of i, we compute

$$l_{ii} = \sqrt{a_{ii} - \sum_{k=1}^{i-1} l_{ik}^2}, \tag{8.9}$$

and, for $j = i+1,\ i+2,\ \ldots,\ n$,

$$l_{ji} = \frac{1}{l_{ii}} \left(a_{ji} - \sum_{k=1}^{i-1} l_{jk} l_{ik} \right). \tag{8.10}$$

These equations constitute the Cholesky algorithm.

Not every matrix has a Cholesky decomposition, but there are some important cases where we can guarantee such a factorization.

Definition 8.2

A square matrix $\mathbf{A}$ is said to be *positive–definite* if

$$\mathbf{x}^T\mathbf{A}\mathbf{x} > 0 \tag{8.11}$$

for all $\mathbf{x} \neq \mathbf{0}$. It can be shown that (8.11) implies the existence of a constant $k > 0$, such that

$$\mathbf{x}^T\mathbf{A}\mathbf{x} \geq k$$

for all $\mathbf{x}$ that satisfy $||\mathbf{x}|| = 1$.

Theorem 8.2

If $\mathbf{A}$ is a symmetric positive–definite matrix, then it has a Cholesky factorization.

Proof: The arguments that show this are outlined in Exercises 5 to 7 at the end of this section. ■

The Cholesky method does not allow for row interchanges, but it is known that for positive–definite matrices this is not necessary. The advantage of the Cholesky method is that it is faster than the LU or GEM methods. It is not hard to show that the total number of arithmetic operations in the Cholesky method is about one-half that of the LU method. Since many applications lead to positive–definite matrices, the Cholesky variation is quite useful.

EXERCISES

1. Justify (8.5).
2. Find the LU decomposition of the matrix

$$\mathbf{A} = \begin{bmatrix} 1 & 2 & 3 \\ 2 & 3 & 5 \\ 6 & 3 & 1 \end{bmatrix}.$$

3. Find the Cholesky decomposition of $\mathbf{A}^T\mathbf{A}$, where $\mathbf{A}$ is the matrix in the previous exercise.

4. Show that the matrix

$$\mathbf{A} = \begin{bmatrix} 1 & 0 \\ -2 & 2 \end{bmatrix}$$

is positive–definite.

5. Show that if $\mathbf{A}$ is positive–definite, then $[\mathbf{A}]_{ii}$ must be positive for all i.

6. A principal k-minor of a matrix $\mathbf{A}$ is the $k \times k$ matrix consisting of the first k rows and first k columns of $\mathbf{A}$. Prove that if $\mathbf{A}$ is positive–definite, then every one of its principal minors must also be positive–definite.

7. Let $\mathbf{A}_n$ be a symmetric positive–definite matrix of the form

$$\mathbf{A}_n = \begin{bmatrix} \mathbf{A}_{n-1} & \mathbf{b}^T \\ \mathbf{b} & a_{nn} \end{bmatrix}.$$

(a) Show that if $\mathbf{A}_{n-1}$ has a Cholesky decomposition

$$\mathbf{A}_{n-1} = \mathbf{L}_{n-1}\mathbf{L}_{n-1}^T,$$

with $\mathbf{L}_{n-1}$ nonsingular, then there exist a vector $\mathbf{c}$ and a real number d such that

$$\mathbf{A}_n = \begin{bmatrix} \mathbf{L}_{n-1} & 0 \\ \mathbf{c} & d \end{bmatrix} \begin{bmatrix} \mathbf{L}_{n-1}^T & \mathbf{c}^T \\ 0 & d \end{bmatrix}.$$

(b) Use the observation in part (a) in an inductive argument to prove that every symmetric, positive–definite matrix has a Cholesky decomposition.

8. Show that if $\mathbf{A}$ is an invertible matrix, then $\mathbf{A}^T\mathbf{A}$ is positive–definite.

9. Use the result of Exercise 8 to prove the converse of Theorem 8.2, namely if a matrix has a Cholesky decomposition then it must be positive–definite.

10. What happens if we apply the Cholesky algorithm to a matrix that does not have a Cholesky decomposition?

11. Show that the Cholesky method requires about half the work of the LU decomposition.

12. Prove that if $\mathbf{P}$ is a permutation matrix, then $\mathbf{AP}$ is derived from $\mathbf{A}$ by permuting its columns.

13. Show that if $\mathbf{P}$ is a permutation matrix, then so is $\mathbf{P}^k$ for any positive integer k.

14. Show that every permutation matrix $\mathbf{P}$ is invertible and that

$$\mathbf{P}^{-1} = \mathbf{P}^T.$$

15. Assume that $\mathbf{A}$ is a positive–definite matrix. Show that for sufficiently small $||\Delta\mathbf{A}||$ the matrix $\mathbf{A} + \Delta\mathbf{A}$ is also positive–definite.

16. Determine whether or not each of the following matrices is positive–definite.

(a) $\mathbf{A} = \begin{bmatrix} 4 & 2 & 6 \\ -2 & 2 & 5 \\ 6 & 5 & 29 \end{bmatrix}$ (b) $\mathbf{B} = \begin{bmatrix} 4 & 4 & 8 \\ 4 & -4 & 1 \\ 8 & 1 & 6 \end{bmatrix}$ (c) $\mathbf{C} = \begin{bmatrix} 1 & 1 & 1 \\ 1 & 2 & 2 \\ 1 & 2 & 3 \end{bmatrix}$

8.2 The QR Decomposition and Least Squares Solutions

The concept of orthogonality is important in many applications, particularly in linear algebra. With vectors and matrices, orthogonality has an intuitive geometrical interpretation. We say that two n-vectors $\mathbf{x}$ and $\mathbf{y}$ are orthogonal if

$$\begin{aligned} \mathbf{x}^T\mathbf{y} &= \sum_{i=1}^{n} [\mathbf{x}]_i[\mathbf{y}]_i \\ &= 0. \end{aligned}$$

An element $\mathbf{x}$ for which $||\mathbf{x}||_2 = 1$ is said to be normalized.[1] The vectors $\{\mathbf{x}_1, \mathbf{x}_2, \ldots, \mathbf{x}_n\}$ form an orthonormal sequence if they are mutually orthogonal and normalized; that is,

$$\begin{aligned} \mathbf{x}_i^T\mathbf{x}_j &= 0, \text{ if } i \neq j, \\ &= 1, \text{ if } i = j. \end{aligned}$$

We can carry this over to matrices by requiring a similar condition for all rows and columns.

Definition 8.3

A square matrix $\mathbf{Q}$ is said to be orthonormal, or *unitary*, if

$$\begin{aligned} \mathbf{Q}^T\mathbf{Q} &= \mathbf{Q}\mathbf{Q}^T \\ &= \mathbf{I} \end{aligned} \tag{8.12}$$

If $\mathbf{Q}$ is a unitary matrix, then $\mathbf{Q}\mathbf{x}$ is called a *unitary transformation* of $\mathbf{x}$.

[1] In this chapter we use the 2-norm; that is, the norm defined by $||\mathbf{x}||^2 = \sum [\mathbf{x}]_i^2 = \mathbf{x}^T\mathbf{x}$.

Unitary matrices have the nice property that they are easily inverted, since the inverse of a unitary matrix is just its transpose. Unitary transformations have another desirable property: they preserve the norm of a vector. We can see this from

$$\begin{aligned}||\mathbf{Qx}||^2 &= (\mathbf{Qx})^T\mathbf{Qx} \\ &= \mathbf{x}^T\mathbf{Q}^T\mathbf{Qx} \\ &= \mathbf{x}^T\mathbf{x} \\ &= ||\mathbf{x}||^2.\end{aligned}$$

Since a unitary transformation does not change the magnitude of $\mathbf{x}$, any error in $\mathbf{x}$ will not be magnified. This indicates that algorithms that involve only unitary transformations tend to be stable numerically.

Definition 8.4

An $m \times n$ matrix $\mathbf{A}$ is said to have a *QR factorization* if

$$\mathbf{A} = \mathbf{QR}, \tag{8.13}$$

where $\mathbf{Q}$ is an $m \times m$ unitary matrix and $\mathbf{R}$ is an $m \times n$ upper triangular matrix.[2]

Before we look at the possibility for a QR factorization, let us see why such a factorization is useful. Suppose we want to solve the $n \times n$ system

$$\mathbf{Ax} = \mathbf{b},$$

where $\mathbf{A}$ is invertible. If $\mathbf{A}$ has a known QR factorization,

$$\mathbf{A} = \mathbf{QR},$$

then

$$\mathbf{QRx} = \mathbf{b}.$$

Multiplying this by $\mathbf{Q}^T$, we get

$$\mathbf{Rx} = \mathbf{Q}^T\mathbf{b}, \tag{8.14}$$

[2]For an $m \times n$ matrix $\mathbf{A}$ with $m > n$, the term "upper triangular" means that $[\mathbf{A}]_{ij} = 0$ for all $i > j$.

which is an easily solved triangular system. Finding a QR factorization of a matrix is therefore an alternative to the LU method.

Actually, one can do more. Suppose that $\mathbf{A}$ is an $m \times n$ matrix with $m > n$ and that we want to find the least squares solution of the system. This means that we want to minimize

$$\rho(\mathbf{x}) = ||\mathbf{Ax} - \mathbf{b}||.$$

But

$$\begin{aligned}\rho(\mathbf{x}) &= ||\mathbf{QRx} - \mathbf{b}|| \\ &= ||\mathbf{Rx} - \mathbf{Q}^T\mathbf{b}\,||,\end{aligned}$$

where the second step follows from the norm-preserving property of unitary transformations. Because the last $m - n$ rows of $\mathbf{R}$ are all zeros, their contribution to the residual does not depend on $\mathbf{x}$, and we can find instead the least squares solution of

$$\mathbf{R}_1\mathbf{x} = \mathbf{z} \tag{8.15}$$

where $\mathbf{R}_1$ is the $n \times n$ matrix made up of the first n rows of $\mathbf{R}$ and $\mathbf{z}$ is the n-vector of the first n rows of $\mathbf{Q}^T\mathbf{b}$. If $\mathbf{R}_1$ is nonsingular, then (8.15) can be solved quickly since the matrix is upper triangular.

Example 8.3 Solve the least squares problem with

$$\mathbf{A} = \begin{bmatrix} 2 & 3 \\ 1 & -1 \\ 2 & 2 \\ 4 & 1 \end{bmatrix}$$

and

$$\mathbf{b} = \begin{bmatrix} 7 \\ 1 \\ 5 \\ 6 \end{bmatrix}.$$

As is easily verified, the matrices

$$\mathbf{Q} = \begin{bmatrix} -0.4000 & 0.6828 & -0.4756 & -0.3842 \\ -0.2000 & -0.5295 & 0.0714 & -0.8213 \\ -0.4000 & 0.3344 & 0.8522 & -0.0441 \\ -0.8000 & -0.3762 & -0.2061 & 0.4195 \end{bmatrix}$$

and

$$\mathbf{R} = \begin{bmatrix} -5.0000 & -2.6000 \\ 0 & 2.8705 \\ 0 & 0 \\ 0 & 0 \end{bmatrix}$$

satisfy the conditions of the QR factorization theorem. If we use this in (8.15) we get the least squares solution

$$\mathbf{x} = \begin{bmatrix} 1.2961 \\ 1.2767 \end{bmatrix}.$$

■

Note that once the QR decomposition has been computed, the least squares problem can be solved efficiently for several right sides. Also, this method for computing the least squares solution tends to be much more stable than the elementary method suggested in Chapter 3.

Example 8.4 The matrix equation

$$\begin{bmatrix} 1 & \frac{1}{2} & \frac{1}{3} & \frac{1}{4} \\ \frac{1}{2} & \frac{1}{3} & \frac{1}{4} & \frac{1}{5} \\ \frac{1}{3} & \frac{1}{4} & \frac{1}{5} & \frac{1}{6} \\ \frac{1}{4} & \frac{1}{5} & \frac{1}{6} & \frac{1}{7} \\ \frac{1}{5} & \frac{1}{6} & \frac{1}{7} & \frac{1}{8} \\ \frac{1}{6} & \frac{1}{7} & \frac{1}{8} & \frac{1}{9} \end{bmatrix} \begin{bmatrix} x_1 \\ x_2 \\ x_3 \\ x_4 \end{bmatrix} = \begin{bmatrix} 4 \\ \frac{163}{60} \\ \frac{21}{10} \\ \frac{241}{140} \\ \frac{307}{210} \\ \frac{641}{504} \end{bmatrix}$$

has an exact solution $\mathbf{x} = [1 \ 2 \ 3 \ 4]$. Since the matrix is a section of the 6×6 Hilbert matrix, we should expect some instability and a solution affected by rounding.

When (3.22) was used, the maximum observed error was 1.5×10^{-9}, but with (8.15) this was reduced to 3.2×10^{-13}. The QR method does as well as one can expect, but the method in (3.22) magnifies any inherent ill-conditioning.

■

Having established the usefulness of a QR factorization, we must now face the more difficult issue of when this is possible and how to find one.

Theorem 8.3 Let $\mathbf{A}$ be an $m \times n$ matrix with $m \geq n$. Then there exists an $m \times m$ unitary matrix $\mathbf{Q}$ and an $m \times n$ upper triangular matrix $\mathbf{R}$ such that $\mathbf{A} = \mathbf{QR}$.

Proof: We prove the result by exhibiting a sequence of $m \times m$ unitary matrices $\mathbf{Q}_1, \ \mathbf{Q}_2, \ \ldots, \ \mathbf{Q}_n$, such that

$$\mathbf{Q}_n \ \ldots \ \mathbf{Q}_2\mathbf{Q}_1\mathbf{A} = \mathbf{R}. \tag{8.16}$$

Suppose that we can find unitary transformations $\mathbf{Q}_i$ that reduce to zero all elements of the ith column below the diagonal, leaving previously reduced columns unchanged. Then, after the kth reduction, we would have

$$\mathbf{Q}_k \ \ldots \ \mathbf{Q}_1\mathbf{A} = \begin{bmatrix} \times & \cdots & \times & \cdots & \times \\ 0 & \ddots & \times & \ldots & \times \\ \vdots & 0 & c_{k+1} & \times & \times \\ & \vdots & \vdots & \vdots & \times \\ 0 & 0 & c_m & \times & \times \end{bmatrix},$$

where $\times$ denotes a possibly nonzero element whose particular value is of no immediate importance. At this point, we want to choose $\mathbf{Q}_{k+1}$ so that the vector

$$\mathbf{c} = \begin{bmatrix} c_{k+1} \\ c_{k+2} \\ \vdots \\ c_m \end{bmatrix} \tag{8.17}$$

is reduced to

$$\mathbf{r} = \begin{bmatrix} \times \\ 0 \\ \vdots \\ 0 \end{bmatrix}, \tag{8.18}$$

while existing zeros in the lower triangle are unchanged. If we can always do this, then clearly the matrix can be reduced to the required form by a sequence of such operations.

A type of transformation that can do this is the *Householder transformation.* We choose

$$\mathbf{Q}_{k+1} = \begin{bmatrix} \mathbf{I}_k & 0 \\ 0 & \mathbf{P}_{m-k} \end{bmatrix}, \tag{8.19}$$

where $\mathbf{I}_k$ denotes the $k \times k$ identity matrix and $\mathbf{P}_{m-k}$ is defined by

$$\mathbf{P}_{m-k} = \mathbf{I}_{m-k} - 2\frac{\mathbf{v}\mathbf{v}^T}{\mathbf{v}^T\mathbf{v}}, \tag{8.20}$$

with

$$\mathbf{v} = \begin{bmatrix} c_{k+1} + \alpha \\ c_{k+2} \\ \vdots \\ c_m \end{bmatrix}$$

and

$$\alpha = \sqrt{\sum_{i=k+1}^{m} c_i^2}.$$

This transformation has the desired effect, so that

$$\mathbf{Q}_n\mathbf{Q}_{n-1} \ \ldots \ \mathbf{Q}_2\mathbf{Q}_1\mathbf{A} = \mathbf{R}.$$

Choosing then

$$\begin{aligned} \mathbf{Q} &= (\mathbf{Q}_n\mathbf{Q}_{n-1} \ \ldots \ \mathbf{Q}_2\mathbf{Q}_1)^T \\ &= \mathbf{Q}_1^T\mathbf{Q}_2^T \ \ldots \ \mathbf{Q}_{n-1}^T\mathbf{Q}_n^T \end{aligned} \tag{8.21}$$

gives the desired factorization.

The loose ends, showing that the transformations are unitary and that they achieve the stated purpose, are left as an exercise to the reader. ■

An operation count shows that, for a square matrix, the total number of additions and multiplications for this transformation is approximately $4n^3/3$. Solving a nonsingular linear system by the QR factorization is therefore more expensive than by the GEM. The advantage of the QR method is that it is more general and for the solution of the least squares problem this method is preferred to the normal equations (3.22).

The solution of the least squares problem by (8.15) requires that $m \geq n$ and that the triangular matrix $\mathbf{R}_1$ is nonsingular. When these conditions do not hold, the least squares problem does not have a unique solution and we need a different approach.

EXERCISES

1. Show that if $\mathbf{Q}$ and $\mathbf{P}$ are unitary matrices of the same dimension, so is $\mathbf{PQ}$.
2. Find the QR factorization of the matrix

$$\mathbf{A} = \begin{bmatrix} 1 & 2 & 3 \\ 2 & 3 & 5 \\ 6 & 3 & 1 \end{bmatrix}.$$

3. Use QR factorization to compute the least squares solution of

$$x_1 + x_2 = 1,$$
$$x_1 - x_2 = 2,$$
$$x_1 + 2x_2 = 4.$$

4. In Theorem 8.3, show that the matrices $\mathbf{Q}_i$ are unitary.
5. In Theorem 8.3, show that the transformations $\mathbf{Q}_i$ achieve the stated purpose.
6. In the QR factorization by Householder transformations, show that the total number of additions and multiplications is approximately $4n^3/3$.

8.3 The Singular Value Decomposition*

Singular value decomposition is another useful factorization. It involves three matrices, two of which are unitary.

Theorem 8.4 Let $\mathbf{A}$ be an $m \times n$ matrix. Then there exists an $m \times m$ unitary matrix $\mathbf{Q}$, an $n \times n$ unitary matrix $\mathbf{P}$, and an $m \times n$ special matrix $\mathbf{S}$ such that

$$\mathbf{A} = \mathbf{QSP}^T. \tag{8.22}$$

The matrix $\mathbf{S}$ has elements

$$\begin{aligned} [\mathbf{S}]_{ij} &= 0, \quad \text{for } i \neq j, \\ &\geq 0, \quad \text{for } i = j. \end{aligned}$$

Proof: From elementary linear algebra, we know that the symmetric matrix $\mathbf{A}^T\mathbf{A}$ has n real and nonnegative eigenvalues. If we write these eigenvalues as $\lambda_1^2,\ \lambda_2^2,\ \ldots,\ \lambda_n^2$ and the corresponding real orthonormal eigenvectors as $\mathbf{u}_1,\ \mathbf{u}_2,\ \ldots,\ \mathbf{u}_n$, then

$$\mathbf{A}^T\mathbf{A}\mathbf{u}_i = \lambda_i^2\mathbf{u}_i, \quad i = 1,\ 2,\ \ldots,\ n. \tag{8.23}$$

Similarly, the symmetric matrix $\mathbf{AA}^T$ has eigenvalues $\mu_1^2, \mu_2^2, \ldots, \mu_m^2$ and orthonormal eigenvectors $\mathbf{v}_1, \mathbf{v}_2, \ldots, \mathbf{v}_m$ such that

$$\mathbf{AA}^T\mathbf{v}_i = \mu_i^2\mathbf{v}_i, \quad i = 1, 2, \ldots, m. \tag{8.24}$$

Since the rank of $\mathbf{A}$ is at most

$$q = \min(m, n),$$

only the first q of the eigenvalues of either $\mathbf{A}^T\mathbf{A}$ or $\mathbf{AA}^T$ can be nonzero. From Exercise 3 of this section, it follows that the λ_i and μ_i can be ordered so that

$$\lambda_i^2 = \mu_i^2, \quad i = 1, 2, \ldots, q.$$

The λ_i, $i = 1, 2 \ldots, q$ are the *singular values*, the $\mathbf{u}_i$, $i = 1, 2, \ldots, n$ are the *right singular* vectors, and the $\mathbf{v}_i$, $i = 1, 2, \ldots, m$ the *left singular* vectors of $\mathbf{A}$.

A simple computation shows that, for $i = 1, 2, \ldots, q$,

$$\mathbf{Au}_i = \lambda_i\mathbf{v}_i \tag{8.25}$$

and

$$\mathbf{A}^T\mathbf{v}_i = \lambda_i\mathbf{u}_i. \tag{8.26}$$

Because the $\mathbf{u}_i$ and $\mathbf{v}_i$ are indeterminate within a multiplicative constant, we can assume that all λ_i are nonnegative. Furthermore, if $m > n$,

$$\mathbf{A}^T\mathbf{v}_i = 0, i = q+1, q+2, \ldots, m,$$

or, if $n > m$,

$$\mathbf{Au}_i = 0, i = q+1, q+2, \ldots n.$$

We now take as $\mathbf{Q}$ the orthonormal matrix whose columns are the left singular vectors, as $\mathbf{P}$ the orthonormal matrix whose columns are the corresponding right singular vectors, and as the diagonal elements of $\mathbf{S}$ the singular values; that is

$$[\mathbf{S}]_{ii} = \lambda_i.$$

Then we can show that

$$\begin{aligned} \mathbf{QSP}^T\mathbf{u}_i &= \lambda_i\mathbf{v}_i, \quad i = 1, 2, \ldots, q, \\ &= 0, \quad i = q+1, q+2, \ldots, n. \end{aligned} \tag{8.27}$$

Since every n-vector can be expanded as a linear combination of the $\mathbf{u}_i$, we have

$$\begin{aligned}\mathbf{QSP}^T\mathbf{x} &= \mathbf{QSP}^T \sum_{i=1}^{n} a_i \mathbf{u}_i \\ &= \sum_{i=1}^{n} a_i \lambda_i \mathbf{v}_i \\ &= \mathbf{Ax}\end{aligned}$$

for all $\mathbf{x}$. This proves (8.22). ■

Methods for finding singular value decompositions are somewhat complicated and we will not go into them here. An extensive treatment can be found in Golub and VanLoan [10].

Example 8.5 The singular value decomposition of

$$\mathbf{A} = \begin{bmatrix} 1 & 2 \\ 3 & 4 \\ 4 & 6 \end{bmatrix}$$

is

$$\mathbf{Q} = \begin{bmatrix} 0.2449 & 0.7789 & -0.5774 \\ 0.5521 & -0.6016 & -0.5774 \\ 0.7970 & 0.1773 & 0.5774 \end{bmatrix},$$

$$\mathbf{S} = \begin{bmatrix} 9.0473 & 0 \\ 0 & 0.3829 \\ 0 & 0 \end{bmatrix},$$

$$\mathbf{P} = \begin{bmatrix} 0.5625 & -0.8268 \\ 0.8268 & 0.5625 \end{bmatrix}.$$

This can be verified simply by multiplying out the matrices. ■

Singular value decomposition can be used to solve linear systems, either in the exact or the least squares sense. For this discussion, we will assume that the singular values are ordered such that $\lambda_1 \geq \lambda_2 \geq \ldots \geq \lambda_n$. This can always be arranged. We will also take $m \geq n$ and, for the moment, assume that $\lambda_i > 0$ for all $i = 1, 2, \ldots, n$; that is, that the rank of $\mathbf{A}$ is n. As we know, a true or a least squares solution $\mathbf{x}_0$ of (3.2) satisfies

$$\mathbf{A}^T\mathbf{A}\mathbf{x}_0 = \mathbf{A}^T\mathbf{b}. \tag{8.28}$$

Suppose we write

$$\mathbf{b} = \sum_{i=1}^{m} b_i \mathbf{v}_i,$$

where $\mathbf{v}_i$ are the orthonomal eigenvectors of $\mathbf{A}^T\mathbf{A}$. Then, because of the orthonormality of the $\mathbf{v}_i$, we see that

$$b_i = \mathbf{v}_i^T\mathbf{b}.$$

If we expand

$$\mathbf{x}_0 = \sum_{i=1}^{n} a_i \mathbf{u}_i,$$

then,

$$\begin{aligned}\mathbf{A}^T\mathbf{A}\mathbf{x}_0 &= \mathbf{A}^T\mathbf{A}\sum_{i=1}^{n} a_i \mathbf{u}_i \\ &= \sum_{i=1}^{n} a_i \lambda_i^2 \mathbf{u}_i.\end{aligned}$$

Also

$$\begin{aligned}\mathbf{A}^T\mathbf{b} &= \mathbf{A}^T\sum_{i=1}^{m} b_i \mathbf{v}_i \\ &= \sum_{i=1}^{n} b_i \lambda_i \mathbf{u}_i,\end{aligned}$$

where the last step comes from $\mathbf{A}^T\mathbf{v}_i = 0,\ i = n+1,\ n+2,\ \ldots,\ m$. It follows then from comparing the expansions for $\mathbf{A}^T\mathbf{b}$ and $\mathbf{A}^T\mathbf{A}\mathbf{x}_0$ that

$$a_i = \frac{\mathbf{v}_i^T\mathbf{b}}{\lambda_i}$$

and

$$\mathbf{x}_0 = \sum_{i=1}^{n} \frac{\mathbf{v}_i^T\mathbf{b}}{\lambda_i}\mathbf{u}_i. \tag{8.29}$$

The least squares solution is unique only if the rank of the matrix $\mathbf{A}$ is the same as the number of unknowns n. While it is traditional to define the rank of $\mathbf{A}$ as the largest number of linearly independent rows, it can be shown that the rank is also equal to the number of nonzero singular values. If some of the singular values are zero, say

$$\lambda_i = 0, \; i = p+1, \; p+2, \; \ldots, \; n,$$

then the coefficients a_i, $i = p+1, \; p+2, \; \ldots, \; n$ are indeterminate. Since they can be chosen arbitrarily, the least squares problem does not have a unique solution. In this case, we can change the problem to finding the *least squares solution of minimum norm*. Since

$$||\mathbf{x}_0||^2 = \sum_{i=1}^{n} a_i^2,$$

we get the smallest norm by setting the indeterminate coefficients to zero. Therefore, the minimum norm least squares solution is given by

$$\mathbf{x}_0^* = \sum_{i=1}^{p} \frac{\mathbf{v}_i^T \mathbf{b}}{\lambda_i} \mathbf{u}_i. \tag{8.30}$$

The least squares solution of a minimum norm is called a *generalized solution*. While not every linear system has a unique least squares solution, it can be shown that a unique generalized solution always exists.

Example 8.6 Examine the solutions of the equation

$$\begin{bmatrix} 1 & 2 & 3 \\ 4 & 5 & 6 \\ 5 & 7 & 9 \end{bmatrix} \begin{bmatrix} x_1 \\ x_2 \\ x_3 \end{bmatrix} = \begin{bmatrix} 1 \\ 1 \\ 1 \end{bmatrix}.$$

A singular value decomposition gives

$$\lambda_1 = 15.6633, \lambda_2 = 0.8126, \lambda_3 = 0.$$

This shows that there is not a unique least squares solution. To compute the minimum norm least squares solution, we use the computed values

$$\begin{aligned}
\mathbf{v}_1 &= (0.2354, \quad 0.5594, \quad 0.7948)^T, \\
\mathbf{v}_2 &= (0.7818, \; -0.5948, \; -0.1870)^T, \\
\mathbf{v}_3 &= (-0.5774, \; -0.5774, \quad 0.5774)^T, \\
\mathbf{u}_1 &= (0.4116, \quad 0.5638, \quad 0.7160)^T, \\
\mathbf{u}_2 &= (-0.8148, \; -0.1243, \quad 0.5662)^T, \\
\mathbf{u}_3 &= (0.4082, \; -0.8165, \quad 0.4082)^T.
\end{aligned}$$

When we substitute these values into (8.30) with $p = 2$, we get

$$\mathbf{x}_0^* = \begin{bmatrix} -0.3333 \\ 0 \\ 0.3333 \end{bmatrix}.$$

These computations yield reasonable results, but are based on the fortuitous assumption that $\lambda_3 = 0$. The actual numerical results were

$$\lambda_3 = 5.48 \times 10^{-16}.$$

If we take this as anything but rounding noise and use $p = 3$ in (8.30), we get the completely useless answer

$$\mathbf{x}_0^* = \begin{bmatrix} -4.30 \times 10^{14} \\ 8.60 \times 10^{14} \\ -4.30 \times 10^{14} \end{bmatrix}.$$

What has happened is that the computational errors have caused a confusion between a singular, rank-deficient matrix and a nonsingular, but highly ill-conditioned one.

To avoid such unpleasantness, we need to distinguish between truly nonzero results and noise. Here the answer is fairly obvious: because λ_3 is just about two ulp, it would not be sensible to mistake it for anything other than zero. A typical strategy is to take some threshold η, say one or two orders of magnitude above machine accuracy, and set every singular value smaller than η to zero.[3] This works well in many cases, but there are also situations where much more sophistication is needed. We will consider this problem in more depth in Chapter 14.

■

EXERCISES

1. Let $\mathbf{A}$ be a square nonsingular matrix with a singular value decomposition

$$\mathbf{A} = \mathbf{QSP}^T.$$

Show that then

$$\mathbf{A}^{-1} = \mathbf{PS}^{-1}\mathbf{Q}^T.$$

[3]More properly, to avoid scaling problems we should look at the ratio between the smallest and the largest singular value.

2. Show that, with the 2-norm, the condition number of a square invertible matrix, as defined in Chapter 3, is the same as the ratio of the largest to the smallest singular values of the matrix.

3. Show that if λ is an eigenvalue of $\mathbf{A}^T\mathbf{A}$ then it is also an eigenvalue of $\mathbf{A}\mathbf{A}^T$.

4. Let $\mathbf{A}$ and $\mathbf{B}$ be two $m \times n$ matrices such that

$$\mathbf{Ax} = \mathbf{Bx}$$

for all n-vectors $\mathbf{x}$. Show that this implies $\mathbf{A} = \mathbf{B}$.

5. Find the minimum norm least squares solution for the system in Example 8.6 for the right side

$$\mathbf{b} = \begin{bmatrix} 0 \\ 1 \\ 2 \end{bmatrix}.$$

6. Use singular value decomposition to find the least squares solution of the system in Example 8.3.

7. Use singular vectors to find an expression for the set of all least squares solutions of a linear system.

8. The pseudo-inverse of a matrix $\mathbf{A}$, denoted by $\mathbf{A}^+$, is the matrix that gives the minimum norm least squares solution of $\mathbf{Ax} = \mathbf{b}$ as $\mathbf{x} = \mathbf{A}^+\mathbf{b}$. Show that an expression for the pseudo-inverse is

$$\mathbf{A}^+ = \sum_{i=1}^{p} \frac{\mathbf{u}_i\mathbf{v}_i^T}{\lambda_i}.$$

9. Show that $\mathbf{A}^+ = \mathbf{P}\mathbf{S}^+\mathbf{Q}$ where $\mathbf{S}^+$ is an $n \times m$ matrix such that

$$\begin{aligned} [\mathbf{S}^+]_{ij} &= \frac{1}{\lambda_i}, \qquad \text{for } i = j,\ i = 1,\ 2,\ \ldots,\ p \\ &= 0, \qquad \text{otherwise.} \end{aligned}$$

10. Prove (8.25) and (8.26).

11. Discuss why a very small value of λ_p can make the numerical evaluation of the minimum norm solution difficult to obtain.

8.4 Sparse Matrices*

The algorithms that we have discussed so far are used primarily in situations where the matrices have no special structure that can be taken advantage of. For matrices of relatively small size, typically $n = 200$ or less, the methods work well and can be carried out even on desktop computers. But resource requirements increase rapidly as the matrix size increases. Storage requirements are of order $O(n^2)$, while execution times typically grow like n^3. For large matrices, say $n = 10{,}000$, this taxes even the most powerful present-day computers. Since such systems occur routinely in practice, dealing with large matrices is an important issue. Fortunately, when very large matrices do occur, they usually have some special properties that can be utilized to make things more manageable.

One characteristic that many large systems have is that they are *sparse*. By this we mean that most elements in the matrices are zero. Large sparse matrices occur in many applications.

Example 8.7

Example 3.1 shows how an electric circuit problem leads to a system of linear equations. The specific case in Example 3.1 is quite small; in practice the systems we get may be very large. Fortunately, there is usually some structure we can use advantageously. If we look at part of such a circuit, as in Figure 8.1, we see that only a few of the unknown currents are connected at any node. When we write down the current balance equation, the corresponding row in the matrix will have mostly zero entries and so will be very sparse. ■

A prime example of sparse matrices occurs in the numerical solution of partial differential equations. In our discussion of differential equations we will see how such matrices arise and what their special structure is. For the moment, we consider only how the sparseness can be used to reduce the resource requirements for the solution of systems of equations.

Schemes for the efficient storage of sparse matrices readily suggest themselves and a variety of well-understood data structures can be used. One of the simplest schemes is to store all the nonzero elements in some type of list that includes values as well as position of the elements.

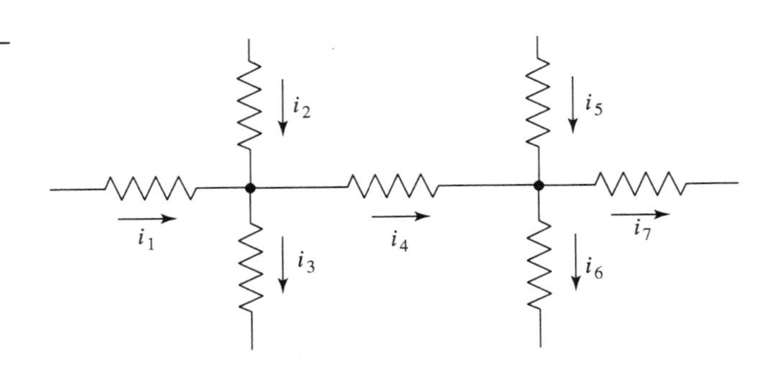

Figure 8.1
Part of an electric circuit.

Example 8.8 The matrix

$$\begin{bmatrix} 0 & 0 & -2 & 0 \\ 2 & 0 & 0 & 1 \\ 0 & 4 & 1 & 0 \\ 0 & 0 & 0 & 1 \end{bmatrix}$$

can be stored as the sequence

$$(1,\ 3,\ -2),\ (2,\ 1,\ 2),\ (2,\ 4,\ 1),\ (3,\ 2,\ 4),\ (3,\ 3,\ 1),\ (4,\ 4,\ 1).$$

The first entry is the row number and the second the column number; the third represents the value of the element. All elements not explicitly shown are assumed to be zero. Although each nonzero element now requires three numbers, for very sparse matrices this saves a great deal of space. An alternative is to use a collection of lists. For example, the above matrix can be represented as

column 1: (2, 2),

column 2: (3, 4),

column 3: (1, −2), (3, 1),

column 4: (2, 1), (4, 1).

Each column is stored as a separate list in which the first number of each entry is the row number, the second the element value. ■

When we implement these storage schemes, we draw on what we know about data structures from computer science. Linked list structures are popular because they make insertion and deletion simple. Figure 8.2 shows the schematic for a linked list representation of the first option in Example 8.8.

There are many other ways for representing sparse matrices and implementing the representations in a computer. What scheme is most suitable is very application-dependent. The efficiency of the structure cannot be judged by the storage requirements only, but by how it affects the computational processes that are to be carried out. These are issues whose

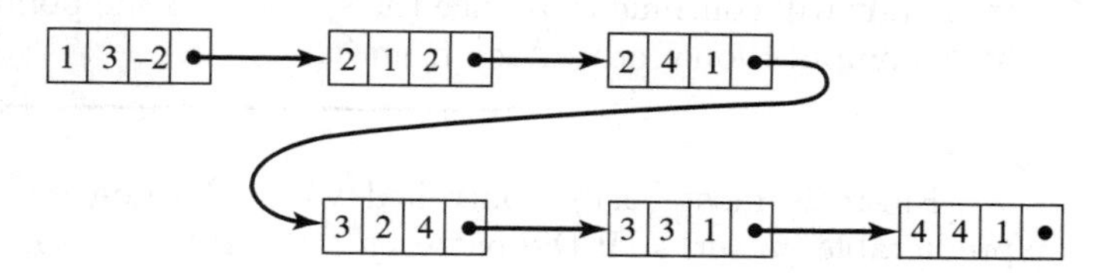

Figure 8.2
A simple linked list representation of a sparse matrix.

effective solution requires knowledge both of numerical methods and data structures. Here we concentrate on the numerical aspects.

For simplicity, we restrict ourselves to discussing the GEM and see how it can be adapted to sparse matrices. The issues that we must address are how to take advantage of the sparsity, and how to preserve sparsity as far as possible throughout the solution process. It is the preservation of the sparsity that gives us the most trouble.

Example 8.9 Consider a 6×6 matrix, represented schematically below, with $\times$ denoting nonzero elements

$$\begin{bmatrix} \times & 0 & \times & 0 & \times & 0 \\ 0 & \times & 0 & 0 & \times & 0 \\ 0 & 0 & \times & \times & 0 & 0 \\ \times & 0 & 0 & \times & 0 & \times \\ 0 & \times & 0 & \times & \times & 0 \\ \times & 0 & 0 & \times & 0 & \times \end{bmatrix}.$$

The first step in the GEM is to reduce to zero all the elements below the diagonal in the first column. We assume that the (1, 1) element is used as the pivot. Since there are only two elements to be eliminated and since there are lots of zeros in the first row, this can be done with just a few operations, and we can easily take advantage of the sparsity. The first step in the reduction gives the result

$$\begin{bmatrix} \times & 0 & \times & 0 & \times & 0 \\ 0 & \times & 0 & 0 & \times & 0 \\ 0 & 0 & \times & \times & 0 & 0 \\ 0 & 0 & * & \times & * & \times \\ 0 & \times & 0 & \times & \times & 0 \\ 0 & 0 & * & \times & * & \times \end{bmatrix}.$$

Note the appearance of new nonzero elements, shown by $*$, which decreases the sparsity and causes what is known as *fill-in*. Unless fill-in can be controlled, it will continue to reduce the sparsity to the point where we can get little computational advantage from it. ■

Fill-in is most easily controlled when the nonzero elements occur in predictable patterns. If the patterns are regular enough we can sometimes solve the system without any fill-in at all. The simplest sparsity pattern

Figure 8.3
A tridiagonal system.

$$\begin{bmatrix} a_1 & c_1 & 0 & \dots & 0 \\ e_2 & a_2 & c_2 & \ddots & \vdots \\ 0 & e_3 & a_3 & \ddots & 0 \\ \vdots & \ddots & \ddots & \ddots & c_{n-1} \\ 0 & \dots & 0 & e_n & a_n \end{bmatrix} \begin{bmatrix} x_1 \\ x_2 \\ \vdots \\ x_{n-1} \\ x_n \end{bmatrix} = \begin{bmatrix} b_1 \\ b_2 \\ \vdots \\ b_{n-1} \\ b_n \end{bmatrix}$$

appears in *tridiagonal* matrices, in which all nonzero elements occur in the two *co-diagonals*; that is, in the two lines next to and parallel to the main diagonal (Figure 8.3).

From Figure 8.3 we can see easily what happens when the GEM is applied to a tridiagonal matrix. The elimination of e_2 changes only a_2 and none of the other elements. A similar thing happens at subsequent stages. If no row interchanges are necessary, the GEM creates no fill-in and the matrix can be reduced to triangular form with $O(n)$ operations. The backsubstitution can also be done in $O(n)$ operations, so the whole process is quite efficient. There is in fact a simple formula for the solution. The solution of the tridiagonal system in Figure 8.3 is given by

$$\begin{aligned} d_1 &= a_1, \\ y_1 &= b_1, \\ d_i &= a_i - \frac{e_i c_{i-1}}{d_{i-1}}, \quad i = 2,\ 3,\ \dots,\ n, \\ y_i &= b_i - \frac{e_i y_{i-1}}{d_{i-1}}, \quad i = 2,\ 3,\ \dots,\ n, \\ x_n &= \frac{y_n}{d_n}, \\ x_{n-i} &= \frac{1}{d_{n-i}}(y_{n-i} - c_{n-i}x_{n-i+1}), \quad i = 1,\ 2,\ \dots,\ n-1. \end{aligned} \tag{8.31}$$

A tridiagonal matrix is a special case of a *banded* matrix. A matrix is said to be banded if all the nonzero elements occur in lines parallel to the main diagonal, as shown in Figure 8.4. The largest number of nonzero elements in any row or column is the *bandwidth* of the matrix. The matrix in Figure 8.4 has a bandwidth of four.

Figure 8.4
A matrix with bandwidth four.

$$\begin{bmatrix} \times & \times & \times & 0 & 0 & 0 & 0 \\ \times & \times & \times & \times & 0 & 0 & 0 \\ 0 & \times & \times & \times & \times & 0 & 0 \\ 0 & 0 & \times & \times & \times & \times & 0 \\ 0 & 0 & 0 & \times & \times & \times & \times \\ 0 & 0 & 0 & 0 & \times & \times & \times \\ 0 & 0 & 0 & 0 & 0 & \times & \times \end{bmatrix}$$

If no row interchanges are needed, the application of the GEM to a banded matrix does not create any fill-in outside the band and the reduction to triangular form can be done efficiently. If the bandwidth is β then, as we can easily show, the amount of work for the whole GEM is $O(n\beta^2)$. If the bandwidth is not too large, this can be quite efficient even for large n. Unfortunately, if row exchanges are necessary there could be significant fill-in outside the band and the efficiency of the process may suffer.

From these considerations we can surmise what an effective attack on the sparse matrix problem might be. First, we want to get the matrix into a banded form of small bandwidth. Next, in the triangularization, we want to eliminate or at least reduce the number of row exchanges. If row exchanges become necessary, we should do them in a way that tends to minimize the fill-in. While it is not known how to achieve all of this in an optimal way, a number of effective heuristic approaches have been developed.

If we have an arbitrary sparse matrix, we can try to reduce the bandwidth by row and column interchanges. Row interchanges have no effect on the solution, while column interchanges simply relabel variables. In any case, neither interchange affects the final solution. Algorithms for reordering rows and columns have been studied. One of the most widely used is the *Reverse Cuthill-McKee* method, which often does a very good job (see Duff, Erisman, and Reid [8], p. 154).

If a matrix is known to be positive–definite, then no row interchanges are necessary, and once we have a nice banded form the worst problem is solved. In the general case, though, we cannot do entirely without row exchanges. A diagonal element may be zero, so we cannot proceed, but even accepting a very small pivot may lead to instability. While maximal pivoting is best for stability, it may lead to excessive fill-in. Usually, one makes some kind of compromise by establishing a threshold above which a pivot is accepted. One might use, for example, a threshold of 10 percent of the maximum pivot. This will give us a set of possible pivots from which we then select one to keep the fill-in small. We can perhaps select that pivot for which the product of the numbers of nonzero elements in its row and its column is minimized.

There are further complications. Much of the efficiency of sparse matrix algorithms depends on the chosen data structures, and what is easy in one structure is not necessarily so in another. Implementing sparse matrix programs is a tricky matter and requires sophistication and much work. Fortunately, as in most matrix problems, there is ready-made software available, and this should be the starting point for anyone who needs to solve sparse systems.

Certain iterative methods are useful in special circumstances. Since these circumstances most often arise in connection with the numerical solution of partial differential equations, we defer this topic until later.

EXERCISES

1. Evaluate the two suggested sparse matrix storage schemes with respect to simple matrix operations, such as finding the element in position (i, j), comparing two matrices, adding and subtracting, and matrix multiplication.

2. Show that the LU decomposition of a tridiagonal matrix yields two matrices $\mathbf{L}$ and $\mathbf{U}$, both with bandwidth two.

3. A five-banded matrix is one for which $a_{ij} = 0$ for all $|i - j| > 2$. Develop formulas for the GEM solution for the solution of five-banded matrices, similar to (8.31).

4. Show that the amount of work needed for applying the GEM without row interchanges to a banded matrix of bandwidth β is of order $O(n\beta^2)$.

5. Show that the following matrix can be reduced to a matrix with bandwidth four by one row interchange followed by a column interchange.

$$\begin{bmatrix} \times & \times & 0 & 0 & 0 & \times & 0 \\ 0 & 0 & \times & 0 & 0 & 0 & \times \\ 0 & \times & 0 & \times & \times & \times & 0 \\ 0 & 0 & \times & \times & \times & \times & 0 \\ 0 & 0 & \times & \times & \times & 0 & \times \\ 0 & 0 & \times & 0 & \times & 0 & \times \\ \times & \times & 0 & \times & 0 & \times & 0 \end{bmatrix}$$

Chapter 9

Eigenvalue Problems

Another important problem in linear algebra is the computation of the *eigenvalues* and *eigenvectors* of a matrix. Eigenvalue problems occur frequently in connection with mechanical vibrations and other cases of periodic motion.

Example 9.1

Three equal masses are connected by a taut string, uniformly spaced, between two supports. (Figure 9.1).

When the masses are pulled from equilibrium in the y-direction, the tension in the string will pull them back toward equilibrium and cause the entire assembly to vibrate. Using a number of simplifying assumptions (e.g.,

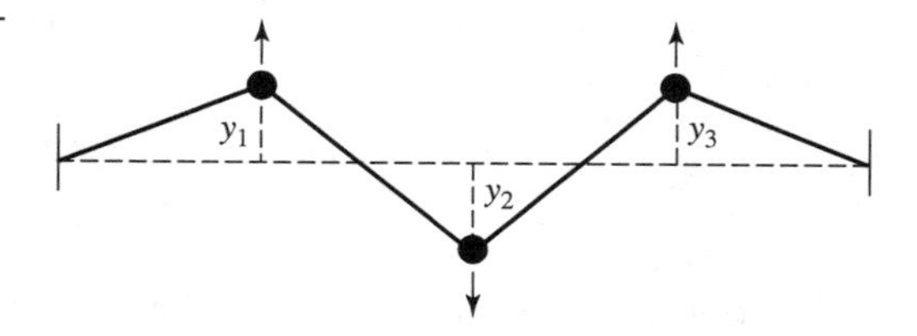

Figure 9.1
A simple vibration problem.

that the extension from equilibrium is small compared with the length of the string) and some suitable scaling, the equations of motion will have the form

$$\frac{d^2y_1}{dt^2} = -2y_1 + y_2,$$
$$\frac{d^2y_2}{dt^2} = y_1 - 2y_2 + y_3,$$
$$\frac{d^2y_3}{dt^2} = y_2 - 2y_3.$$

Under certain conditions, the assembly will settle into a steady state in which all three masses move periodically with the same frequency. If we write

$$y_i = a_i \cos(\omega\, t)$$

and substitute into the equation of motions, we find that the vibrational frequencies are determined by

$$-\omega^2 a_1 = -2a_1 + a_2,$$
$$-\omega^2 a_2 = a_1 - 2a_2 + a_3,$$
$$-\omega^2 a_3 = a_2 - 2a_3.$$

Putting $\lambda = \omega^2$, this can be written in matrix form as

$$\begin{bmatrix} 2 & -1 & 0 \\ -1 & 2 & -1 \\ 0 & -1 & 2 \end{bmatrix} \begin{bmatrix} a_1 \\ a_2 \\ a_3 \end{bmatrix} = \lambda \begin{bmatrix} a_1 \\ a_2 \\ a_3 \end{bmatrix}.$$

The vibrational frequencies are therefore determined by the eigenvalues of a symmetric 3×3 matrix.

■

The preceding is an instance of the *simple* eigenvalue problem

$$\mathbf{Au} = \lambda\,\mathbf{u}; \tag{9.1}$$

that is common in practice. Some physical situations lead to the more general eigenvalue problem

$$\mathbf{Au} = \lambda\,\mathbf{Bu}. \tag{9.2}$$

In both cases, the matrices are square, and we want to find eigenvalues λ for which the equation has a nontrivial eigenvector $\mathbf{u}$. Although the general eigenvalue problem has applications, we will restrict our discussion to the more common simple problem. The set of all eigenvalues of (9.1), called the *spectrum* of $\mathbf{A}$, will be denoted by $\sigma(\mathbf{A})$. The pair $(\lambda, \mathbf{u})$ is an *eigensolution* of (9.1).

As in the case of solutions of linear systems, the algorithms for computing eigenvalues involve a fair amount of manipulative detail. Fortunately, computer programs for them are readily available so the need for implementing them does not often arise. For most users it is not essential to remember all the fine points of the eigenvalue algorithms, but it is important to understand the basic ideas behind these methods, know how to use them, and have an idea of what their limitations are. For this, we first need to review some of the elementary results from the theory of eigenvalues.

9.1 Some Useful Results

The results quoted here can be found in books on linear algebra. We will leave it to the reader to look up the proofs. The first few theorems are simple and can be proved as exercises.

Theorem 9.1

If $(\lambda, \mathbf{u})$ is an eigensolution of the matrix $\mathbf{A}$, then $(\alpha\lambda + \beta, \mathbf{u})$ is an eigensolution of the matrix $\alpha\mathbf{A} + \beta\mathbf{I}$. If $\mathbf{u}$ is an eigenvector, so is $c\mathbf{u}$ for any constant $c \neq 0$.

The second part of this theorem states that the eigenvectors are indeterminate to within a constant. When convenient, we can therefore assume that all eigenvectors are normalized[1] by

$$||\mathbf{u}|| = 1.$$

Theorem 9.2

If $(\lambda,\ \mathbf{u})$ is an eigensolution for the matrix $\mathbf{A}$ and $\mathbf{A}$ is invertible, then $(\frac{1}{\lambda},\ \mathbf{u})$ is an eigensolution of $\mathbf{A}^{-1}$.

Many of the techniques for solving eigenvalue problems involve transforming the original matrix into another one with closely related, but more easily computed, eigensolutions. The primary transformation used in this is given in the following definition.

[1] As in Chapter 8, we use the 2-norm exclusively.

Definition 9.1

Let $\mathbf{T}$ be an invertible matrix. Then the transformation

$$\mathbf{B} = \mathbf{TAT}^{-1}$$

is called a *similarity transformation*. The matrices $\mathbf{A}$ and $\mathbf{B}$ are said to be similar.

Theorem 9.3 Suppose that $(\lambda, \mathbf{u})$ is an eigensolution for the matrix $\mathbf{A}$. Then $(\lambda, \mathbf{Tu})$ is an eigensolution of the similar matrix

$$\mathbf{B} = \mathbf{TAT}^{-1}.$$

An important theorem, called the *Alternative Theorem*, connects the solvability of a system $\mathbf{Ax} = \mathbf{b}$ with the spectrum of $\mathbf{A}$.

Theorem 9.4 The $n \times n$ linear system

$$(\mathbf{A} - \alpha\,\mathbf{I})\mathbf{x} = \mathbf{y}$$

has a unique solution if and only if α is not in the spectrum of $\mathbf{A}$.

The Alternative Theorem gives us a way of analyzing the spectrum of a matrix. If λ is in $\sigma(\mathbf{A})$ then $\mathbf{A} - \lambda\mathbf{I}$ cannot have an inverse; conversely, if $\mathbf{A} - \lambda\mathbf{I}$ is not invertible, then λ must be in the spectrum of $\mathbf{A}$. This implies that the spectrum of $\mathbf{A}$ is the set of all λ for which

$$\det(\mathbf{A} - \lambda\mathbf{I}) = 0.$$

Since the determinant of any matrix can be found with a finite number of multiplications and divisions, say by expansion of minors, it follows that $\det(\mathbf{A} - \lambda\mathbf{I})$ is a polynomial of degree n in λ. This is the *characteristic polynomial* which will be denoted by

$$p_{\mathbf{A}}(\lambda) = \det(\mathbf{A} - \lambda\mathbf{I}). \tag{9.3}$$

The eigenvalue problem is therefore reducible to a polynomial root-finding problem which, at least in theory, is well understood. While we usually do not solve eigenvalue problems quite in this way, the observation is used in many of the eigenvalue algorithms. For instance, we can draw on our extensive knowledge of polynomials and their roots to characterize the spectrum.

Theorem 9.5 An $n \times n$ matrix has exactly n eigenvalues in the complex plane.

Proof: This follows from the fact that a polynomial of degree n has exactly n roots in the complex plane. ■

In this theorem we must account for multiple roots of $p_{\mathbf{A}}(\lambda)$. If λ_0 is a root of $p_{\mathbf{A}}(\lambda)$ with multiplicity k, then λ_0 is an eigenvalue with the same multiplicity.

In practice, one way of finding eigenvalues is to transform the original matrix by a sequence of similarity transformations until it becomes more manageable, for example, so that its determinant can be evaluated quickly or so that its eigenvalues can be immediately read off. The latter is certainly true if the matrix is diagonal. When a matrix is not diagonal, but in some way nearly so, we can use the next result to locate its eigenvalues approximately.

Theorem 9.6 The spectrum of a matrix $\mathbf{A}$ is contained entirely in the union of the disks (in the complex plane)

$$d_i(\alpha) = \{\alpha \ : \ |[\mathbf{A}]_{ii} - \alpha| \leq \sum_{j \neq i} |[\mathbf{A}]_{ij}|\}, \quad i = 1,\ 2,\ \ldots,\ n. \tag{9.4}$$

Proof: Assume that λ is in $\sigma(\mathbf{A})$ and let $\mathbf{u}$ be the corresponding eigenvector. Then, for every i,

$$[\mathbf{A}]_{ii}[\mathbf{u}]_i - \lambda[\mathbf{u}]_i = \sum_{j \neq i} [\mathbf{A}]_{ij}\, [\mathbf{u}]_j.$$

Suppose now that the largest component of $\mathbf{u}$ in magnitude is $[\mathbf{u}]_k$. Then

$$([\mathbf{A}]_{kk} - \lambda)[\mathbf{u}]_k = \sum_{j \neq k} [\mathbf{A}]_{kj}\, [\mathbf{u}]_j$$

and

$$\begin{aligned} |[\mathbf{A}]_{kk} - \lambda| &= \left| \sum_{j \neq k} [\mathbf{A}]_{kj}\, [\mathbf{u}]_j / [\mathbf{u}]_k \right| \\ &\leq \sum_{j \neq k} |[\mathbf{A}]_{kj}|. \end{aligned}$$

Therefore, λ is in $d_k(\alpha)$, proving (9.4). ■

This is a simplified version of the *Gerschgorin Theorem.* The more general version states that each disjoint region, which is made up from the union of one or more disk components, contains exactly as many eigenvalues as it has components. For a proof of the more general result, see Atkinson [4].

Example 9.2 Consider the matrix

$$\mathbf{A} = \begin{bmatrix} 1.0 & 0.1 & 0.2 \\ 0.2 & 1.4 & 0 \\ 0.1 & 0.1 & -3.0 \end{bmatrix}.$$

Then Theorem 9.6 shows that all the eigenvalues lie in the shaded disks shown in Figure 9.2. It also proves that $0 \notin \sigma(\mathbf{A})$, so that the matrix is invertible. The extension of Gerschgorin's theorem tells us that there are two eigenvalues in the right half-plane and one in the left half-plane. ■

Theorem 9.6 is useful because it tells us exactly how many eigenvalues there are in a region. Unfortunately, we may have to go into the complex plane to find them and this complicates matters. There is also the issue of the eigenvectors that is less readily resolved. All of this becomes a lot easier when we deal with symmetric matrices.

Theorem 9.7 The eigenvalues of a symmetric matrix are all real. To each eigenvalue there corresponds a distinct, real eigenvector.

Figure 9.2 Gerschgorin disks for locating eigenvalues.

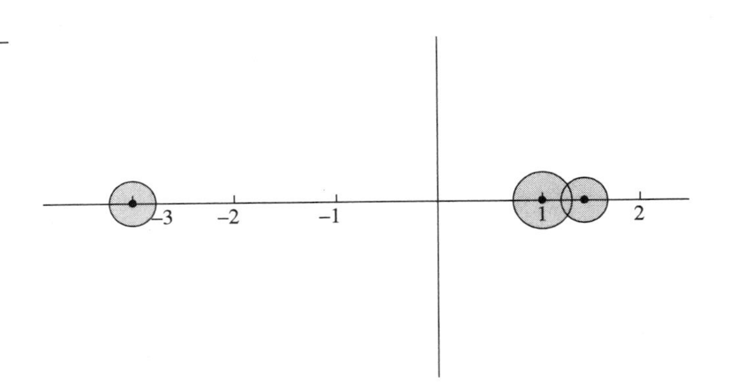

Theorem 9.8 If $\mathbf{A}$ is symmetric, then the eigenvectors associated with two different eigenvalues are orthogonal to each other. This implies that, even if there are multiple eigenvalues, there are eigenvectors $\{\mathbf{u}_1, \mathbf{u}_2, \ldots, \mathbf{u}_n\}$ such that

$$\mathbf{u}_i^T \mathbf{u}_j = 0$$

for all $i \neq j$.

Since the orthogonality of the eigenvectors makes them linearly independent, it follows from this theorem that every n-vector $\mathbf{x}$ can be expanded as a linear combination of the eigenvectors of a symmetric $n \times n$ matrix. In particular, because we can always normalize the eigenvectors such that $\mathbf{u}_i^T \mathbf{u}_i = 1$, we have that for every n-vector $\mathbf{x}$,

$$\mathbf{x} = \sum_{i=1}^{n} (\mathbf{x}^T \mathbf{u}_i)\mathbf{u}_i.$$

Such simple results normally do not hold for nonsymmetric matrices, but they are of great help in working with symmetric cases. Since in practice symmetric matrices are very common, the symmetric eigenvalue problem has received a great deal of attention.

EXERCISES

1. Prove Theorem 9.1.
2. Prove Theorem 9.2.
3. Prove Theorem 9.3.
4. Show that if the matrix $\mathbf{B}$ is invertible, then the eigensolutions of the generalized eigenvalue problem (9.2) are the eigensolutions of $\mathbf{B}^{-1}\mathbf{A}$.
5. Use the Gerschgorin theorem to locate the eigenvalues of

$$\mathbf{A} = \begin{bmatrix} 1 & 0.2 & 0.2 & 0 \\ 0 & 1 & 0.2 & 0.1 \\ 0.3 & 0 & 2 & 0.1 \\ 0.2 & 0.2 & 0 & -1 \end{bmatrix}.$$

6. Let $\mathbf{A}$ be a symmetric matrix with two eigensolutions $(\lambda_1, \mathbf{u}_1)$ and $(\lambda_2, \mathbf{u}_2)$. Show that if $\lambda_1 \neq \lambda_2$, then $\mathbf{u}_1$ and $\mathbf{u}_2$ must be orthogonal to each other.
7. For the matrix

$$\mathbf{A} = \begin{bmatrix} 1 & 0 & 2 \\ 2 & 3 & 1 \\ 0 & 2 & 6 \end{bmatrix},$$

find $p_{\mathbf{A}}(\lambda)$ and use it to find all the eigenvalues of $\mathbf{A}$.

8. Find the eigenvalues of the general eigenvalue problem with $\mathbf{A}$ as in Exercise 7 and

$$\mathbf{B} = \begin{bmatrix} 2 & 0 & 1 \\ 0 & 2 & 0 \\ 1 & 0 & 2 \end{bmatrix}.$$

9. Prove that $\det(\mathbf{A} - \lambda\mathbf{I})$ is a polynomial in λ.

10. Show that if $\lambda \in \sigma(\mathbf{A})$ then $\lambda^k \in \sigma(\mathbf{A}^k)$ for all positive integers k.

11. Show that a symmetric matrix is positive–definite if and only if its smallest eigenvalue is positive.

12. Two masses m_1 and m_2 are attached to three springs, with spring constants k_1, k_2, and k_3. The masses move on a frictionless plane, as shown in the diagram below. Write the equations of motion for this system and derive the eigenvalue problem that can be used to find the frequencies of periodic motion.

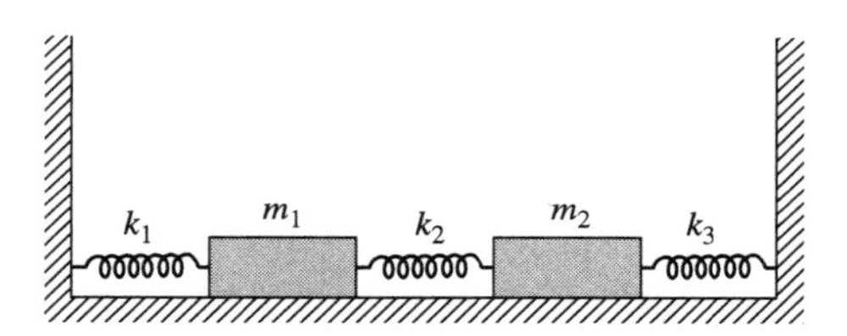

13. Two masses m_1 and m_2 are attached to springs, as shown below. In addition to the forces from the springs, the masses are also acted on by gravity. Write the equations of motion for this system and derive the eigenvalue problem that can be used to find the frequencies of periodic motion.

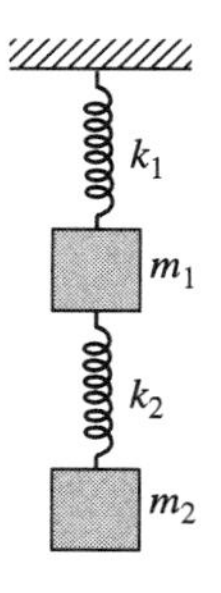

14. Reformulate Example 9.1 for n masses, uniformly spaced along the string. What system is approached as $n \to \infty$? Can you guess what equation might describe such a system?

9.2 The Power Method

The power method is conceptually the simplest way of getting eigenvalues of a symmetric matrix. It is an iterative method that, in many cases, converges to the eigenvalue of largest magnitude. For simplicity, we will assume that the eigenvalues of the symmetric matrix $\mathbf{A}$ are all simple and have been labeled so that $|\lambda_1| > |\lambda_2| \geq \ldots$. The corresponding eigenvectors will be $\mathbf{u}_1, \mathbf{u}_2, \ldots$.

Take any vector $\mathbf{x}$; by Theorem 9.8, this vector can be expanded as

$$\mathbf{x} = \sum_{i=1}^{n} a_i \mathbf{u}_i.$$

Then

$$\mathbf{A}\mathbf{x} = \sum_{i=1}^{n} a_i \lambda_i \mathbf{u}_i,$$

and generally,

$$\begin{aligned} \mathbf{A}^k\mathbf{x} &= \sum_{i=1}^{n} a_i \lambda_i^k \mathbf{u}_i \\ &= \lambda_1^k \sum_{i=1}^{n} a_i \left(\frac{\lambda_i}{\lambda_1}\right)^k \mathbf{u}_i. \end{aligned} \tag{9.5}$$

As k becomes large all the terms in this sum will become small except the first, so that eventually

$$\mathbf{A}^k\mathbf{x} \cong \lambda_1^k a_1 \mathbf{u}_1. \tag{9.6}$$

If we take any component of this vector, say the ith one, and compare it for successive iterates, we find that the *dominant* eigenvalue λ_1 is approximated by

$$\lambda_1 \cong \frac{[\mathbf{A}^{k+1}\mathbf{x}]_i}{[\mathbf{A}^k\mathbf{x}]_i}. \tag{9.7}$$

For stability reasons, we usually take i so that we get the largest component of $[\mathbf{A}^k\mathbf{x}]_i$, but in principle any i will do. An approximation to the corresponding eigenvector can be obtained by

$$\mathbf{u}_1 \cong c\mathbf{A}^k\mathbf{x}, \tag{9.8}$$

where the constant c can be chosen for normalization.

In practice, we do not evaluate (9.7) by computing the powers of $\mathbf{A}$. Instead, we iteratively compute

$$\mathbf{x}_{k+1} = \mathbf{A}\mathbf{x}_k,$$

with initial guess $\mathbf{x}_0 = \mathbf{x}$. Then

$$\lambda_1 \cong \frac{[\mathbf{x}_{k+1}]_i}{[\mathbf{x}_k]_i}.$$

Example 9.3 The power method was applied to

$$\mathbf{A} = \begin{bmatrix} 1 & 0 & 0.2 & 0.1 \\ 0 & 3 & 0.1 & 0.2 \\ 0.2 & 0.1 & 0 & 0.1 \\ 0.1 & 0.2 & 0.1 & -2 \end{bmatrix},$$

with an initial guess $\mathbf{x} = (1,\ 1,\ 1,\ 1)^T$. After fourteen iterations, we obtained the estimates

$$\lambda_1 = 3.0108, \qquad \mathbf{u}_1 = (0.0054,\ 1.0000,\ 0.0348,\ 0.0437)^T.$$

After an additional seven iterations, the results were

$$\lambda_1 = 3.0117, \qquad \mathbf{u}_1 = (0.0055,\ 1.0000,\ 0.0349,\ 0.0405)^T.$$

The last approximate eigenvalue can be expected to be accurate to at least three significant digits. A similar accuracy can be expected for the eigenvector.

■

The power method is simple, but it has some obvious shortcomings. First, note that the argument leading to (9.7) works only if $a_1 \neq 0$; that is, the starting guess $\mathbf{x}$ must have a component in the direction of the eigenvector $\mathbf{u}_1$. Since we do not know $\mathbf{u}_1$, this is hard to enforce. In practice, though, rounding will eventually introduce a small component in this direction, so the power method should work in any case. But it may be quite slow and it is a good idea to use the best guess for $\mathbf{u}_1$ as an initial guess. If no reasonable value for $\mathbf{u}_1$ is available, we can simply use a random number generator to choose a starting value.

The power method is iterative, so its rate of convergence is of concern. It is not hard to see that it has an iterative order of convergence one and that each step reduces the error roughly by a factor of

$$c = \left| \frac{\lambda_2}{\lambda_1} \right|. \tag{9.9}$$

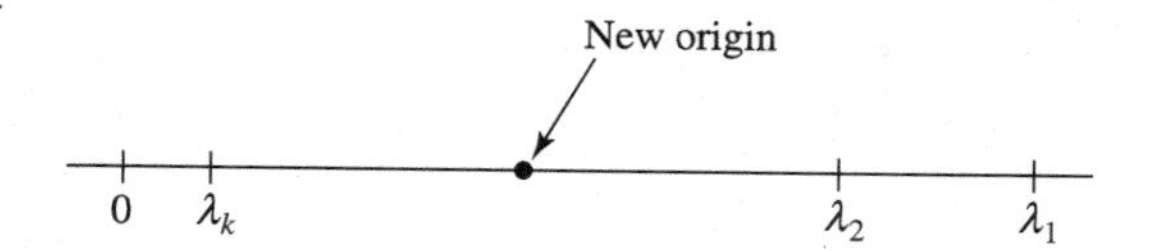

Figure 9.3 Shifting the origin to accelerate the power method.

The method, as described, works reasonably well only if the dominant eigenvalue is simple and significantly separated from the next largest eigenvalue. We can sometimes make an improvement by shifting the origin, as indicated by Theorem 9.1. If we know the approximate positions of the eigenvalue closest to λ_1, say λ_2, and the eigenvalue farthest from it, say λ_k, we can shift the origin so that

$$\left|\frac{\lambda_2}{\lambda_1}\right| = \left|\frac{\lambda_k}{\lambda_1}\right|,$$

which minimizes the ratio of largest to second largest eigenvalue. This can be achieved by shifting the origin by an amount β, where β is halfway between λ_2 and λ_k, as shown in Figure 9.3. When we now apply the power method to the matrix $\mathbf{A} - \beta\mathbf{I}$, we can expect faster convergence.

Example 9.4 Consider the matrix in Example 9.3

$$\mathbf{A} = \begin{bmatrix} 1 & 0 & 0.2 & 0.1 \\ 0 & 3 & 0.1 & 0.2 \\ 0.2 & 0 & 0 & 0.1 \\ 0.1 & 0.2 & 0.1 & -2 \end{bmatrix}.$$

Gerschgorin's theorem tells us that the approximate location of the eigenvalues are $-2,\ 0,\ 1,\ 3$. We therefore expect that in the computations for Example 9.3, the error is reduced approximately by a factor of $2/3$ on each iteration. If we shift the origin with $\beta = -0.5$, we can expect an error attenuation of about $1.5/3.5 = 0.4286$ on each iteration, about twice as fast as the original computation. Using the same starting value as in Example 9.3, the power method with this shift gave

$$\lambda_1 = 3.0117$$

after eleven iterations. ■

Other adjustments can be made to increase the usefulness of the power method. For example, we can get the smallest eigenvalue by inverting the matrix and applying the power method to $\mathbf{A}^{-1}$. By Theorem 9.2 this will give the reciprocal of the smallest eigenvalue of $\mathbf{A}$. Once the eigenvector $\mathbf{u}_1$ has been found we can *deflate* the original matrix by

$$\mathbf{A}_1 = \mathbf{A} - \lambda_1 \mathbf{u}_1 \mathbf{u}_1^T. \tag{9.10}$$

As is easily shown, the spectrum of $\mathbf{A}_1$ is $\{\lambda_2, \lambda_3, \ldots, \lambda_n, 0\}$, so that we can use this observation to compute the second largest eigenvalue, and so on.

Example 9.5 In Example 9.4, the eigenvalue λ_1 and the corresponding eigenvector $\mathbf{u}_1$ were computed by iteration to an estimated accuracy of 10^{-5}. These were then used in (9.10) and the power method applied. After fourteen iterations, we obtained

$$\lambda_2 = -2.0153.$$

The accuracy of this result is hard to judge since it depends not only on the number of iterations, but also on the limited accuracy with which we know λ_1 and $\mathbf{u}_1$.

■

In principle, we can carry this idea further and look at

$$\mathbf{A}_2 = \mathbf{A} - \lambda_1 \mathbf{u}_1 \mathbf{u}_1^T - \lambda_2 \mathbf{u}_2 \mathbf{u}_2^T,$$

whose dominant eigenvalue is λ_3, but the process becomes increasingly more cumbersome and inaccurate.

The main advantage of the power method is simplicity. It is easily programmed and can take full advantage of special properties of the matrix, such as sparsity. As long as we are interested only in the dominant eigenvalue, it can be quite effective. But while one can modify the power method in various ways to get other eigenvalues and eigenvectors, the process quickly loses its attraction. If we need more than just the dominant eigenvalue, it is probably better to use a method that gives all eigenvalues simultaneously.

EXERCISES

1. In arguments about the power method we have assumed the matrix $\mathbf{A}$ is symmetric. The method often also works for nonsymmetric matrices. What simple assumption can we make so that (9.7) and (9.8) will hold for all matrices?

2. Prove the statement made about the spectrum of $\mathbf{A}_1$ in (9.10).

3. Suppose we know that a matrix has an eigenvalue near $\lambda = 1$, but we do not know anything about the location of the other eigenvalues. How could you use the power method to refine the accuracy of the eigenvalue near 1?

4. Suppose that $\mathbf{A}$ has all positive eigenvalues, with the maximal eigenvalue larger than one. If we iterate with $\mathbf{A}^2$ we could expect faster convergence, since its eigenvalues are better separated than those of $\mathbf{A}$. Do you think this observation has any practical significance?

5. Suppose the dominant eigenvalue of a matrix is double; that is, $\lambda_1 = \lambda_2$. How will this affect the convergence of the power method?

6. What happens in the power method when $\lambda_1 = -\lambda_2$?

7. It is often recommended that in implementation of the power method the iterates are normalized on each step (or at least every few steps) by computing

$$\mathbf{z} = \frac{\mathbf{A}^k\mathbf{x}}{||\mathbf{A}^k\mathbf{x}||}$$

and restarting the iteration with this normalized vector. What is the practical reason behind this suggestion?

8. Can the dominant eigenvalue be estimated by

$$\lambda_1 = \frac{||\mathbf{A}^{k+1}\mathbf{x}||}{||\mathbf{A}^k\mathbf{x}||}$$

instead of (9.7)?

9. Compute the smallest eigenvalue of

$$\mathbf{A} = \begin{bmatrix} -4 & 14 & 0 \\ -5 & 13 & 0 \\ -1 & 0 & 2 \end{bmatrix}$$

to an accuracy of 10^{-3}, using the power method (a) on $\mathbf{A} - 13\mathbf{I}$, and (b) on $\mathbf{A}^{-1}$.

9.3 The Jacobi and Householder Methods for Symmetric Matrices*

Most of the methods for finding eigenvalues of general matrices rely on a sequence of similarity transformations that bring the matrix into a suitable form. The easiest form to deal with is a diagonal matrix, as its eigenvalues are explicitly exhibited. Reduction to diagonal form is the idea behind the *Jacobi method.* Because a matrix with complex eigenvalues cannot be reduced to diagonal form with real arithmetic, the Jacobi method can be guaranteed to work only for symmetric matrices.

Figure 9.4
A Jacobi transformation matrix.

$$
\mathbf{P}_{rs} = \begin{bmatrix}
1 & 0 & & & & & & & & 0 \\
0 & \ddots & & & & & & & & \\
 & & 1 & & & & & & & \\
 & & & \cos\theta & 0 & \cdots & 0 & \sin\theta & & \\
 & & & 0 & 1 & & & 0 & & \\
 & & & \vdots & & \ddots & & \vdots & & \\
 & & & 0 & & & 1 & 0 & & \\
 & & & -\sin\theta & 0 & \cdots & 0 & \cos\theta & & \\
 & & & & & & & & 1 & \\
 & & & & & & & & \ddots & 0 \\
0 & & & & & & & & 0 & 1
\end{bmatrix}
\begin{matrix} \\ \\ \\ \leftarrow r \\ \\ \\ \\ \leftarrow s \\ \\ \\ \\ \end{matrix}
$$

$$\uparrow r \qquad \uparrow s$$

The Jacobi method uses a sequence of simple transformation matrices whose form is shown in Figure 9.4. The transformation represented by $\mathbf{P}_{rs}$ is said to be a *rotation pivoted* at $(r,\ s)$.

It is easy to show that P_{rs} is unitary for any θ, so

$$\mathbf{A}_1 = \mathbf{P}_{rs}\mathbf{A}\mathbf{P}_{rs}^T \tag{9.11}$$

is an eigenvalue preserving similarity transformation. It can also be shown, with a little work, that if we choose θ such that

$$\tan 2\theta = \frac{2\,[\mathbf{A}]_{rs}}{[\mathbf{A}]_{rr} - [\mathbf{A}]_{ss}}, \tag{9.12}$$

then the transformation (9.11) annihilates the $(r,\ s)$ elements of $\mathbf{A}$; that is,

$$[\mathbf{P}_{rs}\mathbf{A}\mathbf{P}_{rs}^T]_{rs} = [\mathbf{P}_{rs}\mathbf{A}\mathbf{P}_{rs}^T]_{sr} = 0. \tag{9.13}$$

The transformation with pivot $(r,\ s)$ changes the elements in rows r and s, and columns r and s, but does not affect the rest of the matrix.

After reducing the elements $[\mathbf{A}]_{rs}$ and $[\mathbf{A}]_{sr}$ to zero, we shift the pivot to another point and eliminate other elements. Unfortunately, it is not possible to do this in a way that the previously eliminated elements stay zero; consequently the process does not usually terminate after a finite number of rotations. But it is known that, if the pivots are selected carefully, the sum of squares of the off-diagonal elements get smaller on each rotation.

Example 9.6 Consider the 4×4 symmetric matrix

$$\mathbf{A} = \begin{bmatrix} 3 & 9 & 4 & -7 \\ 9 & -9 & 8 & 4 \\ 4 & 8 & 5 & -4 \\ -7 & 4 & -4 & 2 \end{bmatrix}.$$

The sum of squares of the off-diagonal elements is 484. After one Jacobi transformation with pivot (1, 2), the matrix

$$\mathbf{A}_1 = \mathbf{P}_{12}\mathbf{A}\mathbf{P}_{12}^T = \begin{bmatrix} 7.8167 & 0.0000 & 7.3016 & -4.2843 \\ 0.0000 & -13.8167 & 5.1660 & 6.8297 \\ 7.3016 & 5.1660 & 5.0000 & -4.0000 \\ -4.2843 & 6.8297 & -4.0000 & 2.0000 \end{bmatrix}$$

has the sum of squares of the off-diagonal elements reduced to 322. Another transformation with pivot (1, 3) results in the matrix

$$\mathbf{A}_2 = \mathbf{P}_{13}\mathbf{A}_1\mathbf{P}_{13}^T = \begin{bmatrix} 13.8445 & 3.2888 & 0.0000 & -5.8504 \\ 3.2888 & -13.8167 & 3.9838 & 6.8297 \\ 0.0000 & 3.9838 & -1.0278 & -0.3571 \\ -5.8504 & 6.8297 & -0.3571 & 2.0000 \end{bmatrix}$$

and has reduced the sum of squares to about 215. ■

In general, we apply rotations until the off-diagonal elements are sufficiently small. Gerschgorin's theorem can be used to determine how small this should be with respect to the required accuracy. Since each rotation involves choosing a pivot, and because this choice affects the efficiency of the method, there is a question of a good pivoting strategy. One successful method for pivot selection is to use (r, s) so that $[\mathbf{A}]_{rs}$ has the largest magnitude of all the off-diagonal elements. Unfortunately, it is expensive to find this maximal element at each step. An alternative strategy is to cycle through all the off-diagonal positions in some order, but rotating only if the element in the pivot position is larger than some threshold (say, 10 percent of the initial maximum element). When the cycle is completed, a new threshold is computed and the process continued.

The eigenvectors are also available from the Jacobi transformations. Since the eigenvectors of a diagonal matrix are just the unit vectors

$$\begin{aligned} [\mathbf{e}_i]_j &= 1, \ i = j, \\ &= 0, \ i \neq j, \end{aligned}$$

we can apply Theorem 9.3 to get the eigenvectors of the original matrix. If the diagonal matrix is obtained by the sequence of rotations $\mathbf{P}_1, \mathbf{P}_2, \ldots, \mathbf{P}_k$, then Theorem 9.3 tells us that the columns of the matrix

$$\mathbf{Q} = \mathbf{P}_1^T \mathbf{P}_2^T \ \ldots \ \mathbf{P}_k^T \tag{9.14}$$

are the eigenvectors of $\mathbf{A}$, with the ith column corresponding to the eigenvalue in the ith row and column of the diagonal matrix $\mathbf{P}_k \ \ldots \ \mathbf{P}_2\mathbf{P}_1\mathbf{A}$. However, because we can never iterate long enough to get a diagonal form, both the eigenvalues and eigenvectors obtained by a finite number of Jacobi transformations are only approximate.

Each rotation of the Jacobi method changes only about $4n$ elements of which, because of symmetry, only $2n$ have to be computed. Each single Jacobi transformation can be done quickly, but when high accuracy is required the whole computation can take many individual transformations and thus become time-consuming. The Jacobi method is most suitable when the original matrix already has small off-diagonal elements.

An alternative is to use Householder transformations. The unitary transformations in Theorem 8.3 reduce the matrix to triangular form, but they are not similarity transformations and so cannot be used for the eigenvalue problem. If we try to fix this by post-multiplying by $\mathbf{Q}_i$ at each step, we fail also because this post-multiplication destroys the just-generated zeros. The right thing to do is not to try to diagonalize the matrix, but rather to convert it to tridiagonal form, which requires only that we "shorten" the column–vector $\mathbf{c}$ in (8.17) that is to be reduced. After making suitable changes to the transformations in Theorem 8.3, we get a method for reducing a symmetric matrix to tridiagonal form.

Theorem 9.9 Let $\mathbf{A}$ be a symmetric $n \times n$ matrix. Define a sequence of matrices by

$$\mathbf{A}_k = \mathbf{Q}_k \mathbf{A}_{k-1} \mathbf{Q}_k,$$

with

$$\mathbf{A}_0 = \mathbf{A},$$

$$\mathbf{Q}_k = \begin{bmatrix} \mathbf{I}_k & 0 \\ 0 & \mathbf{P}_{n-k} \end{bmatrix},$$

$$\mathbf{P}_{n-k} = \mathbf{I}_{n-k} - 2\frac{\mathbf{v}\mathbf{v}^T}{\mathbf{v}^T\mathbf{v}},$$

$$\mathbf{v} = \begin{bmatrix} c_{k+1} + \alpha \\ c_{k+2} \\ \vdots \\ c_n \end{bmatrix},$$

$$\alpha = \sqrt{\sum_{i=k+1}^{n} c_i^2},$$

where $c_{k+1}, c_{k+2}, \ldots, c_n$ denote the elements below the main diagonal in the kth column of $\mathbf{A}_{k-1}$. Then $\mathbf{A}_{n-2}$ is a tridiagonal matrix similar to $\mathbf{A}$.

Proof: This requires some matrix manipulations which are suitable as an exercise.[2] ■

Once a matrix has been reduced to tridiagonal form, much of the work is done. To find the eigenvalues we can construct the characteristic polynomial $p_{\mathbf{A}}(\lambda)$ and find its roots. For general matrices this is not feasible, but it is quite suitable for tridiagonal matrices. Suppose we have the symmetric tridiagonal matrix

$$\mathbf{A} = \begin{bmatrix} a_1 & b_1 & 0 & \ldots & 0 \\ b_1 & a_2 & b_2 & \ddots & \vdots \\ 0 & b_2 & a_3 & \ddots & 0 \\ \vdots & \ddots & \ddots & \ddots & b_{n-1} \\ 0 & \ldots & 0 & b_{n-1} & a_n \end{bmatrix}.$$

Let

$$p_i(\lambda) = \det(\mathbf{M}_i - \lambda \mathbf{I}),$$

where $\mathbf{M}_i$ is the principal minor of $\mathbf{A} - \lambda\mathbf{I}$, consisting of rows i to n and columns i to n of $\mathbf{A} - \lambda\mathbf{I}$. Then

$$p_n(\lambda) = a_n - \lambda,$$
$$p_{n-1}(\lambda) = (a_{n-1} - \lambda)(a_n - \lambda) - b_{n-1}^2.$$

An expansion by minors then gives

$$p_{n-i}(\lambda) = (a_{n-i} - \lambda)p_{n-i+1}(\lambda) - b_{n-i}^2 p_{n-i+2}(\lambda), \; i = 2, 3, \ldots, n-1. \tag{9.15}$$

[2]Note that the transformation here is not identical to that defined in Theorem 8.3. The difference is that the $\mathbf{Q}_k$ here is $\mathbf{Q}_{k+1}$ in Theorem 8.3.

Since $p_{\mathbf{A}}(\lambda) = p_1(\lambda)$, this recursive process can be used to find the nth characteristic polynomial quickly. After that, we can use polynomial root-finders to get the eigenvalues.

Example 9.7 Find the eigenvalues of

$$\mathbf{A} = \begin{bmatrix} 2 & 1 & 0 & 1 \\ 1 & 3 & 2 & 1 \\ 0 & 2 & 4 & 1 \\ 1 & 1 & 1 & 1 \end{bmatrix}.$$

Two Householder transformations, eliminating $[\mathbf{A}]_{41}$ and $[\mathbf{A}]_{42}$, give

$$\mathbf{A}_2 = \begin{bmatrix} 2 & -1.4142 & 0 & 0 \\ -1.4142 & 3 & -2.3452 & 0 \\ 0 & -2.3452 & 4 & -0.7071 \\ 0 & 0 & -0.7071 & 1 \end{bmatrix},$$

so that

$$\begin{aligned} p_4(\lambda) &= 1 - \lambda, \\ p_3(\lambda) &= (1-\lambda)(4-\lambda) - 0.5 \\ &= 3.5 - 5\lambda + \lambda^2, \\ p_2(\lambda) &= (3-\lambda)p_3(\lambda) - 5.5p_4(\lambda) \\ &= 5 - 13\lambda + 8\lambda^2 - \lambda^3, \\ p_1(\lambda) &= (2-\lambda)p_2(\lambda) - 2p_3(\lambda) \\ &= 3 - 21\lambda + 27\lambda^2 - 10\lambda^3 + \lambda^4 \\ &= p_{\mathbf{A}}(\lambda). \end{aligned}$$

Using a polynomial root-finding method, we find that the roots of $p_{\mathbf{A}}(\lambda)$ are

$$\begin{aligned} \lambda_1 &= 6.1545, \\ \lambda_2 &= 2.6624, \\ \lambda_3 &= 1.0000, \\ \lambda_4 &= 0.1831, \end{aligned}$$

which are the eigenvalues of $\mathbf{A}$ to the accuracy shown. ■

The characteristic polynomial $p_{\mathbf{A}}(\lambda)$ has some special properties that can be exploited in actual implementation of this method, so we do not always use standard polynomial root-finders. But this is a technical matter that we need not pursue here; we only need to understand that methods based on Householder transformations are efficient and stable. For details see Atkinson [4], p.533.

EXERCISES

1. Show that the Jacobi rotation matrices are unitary.
2. Show that the choice of θ in (9.12) annihilates the matrix elements in (9.13).
3. Substantiate the claim that $\mathbf{Q}$ in (9.14) contains all the eigenvectors.
4. Prove Theorem 9.9.
5. Apply two Jacobi rotations, pivoted at (2, 1) and (4, 1), to the matrix

$$\mathbf{A} = \begin{bmatrix} 4 & 2 & 0 & 0 \\ 2 & 3 & 6 & 0 \\ 0 & 6 & 1 & 2 \\ 1 & 0 & 2 & 2 \end{bmatrix}.$$

   Compute the off-diagonal elements before and after each rotation.
6. Find the eigenvalues of

$$\mathbf{A} = \begin{bmatrix} 4 & 2 & 0 & 0 \\ 2 & 3 & 6 & 0 \\ 0 & 6 & 1 & 2 \\ 0 & 0 & 2 & 2 \end{bmatrix}.$$

7. Make a rough count of the number of operations needed to tridiagonalize a symmetric matrix by Householder reductions.
8. Carry out the expansion by minors to show that (9.15) is correct.
9. A *persymmetric* matrix is a matrix that is symmetric about both diagonals. Consider the 4×4 persymmetric matrix

$$\mathbf{A} = \begin{bmatrix} 2 & -1 & 0 & 0 \\ -1 & 2 & -1 & 0 \\ 0 & -1 & 2 & -1 \\ 0 & 0 & -1 & 2 \end{bmatrix}.$$

(a) Show that all the eigenvalues are real and positive.

(b) Form the characteristic polynomial for **A**.

(c) Using the bisection method, compute the minimal eigenvalue to four significant digits by finding the smallest root of the corresponding characteristic polynomial of **A**.

(d) Verify your result in (c) by applying the power method to $\mathbf{A}^{-1}$.

10. Find explicit expressions for the eigenvalues of the Jacobi rotation matrix $\mathbf{P}_{rs}$.

9.4 Eigenvalues of Nonsymmetric Matrices*

The computation of the eigenvalues of nonsymmetric matrices is a difficult topic, much of which is beyond the scope of this text. We give a brief outline here to indicate how one deals with this problem in practice.

The obvious complication for nonsymmetric matrices is that their spectrum can contain complex values. Consequently, the Jacobi method cannot work in the way we have described it. Another complication is that although an $n \times n$ matrix always has n eigenvalues, it does not necessarily have n eigenvectors. A simple example is the matrix

$$\mathbf{A} = \begin{bmatrix} 1 & 1 \\ 0 & 1 \end{bmatrix}, \tag{9.16}$$

which has a double eigenvalue at $\lambda = 1$, but only one eigenvector proportional to $\mathbf{u} = (1,\ 0)$. This potential lack of a complete set of eigenvectors invalidates the argument for the power method, so it cannot always be used for general matrices.

The most successful approach starts with an application of the Householder transformation in Theorem 9.9. We can still eliminate the same elements in the lower triangular part but, because of lack of symmetry, this will not simultaneously get rid of the corresponding elements in the upper triangular part. A complete Householder reduction yields a *Hessenberg* matrix. The general form of a matrix in Hessenberg form is shown in Figure 9.5. All the elements in the lower triangular part below the co-diagonal are zero, but elements on the co-diagonal and in the upper triangular part can be anything.

We can try to evaluate the characteristic polynomial and find its roots. While finding $\det(\mathbf{A} - \lambda\mathbf{I})$ is not as easy as the simple recursion (9.15)

Figure 9.5 The Hessenberg form. An $\times$ represents a nonzero entry.

$$\begin{bmatrix} \mathrm{x} & \mathrm{x} & \mathrm{x} & \cdots & \mathrm{x} \\ \mathrm{x} & \mathrm{x} & \mathrm{x} & \cdots & \mathrm{x} \\ 0 & \mathrm{x} & \mathrm{x} & \cdots & \mathrm{x} \\ \vdots & 0 & \mathrm{x} & \ddots & \mathrm{x} \\ 0 & \cdots & 0 & \mathrm{x} & \mathrm{x} \end{bmatrix}$$

for tridiagonal matrices, it is feasible for Hessenberg matrices. A preferred approach, called the QR method for eigenvalues, computes a sequence of matrices by the iteration

$$\begin{aligned} \mathbf{A}_0 &= \mathbf{A}, \\ \mathbf{A}_k &= \mathbf{Q}_k \mathbf{R}_k, \\ \mathbf{A}_{k+1} &= \mathbf{R}_k \mathbf{Q}_k, \ k = 0,\ 1,\ 2,\ \ldots . \end{aligned} \tag{9.17}$$

In words, in each iteration the matrix $\mathbf{A}_k$ is decomposed by QR factorization into factors $\mathbf{Q}_k$ and $\mathbf{R}_k$. These are then multiplied in reverse order to give the next iterate $\mathbf{A}_{k+1}$. It can be shown that under most conditions this process converges to a matrix that is similar to $\mathbf{A}$ and is such that the eigenvalues can be obtained easily. More specifically, when the eigenvalues are all real the limiting matrix is triangular, so the eigenvalues are explicit on the diagonal. When there are complex eigenvalues, the limiting matrix is triangular except for 2×2 blocks along the diagonal from which the complex eigenvalues can be obtained (Figure 9.6).

While there are other methods for computing the eigenvalues of nonsymmetric matrices, the QR method is probably the most popular. For details, see the classic treatise of Wilkinson [28] and the more recent work by Golub and Van Loan [10].

Incidentally, the QR method can also be used for symmetric matrices. The QR factorization of a tridiagonal matrix can be found quickly as can be the reverse product. This provides an alternative to finding the roots of the characteristic polynomial (9.15).

Figure 9.6 Limit form of a matrix in the QR eigenvalue method.

$$\begin{bmatrix} \mathrm{x} & \mathrm{x} & \mathrm{x} & \mathrm{x} & \mathrm{x} & \mathrm{x} \\ \mathrm{x} & \mathrm{x} & \mathrm{x} & \mathrm{x} & \mathrm{x} & \mathrm{x} \\ & & \mathrm{x} & \mathrm{x} & \mathrm{x} & \mathrm{x} \\ & & \mathrm{x} & \mathrm{x} & \mathrm{x} & \mathrm{x} \\ & \mathbf{0} & & & \mathrm{x} & \mathrm{x} \\ & & & & & \mathrm{x} \end{bmatrix}$$

EXERCISES

1. Show why the matrix in (9.16) has only one eigenvector.
2. Show that the QR method produces a sequence of similar matrices.
3. Explain why the complex eigenvalues of a nonsymmetric matrices must occur in complex conjugate pairs $a + ib$ and $a - ib$.
4. Discuss how one finds the eigenvalues of a matrix in the form shown in Figure 9.6.
5. Find the eigenvalues of

$$\begin{bmatrix} 1 & 4 & 0 & 0 \\ 2 & 3 & 0 & 0 \\ 0 & 0 & 5 & 0 \\ 0 & 0 & 0 & 6 \end{bmatrix}.$$

6. Find an example of a 4×4 matrix that has only two linearly independent eigenvectors.
7. Investigate what could happen when the power method is applied to the matrix in (9.16).

9.5 Finding Eigenvectors by Inverse Iteration*

The power method and the Jacobi method both yield eigenvalues and eigenvectors, but the methods that involve finding the roots of the characteristic polynomial give only eigenvalues. We therefore need a way of getting eigenvectors from the computed eigenvalues. This can be done by the technique of *inverse iteration*, a variant of the power method.

Suppose that we have an $n \times n$ matrix $\mathbf{A}$ with eigenvalues $\lambda_1, \lambda_2, \ldots, \lambda_n$ and corresponding n orthonormal eigenvectors $\mathbf{u}_1, \mathbf{u}_2, \ldots, \mathbf{u}_n$.[3] Suppose also that α is not in the spectrum of $\mathbf{A}$. Then the system

$$(\mathbf{A} - \alpha \mathbf{I})\mathbf{x} = \mathbf{b} \tag{9.18}$$

has a unique solution for all n-vectors $\mathbf{b}$. If we expand $\mathbf{b}$ as

$$\mathbf{b} = \sum_{i=1}^{n} a_i \mathbf{u}_i$$

[3]The arguments as given only apply when $\mathbf{A}$ is symmetric or has a full set of eigenvectors. However, the result can be shown to hold in general.

then

$$\begin{aligned}\mathbf{x} &= (\mathbf{A} - \alpha\mathbf{I})^{-1}\mathbf{b} \\ &= (\mathbf{A} - \alpha\mathbf{I})^{-1}\sum_{i=1}^{n} a_i\mathbf{u}_i \\ &= \sum_{i=1}^{n} \frac{1}{\lambda_i - \alpha} a_i\mathbf{u}_i.\end{aligned}$$

If α is close to one of the eigenvalues, say λ_1, that is simple and well separated from the other eigenvalues, then in

$$\mathbf{x} = \frac{a_1\mathbf{u}_1}{\lambda_1 - \alpha} + \sum_{i=2}^{n} \frac{1}{\lambda_i - \alpha} a_i\mathbf{u}_i$$

the first term on the right will be much larger than the rest. Therefore

$$\mathbf{x} = (\mathbf{A} - \alpha\,\mathbf{I})^{-1}\mathbf{b}$$

is approximately a multiple of $\mathbf{u}_1$ for any $\mathbf{b}$. The closer α is to λ_1, the better the approximation. All we have to do then is substitute our best approximation for λ_1 as α and solve (9.18).

At first, this looks like a poor suggestion. If α is close to λ_1, the matrix $\mathbf{A} - \alpha\mathbf{I}$ is very ill-conditioned and the solution of (9.18) subject to large errors. However, it turns out, perhaps somewhat surprisingly, that this is not a valid objection. If λ_1 is a simple eigenvalue, then $\mathbf{A} - \alpha\mathbf{I}$ has rank $n - 1$. When we apply the GEM, the last diagonal element of the reduced matrix will be very small and seriously affected by rounding. The value of $[\mathbf{x}]_n$, computed in the first step of the backsubstitution (3.10), will then be largely meaningless, reflecting the ill-conditioning of the equations. But for the computation of the eigenvector, the value of $[\mathbf{x}]_n$ does not matter as it just gives a scaling for the eigenvector that is removed by normalization. For any given value of $[\mathbf{x}]_n$ the computation of $[\mathbf{x}]_{n-1}$, $[\mathbf{x}]_{n-2}$, ... is completely stable. This is one of the rare instances in numerical work where an ill-conditioned problem yields well-conditioned results!

Example 9.8 The largest eigenvalue of the matrix in Example 9.7, to thirteen significant digits, is

$$\lambda_1 = 6.154523008670.$$

When this is used as α in (9.18), with $b = (1,\ 1,\ 1,\ 1)^T$, the result, after normalization, is

$$\mathbf{x} = \begin{bmatrix} 0.2179424255587 \\ 0.6087256857810 \\ 0.7027878102191 \\ 0.2967211357816 \end{bmatrix}. \qquad (9.19)$$

This approximation to the eigenvector appears to be accurate to all digits shown.

■

In practice, using the most accurate approximation of λ for α may not work, because $\mathbf{A} - \alpha\mathbf{I}$ could be so close to singular that our linear system solver refuses to give an answer. In that case it is perhaps wiser to use a less accurate value of λ, say by dropping some significant digits, and replacing (9.18) by the iteration

$$(\mathbf{A} - \alpha\mathbf{I})\mathbf{x}^{[k+1]} = \mathbf{x}^{[k]}, \qquad (9.20)$$

with an arbitrary starting vector $\mathbf{x}^{[0]}$. Usually, one or two iterations suffice to get full accuracy.

The iteration (9.20) is just the power method with the inverse of $\mathbf{A} - \alpha\mathbf{I}$. This is why the method is referred to as inverse iteration.

Example 9.9 For the matrix in Example 9.8 we used

$$\alpha = 6.1545$$

instead of the thirteen-digit approximation. With $\mathbf{x}^{[0]} = (1,\ 1,\ 1,\ 1)$, the first iteration gives

$$\mathbf{x}^{[1]} = \begin{bmatrix} 0.2179403078950 \\ 0.6087257678091 \\ 0.7027889426590 \\ 0.2967198407188 \end{bmatrix}$$

which differs from $\mathbf{x}$ in (9.19) by less than 10^{-5}. After two more iterations, the result agrees with (9.19) to all digits shown.

■

EXERCISES

1. Use inverse iteration to find the eigenvector corresponding to the second largest eigenvalue of the matrix in Example 9.7.
2. The largest eigenvalue of

$$\mathbf{A} = \begin{bmatrix} 1 & 2 & 4 & 6 \\ 2 & 0 & 3 & 1 \\ 4 & 3 & 0 & 6 \\ 6 & 1 & 6 & 1 \end{bmatrix}$$

is $\lambda \cong 12.3464$. Find the corresponding eigenvector to three significant digits.
3. Can you defend the claim that, under the conditions stated, the computation of $[\mathbf{x}]_{n-1}$, $[\mathbf{x}]_{n-2}$, ... is stable?
4. What can you expect if inverse iteration is applied near a double eigenvalue that is well separated from the rest of the spectrum?
5. In Example 9.8 each iteration improves the accuracy by about five significant digits. Is there an explanation for this? How can you increase the speed of convergence?
6. Suppose that in (9.18) α is close to the eigenvalue λ_1, which is simple and well separated from the other eigenvalues, but the n-vector is accidentally chosen so that $\mathbf{b} = \sum_{i=2}^{n} a_i \mathbf{u}_i$. What can you expect from the inverse iteration method in this case?

9.6 Error Estimates for Symmetric Matrices*

There are two sources of error in computing eigenvalues and eigenvectors. The first is the inevitable rounding. The second lies in the fact that eigenvalue algorithms are iterative, so we incur an error when we terminate the iteration. While it is possible to analyze the situation and produce rigorous error bounds, most of what we get is not practically useful. A more productive approach is to use *a posteriori* estimates in which we use the computed solution to get some idea of the error. The analysis is much easier in the symmetric case, so we will confine our attention to it.

Many a posteriori methods take a trial solution and substitute it into the original equation to find the discrepancy, or *residual.* Here we take as a trial element the approximate eigensolution $(\mu, \mathbf{x})$, computed by whatever means, and define the residual as

$$\rho(\mu, \mathbf{x}) = \mathbf{A}\mathbf{x} - \mu\mathbf{x}. \tag{9.21}$$

We can assume without loss of generality that $\mathbf{x}$ is normalized by $||\mathbf{x}|| = 1$.

Theorem 9.10 Assume that $\mathbf{A}$ is a symmetric $n \times n$ matrix. Let $(\mu, \mathbf{x})$ be an approximate solution of the eigenvalue problem (9.1) and let $\rho(\mu, \mathbf{x})$ be its residual as defined by (9.21). Then the spectrum $\sigma(\mathbf{A})$ contains at least one point λ such that

$$|\lambda - \mu| \leq ||\rho(\mu, \mathbf{x})||. \tag{9.22}$$

Proof: Let $\lambda_1, \lambda_2, \ldots, \lambda_n$ denote the eigenvalues of $\mathbf{A}$. Since $\mathbf{A}$ is symmetric, it has n independent and orthonormal eigenvectors $\mathbf{u}_i$, so we can write

$$\mathbf{x} = \sum_{i=1}^{n} a_i \mathbf{u}_i.$$

Because of the normalization of $\mathbf{x}$ we also have

$$\sum_{i=1}^{n} a_i^2 = 1. \tag{9.23}$$

Now

$$\begin{aligned} \rho(\mu, \mathbf{x}) &= \mathbf{A}\mathbf{x} - \mu\mathbf{x} \\ &= \sum_{i=1}^{n} a_i(\lambda_i - \mu)\mathbf{u}_i \end{aligned}$$

and

$$||\rho(\mu, \mathbf{x})||^2 = \sum_{i=1}^{m} a_i^2(\lambda_i - \mu)^2.$$

If λ is the eigenvalue closest to μ, then

$$||\rho(\mu, \mathbf{x})||^2 \geq (\lambda - \mu)^2 \sum_{i=1}^{n} a_i^2.$$

Using (9.23), the result (9.22) follows. ■

This theorem tells us that when the residual of a computed eigensolution is small, the computed eigenvalue must be close to a true eigenvalue. If we can do this for every one of the n computed eigenvalues, we have demonstrated that the computed spectrum and the actual spectrum are very closely related. We can establish a similar result for eigenvectors, but here we must be a little careful as eigenvectors of simple eigenvalues are indeterminate within a constant, and eigenvectors associated with multiple eigenvectors are indeterminate within a linear combination. We state and prove only the result for eigenvectors associated with simple eigenvalues, that are well separated from the rest of the spectrum.

Theorem 9.11 Let $\mathbf{A}$ be a symmetric $n \times n$ matrix with eigenvalues $\lambda_1, \lambda_2, \ldots, \lambda_n$ and corresponding eigenvectors $\mathbf{u}_1, \mathbf{u}_2, \ldots, \mathbf{u}_n$ and let $(\mu, \mathbf{x})$ be an approximate eigensolution of (9.1). Suppose that an eigenvalue λ_i satisfies

$$|\lambda_i - \mu| \leq ||\rho(\mu, \mathbf{x})||.$$

According to Theorem 9.10, such a λ_i must exist. Assume now that λ_i is separated from the rest of the spectrum of $\mathbf{A}$, in particular that

$$\min_{i \neq j} |\lambda_i - \lambda_j| = d.$$

Then there exists a constant a such that

$$||\mathbf{x} - a\mathbf{u}_i|| \leq \frac{||\rho(\mu, \mathbf{x})||}{d - ||\rho(\mu, \mathbf{x})||}, \tag{9.24}$$

provided $||\rho(\mu, \mathbf{x})|| < d$.

Proof: Using the expansion for $\mathbf{x}$ from Theorem 9.10, we see that

$$||\mathbf{x} - a_i\mathbf{u}_i||^2 = \sum_{j \neq i} a_j^2 \tag{9.25}$$

and

$$\sum_{j=1}^{n} a_j^2 (\lambda_j - \mu)^2 = ||\rho(\mu, \mathbf{x})||^2,$$

so that

$$\sum_{j \neq i} a_j^2 (\lambda_j - \mu)^2 \leq ||\rho(\mu, \mathbf{x})||^2.$$

As we can show,

$$|\lambda_j - \mu| \geq d - |\lambda_i - \mu| \tag{9.26}$$

for all $j \neq i$, from which it follows that

$$(d - |\lambda_i - \mu|)^2 \sum_{j \neq i} a_j^2 \leq ||\rho(\mu, \mathbf{x})||^2.$$

Putting this into (9.25) gives

$$||\mathbf{x} - a_i\mathbf{u}|| \leq \frac{||\rho(\mu, \mathbf{x})||}{d - |\lambda_i - \mu|}$$

$$\leq \frac{||\rho(\mu, \mathbf{x})||}{d - ||\rho(\mu, \mathbf{x})||},$$

completing the proof. ■

This shows that if the residual is small and μ approximates a simple eigenvalue that is well separated from the other eigenvalues, then the computed eigenvector approximates the true eigenvector closely as well.

EXERCISES

1. Modify the statement of Theorem 9.11 so that it can be applied when λ_i is a multiple eigenvalue.
2. Use Theorems 9.10 and 9.11 to estimate the accuracy of the eigenvalues and eigenvectors computed in Example 9.3.
3. Use Theorem 9.10 to compare the spectra of two symmetric matrices $\mathbf{A}$ and $\mathbf{A} + \Delta\mathbf{A}$, where $||\Delta\mathbf{A}||$ is small.
4. Suppose that we believe that one of the eigenvalues of

$$\mathbf{A} = \begin{bmatrix} 1 & 2 & 3 & 4 \\ 2 & 3 & 4 & 5 \\ 3 & 4 & 5 & 6 \\ 4 & 5 & 6 & 7 \end{bmatrix}$$

is near $\lambda = 17$. Use Theorem 9.10 and the results of Section 9.5 to find an approximation of the corresponding eigenvector and evaluate the accuracy of this guess.

5. Suppose that we know that the largest and the second largest eigenvalues of

$$\mathbf{B} = \begin{bmatrix} 3 & -1 & -1 & 1 \\ -1 & 3 & -1 & -1 \\ -1 & -1 & 3 & -1 \\ 1 & -1 & -1 & 3 \end{bmatrix}$$

are near $\lambda_1 = 5$ and $\lambda_2 = 4$. Use the results of Section 9.5 to find approximate eigenvectors corresponding to λ_1 and λ_2, then

(a) use Theorem 9.10 to evaluate the accuracy of both λ_1 and λ_2,

(b) use the result of (a) and Theorem 9.11 to estimate the accuracy for the computed eigenvector corresponding to λ_1.

6. Use Theorem 9.10 to prove that the eigenvalues of a symmetric matrix **A** are continuous functions of the elements $[\mathbf{A}]_{ij}$.

7. Show why (9.26) is true.

Chapter 10

Initial Value Problems for Ordinary Differential Equations

A great many mathematical models of physical processes lead to differential equations. As a consequence, the numerical solution of ordinary and partial differential equations is a central topic in scientific computation. Numerical methods are necessary because even very simple situations lead to intractable equations.

Example 10.1 At time $t = 0$ a rocket with mass m is launched from rest at (0, 0)at an angle α. After launch, the rocket is subject to two forces: gravity, which exerts a pull in the negative y direction, and the rocket's variable thrust $F(t)$, which acts parallel to the rocket's instantaneous direction (Figure 10.1). The problem is to determine the path of the rocket.

If we make certain simplifying assumptions (for example, that there is no air resistance), we can use Newton's laws of motion to construct a simple model for this situation. Letting $x(t)$ and $y(t)$ denote the coordinates of the rocket at time t, we get

$$\frac{d^2x(t)}{dt^2} = \frac{1}{m}F(t)\cos\theta(t),$$

$$\frac{d^2y(t)}{dt^2} = -g + \frac{1}{m}F(t)\sin\theta(t)$$

where

$$\tan\theta(t) = \frac{dy(t)}{dx(t)}.$$

Figure 10.1
Path of a rocket.

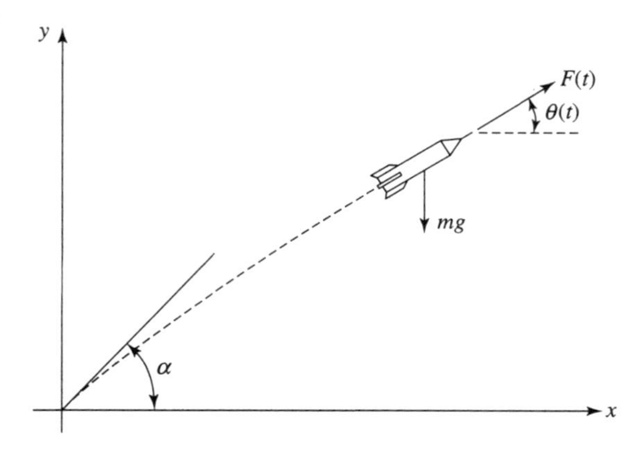

From this

$$\cos\theta = \frac{\frac{dx}{dt}}{\sqrt{\left(\frac{dx}{dt}\right)^2 + \left(\frac{dy}{dt}\right)^2}},$$

and

$$\sin\theta = \frac{\frac{dy}{dt}}{\sqrt{\left(\frac{dx}{dt}\right)^2 + \left(\frac{dy}{dt}\right)^2}}.$$

Putting everything together, we see that the rocket's motion can be described by a system of two second-order ordinary differential equations

$$\frac{d^2x}{dt^2} = \frac{1}{m}F(t)\frac{\frac{dx}{dt}}{\sqrt{\left(\frac{dx}{dt}\right)^2 + \left(\frac{dy}{dt}\right)^2}},$$

$$\frac{d^2y}{dt^2} = -g + \frac{1}{m}F(t)\frac{\frac{dy}{dt}}{\sqrt{\left(\frac{dx}{dt}\right)^2 + \left(\frac{dy}{dt}\right)^2}}.$$

To complete the model, we also need to give the initial conditions $x(0) = y(0) = x'(0) = y'(0) = 0$ and $\theta(0) = \alpha$.

■

In this chapter we concentrate on the solution of the single differential equation

$$y'(x) = f(x,\ y(x)), \tag{10.1}$$

with the given initial condition

$$y(a) = y_0. \tag{10.2}$$

The unknown here is $y(x)$, whose value is to be determined in an interval $a \le x \le b$. We will also consider, somewhat briefly, systems of such equations

$$\begin{aligned} y_1'(x) &= f_1(x,\ y_1(x),\ y_2(x),\ \dots,\ y_m(x)),\\ y_2'(x) &= f_2(x,\ y_1(x),\ y_2(x),\ \dots,\ y_m(x)),\\ &\vdots\\ y_m'(x) &= f_m(x,\ y_1(x),\ y_2(x),\ \dots,\ y_m(x)), \end{aligned} \tag{10.3}$$

with conditions

$$y_i(a) = y_i, \quad i = 1,\ 2,\ \dots,\ m. \tag{10.4}$$

Systems of this type are classified as *initial value problems* because all the subsidiary conditions (10.4) are given at the same point $x = a$. The solution is developed from that point in a stepwise fashion, meaning that if the solution is known in the interval $a \le x \le a_1$, it can be carried forward from there to the next interval $a_1 \le x \le a_2$, and so on.

As we see from Example 10.1, equations of higher order are also encountered in practice. They are often treated by converting them into systems. Consider, for example, the second order equation

$$y''(x) = f(x,\ y(x),\ y'(x)).$$

By making the change of variables $y_1(x) = y(x),\ \ y_2(x) = y'(x)$, this equation becomes the equivalent system

$$\begin{aligned} y_1'(x) &= y_2(x),\\ y_2'(x) &= f(x,\ y_1(x),\ y_2(x)). \end{aligned}$$

If we know how to solve systems of first order equations, we can deal with many of the complicated ordinary differential equations encountered in practice.

10.1 Existence and Uniqueness of the Solution

Before tackling any problem numerically, it is advisable to find out as much as possible about the nature of the solution. Trying to solve a poorly understood problem numerically may lead not only to ineffective algorithms, but to unreliable or entirely misleading results. It is not always easy to reach a good understanding of what to expect, particularly for nonlinear problems. However, for the initial value problem of ordinary differential equations the theory is well developed. This gives us the opportunity to construct many effective numerical algorithms, even when the equations are nonlinear. As we will see, certain types of nonlinearities in (10.3) are only a minor complication.

Definition 10.1

A function $g(x)$ is said to be *Lipschitz-continuous* in some interval $[a, b]$, if there exists a constant c such that

$$|g(x_1) - g(x_2)| \leq c\,|x_1 - x_2| \tag{10.5}$$

for all x_1 and x_2 in $[a, b]$. The c is the *Lipschitz constant* and is to be independent of x_1 and x_2.

A Lipschitz-continuous function is always continuous, but simple continuity is weaker. The function $\sqrt{|x|}$ is continuous everywhere, but not Lipschitz-continuous in any interval that includes $x = 0$.

The standard existence and uniqueness result for ordinary differential equations is stated in terms of Lipschitz-continuity.

Theorem 10.1

Let $f(x, y)$ be bounded for all x and y, continuous for all $a \leq x \leq b$, and *uniformly* Lipschitz-continuous with respect to y; that is, there is a constant c such that

$$|f(x, y_1) - f(x, y_2)| \leq c|y_1 - y_2| \tag{10.6}$$

for all $a \leq x \leq b$ and all $-\infty < y_1, y_2 < \infty$. Then (10.1) has a unique continuously differentiable solution in $a \leq x \leq b$. Furthermore, the solution $y(x)$at any x is a continuous function of the initial value y_0.

Proof: See any book on differential equations. ■

There are many differential equations that do not satisfy the required Lipschitz condition. In such cases, the theoretical situation is more complicated and we may not be able to guarantee the existence or the uniqueness of a solution.

Example 10.2 Consider the differential equation

$$y'(x) = y^2(x)$$

in the interval $0 \leq x \leq 2$, with initial condition

$$y(0) = 1.$$

The function on the right side of the equation does not satisfy the Lipschitz condition (10.6), so Theorem 10.1 does not hold and we cannot claim that there exists a unique solution.

By integrating the equation we find that

$$y(x) = \frac{1}{1-x}$$

which goes to infinity at $x = 1$, so that there is no continuously differentiable function that satisfies the equation in the whole interval $[0,\ 2]$.

■

This example illustrates the typical situation that occurs when the Lipschitz condition fails; while we can no longer guarantee a solution over an arbitrarily large interval, there is still a solution near the starting point. It is possible to restate Theorem 10.1 to guarantee such local solutions for a variety of conditions and stronger nonlinearities, but we will not pursue this here.

EXERCISES

1. Show that if a function is differentiable in a neighborhood of some point x, then it is Lipschitz-continuous near x.
2. Which of the following functions are Lipschitz-continuous in the interval $[-1,\ 1]$?
   (a) $|x|^{1/3}$
   (b) $|x|^{3/2}$
   (c) $\sin(x)/x$
   (d) $\sin(x)/\sqrt{x}$.
3. Give an example of a function that is Lipschitz-continuous but not differentiable.
4. Find the solution of

$$\begin{aligned} y'(x) &= 1 + y(x), \\ y(0) &= 0, \end{aligned}$$

in the interval $0 \leq x \leq 1$. Prove that the solution is unique.

5. Use Theorem 10.1 to show that the linear initial value problem

$$y'(x) = c_0(x) + c_1(x)y(x)$$

has a unique solution for all bounded functions $c_0(x)$ and $c_1(x)$.

6. Can the equation

$$y'(x) = y^3(x) + y^2(x),$$

with $y(0) = 1$, have a continuous and differentiable solution in $0 \leq x \leq 1$?

7. The following simple electric circuit contains a generator producing a voltage of $E(t)$ and a current of $I(t)$ at time t. The circuit also contains a resistor with a resistance R and an inductor with an inductance of L.

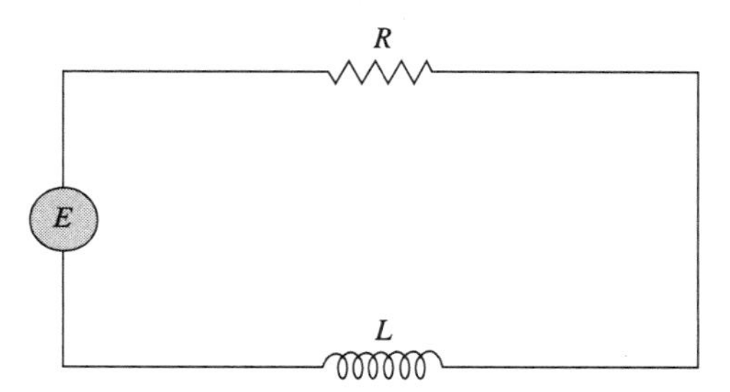

The laws governing electric circuits tell us that the drop in voltage due to the resistor is RI, and that the voltage drop due to the inductor is $L\frac{dI}{dt}$. One of Kirchhoff's Laws says that the sum of the voltage drops is equal to the supplied voltage $E(t)$.Thus we have

$$L\frac{dI}{dt} + RI = E(t),$$

which models the current I at time t.

A particular circuit of this type, after proper scaling, leads to the differential equation

$$\frac{dI}{dt} + 4I = 5e^t$$

with an initial condition $I(0)$=0 for $0 \leq t \leq 3$. Find an I that satisfies the differential equation and show that it is the only solution.

8. A seasonal-growth model that accounts for variations in the rate of growth of certain species is given as

$$\frac{dP}{dt} = 2P\cos^3(12t - \frac{7}{12})$$

with $P(0)$= 1200. Show that a unique solution exists for $0 \leq t \leq 5$.

10.2 Some Simple Numerical Methods

There are many methods for solving (10.1) numerically. The most commonly used ones, to which we will restrict ourselves here, approximate $y(x)$ by a *mesh function* $\{Y_0,\ Y_1,\ \ldots\}$ such that Y_i approximates $y(x_i)$ on the mesh $\{a = x_0 < x_1 < \ \ldots\ < x_n = b\}$. For the moment, assume that we have a uniform mesh with step size

$$h = x_{i+1} - x_i.$$

A simple and intuitive numerical method starts from the interpretation of the derivative as the slope of a line. Suppose at $x = x_i$ we know the value $y(x_i)$. We can then compute $y'(x_i)$ from (10.1) and use this computed slope to step forward to x_{i+1} (Figure 10.2). As is obvious, this will not give the exact value for $y(x_{i+1})$ but perhaps a good enough approximation. Since at x_0 we have the starting value $Y_0 = y_0$, we can compute successive approximations by

$$Y_{i+1} = Y_i + hf(x_i,\ Y_i), \quad i = 0,\ 1,\ \ldots. \tag{10.7}$$

This simple scheme is called *Euler's Method.*

Euler's method is not very accurate unless the step size is very small. The reason for this inaccuracy is easy to see. To get the approximate solution at x_{i+1} we use only the slope at x_i, but the change in the true solution depends on the slope of all the points in the interval $[x_i,\ x_{i+1}]$. We can also see from Figure 10.2 that the individual small errors accumulate and we quickly move away from the true solution. We might suspect that a better way is to average the slope at x_i and x_{i+1} and compute the approximation by

$$Y_{i+1} = Y_i + \frac{h}{2}\left\{f(x_i,\ Y_i) + f(x_{i+1},\ Y_{i+1})\right\}. \tag{10.8}$$

This is known as the *trapezoidal method.* Not surprisingly, the trapezoidal method gives much better results than (10.7), but we notice a difficulty. The

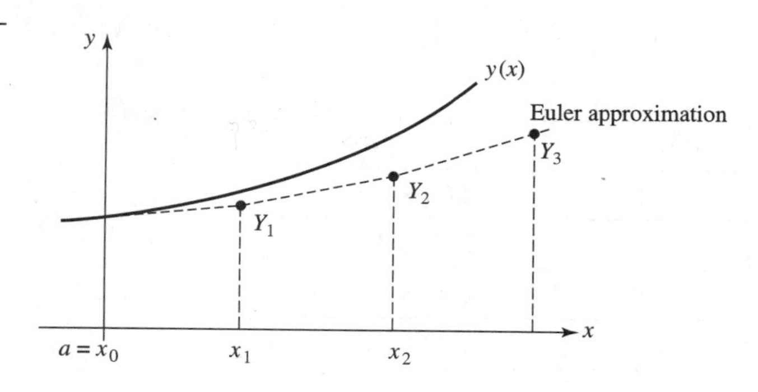

Figure 10.2 Computing an approximate solution by Euler's method.

new unknown Y_{i+1} is defined implicitly and, unless the differential equation is linear, it cannot be immediately obtained from (10.8). Here we have a root-finding problem that can be addressed with the methods developed in Chapter 7. When h is sufficiently small, successive substitution will work and we can use

$$Y_{i+1}^{[k+1]} = Y_i + \frac{h}{2}\left\{f(x_i,\ Y_i) + f(x_{i+1},\ Y_{i+1}^{[k]})\right\} \tag{10.9}$$

for an initial guess $Y_{i+1}^{[0]} = Y_i$. We iterate until we have sufficient accuracy and accept the last iterate as Y_{i+1}.

We also need to consider the efficiency of this process. Since most of the work in solving differential equations is the evaluation of $f(x_{i+1},\ Y_{i+1})$, iterating many times can slow the algorithm considerably. It can be shown that in this case iterating just twice; that is, stopping with $Y_{i+1}^{[2]}$, gives just about as much accuracy as can be obtained with this method.

Example 10.3 Examine the role of iteration of (10.9) on the numerical solution of

$$\begin{aligned} y'(x) &= y^2(x), \\ y(0) &= 1, \end{aligned}$$

in the interval $0 \le x \le 0.5$. In Table 10.1 we show the respective results for $h = 0.01$, using (10.9) with varying numbers of iterations. As we can see, there is a significant difference between one and two iterations, but further iterations do not reduce the error by any appreciable amount. This behavior is typical of most situations. ■

Another way to deal with the nonlinearity in (10.8) is to use Euler's method to predict the value of Y_{i+1} and then use this prediction in (10.9). This gives the approximate integration formula

$$Y_{i+1} = Y_i + \frac{h}{2}\left\{f(x_i,\ Y_i) + f(x_{i+1},\ Y_i + hf(x_i,\ Y_i))\right\}. \tag{10.10}$$

Table 10.1
Effect of iterations in method (10.9).

Iterations	Max. error
1	2.66×10^{-2}
2	1.97×10^{-4}
3	1.94×10^{-4}
4	2.00×10^{-4}

Table 10.2
Error in Example 10.4.

h	Max. error
0.02	1.88×10^{-5}
0.01	4.69×10^{-6}
0.005	1.17×10^{-6}

Although motivated differently, this is exactly what we get if we use (10.9) with $k = 1$.

Example 10.4 Table 10.2 shows the observed error when (10.10) is applied with several step sizes to

$$y'(x) = \sqrt{2y - y^2},$$
$$y(0) = 1,$$

in $0 \leq x \leq 0.5$. The true solution is

$$y(x) = 1 + \sin(x).$$

The error is reduced by a factor of about four with each halving of the step size. We conjecture from this that the method is of second order.[1] ∎

There are still several more ways in which we can rearrange these simple ideas. For example, instead of using Euler's method to estimate the slope at the right end of the interval, we can use it to estimate the slope at the center of the interval. This suggests the scheme

$$Y_{i+1} = Y_i + hf(x_i + \frac{h}{2},\ Y_i + \frac{h}{2} f(x_i,\ Y_i)). \tag{10.11}$$

The accuracy of this scheme is comparable to that of (10.10).

Example 10.5 Table 10.3 shows the observed error when (10.11) is applied with several step sizes to the same problem in Example 10.4.

The errors in Table 10.3 are smaller than those in Table 10.2, but roughly of the same order of magnitude. Also, the error is reduced by about

[1]Since the error is reduced by a factor of four when the step size is halved, we suspect that the error is proportional to h^2. As in the case of quadrature, we call this second order convergence.

Table 10.3
Error in Example 10.5.

h	Max. error
0.02	5.34×10^{-6}
0.01	1.33×10^{-6}
0.005	3.32×10^{-7}

a factor of four when the step size is halved, so its order of convergence appears to be two.

■

EXERCISES

1. Another way of approximating solutions of differential equations is to use two terms in the Taylor expansion

$$y(x+h) \cong y(x) + hy'(x) + \frac{h^2}{2}y''(x),$$

and using the derivative of f to compute y''. Derive a numerical method based on this approximation and discuss advantages and disadvantages of this approach.

2. How is the method (10.8) simplified if f is linear in y?

3. Apply Euler's method with step sizes 0.2 and 0.1 to the equation

$$y'(x) = 1 - xy^2(x),$$

with $y(0) = 0$ in $0 \leq x \leq 1$. Use the two sets of computed results to estimate the accuracy of the computed solution.

4. Apply the method (10.10) to the equation

$$y'(x) = \tfrac{1}{2}xy(x),$$
$$y(0) = 1$$

in $0 \leq x \leq 1$. Find the exact solution of the equation and compare it with the approximation. Try various step sizes to establish a relation between the error and the step size.

5. Using a forward difference for the derivative

$$y'(x_i) \cong \frac{y(x_{i+1}) - y(x_i)}{h}$$

and solving for $y(x_{i+1})$ leads to Euler's method. If we start instead from a backward difference

$$y'(x_{i+1}) \cong \frac{y(x_{i+1}) - y(x_i)}{h},$$

we get the *Backward Euler's formula*

$$Y_{i+1} = Y_i + hf(x_{i+1},\ Y_{i+1}).$$

Under what conditions would it be reasonable to consider this as an alternative to the normal Euler's formula?

6. Explain why one should expect little improvement after the first two iterations in Example 10.3.

7. What is a sufficient condition on h for the iteration in (10.9) to converge?

8. Determine if (10.10) is identical with (10.9) for $k = 1$.

9. Apply the method (10.11) to the equation in Exercise 8, Section 10.1. Use $h = 0.2$ and $h = 0.1$. Estimate the accuracy of your best answer.

10.3 Multistep and Predictor-Corrector Methods

The methods described in the last section are simple and intuitive, but of relatively low accuracy. For initial value problems in ordinary differential equations it is possible to find much more accurate approaches.

A class of methods can be developed by integrating (10.1) over several mesh intervals,

$$y(x_{i+k}) = y(x_i) + \int_{x_i}^{x_{i+k}} f(x,\ y(x))dx. \tag{10.12}$$

This is exact, but for computational purposes we need to replace the integral by a quadrature and the unknown function y by its approximations. This means that the quadrature points must be located at the mesh points, as these are the only places where y is approximated. If we use quadrature points $x_i,\ x_{i+1},\ \ldots,\ x_{i+k}$ and denote the corresponding quadrature weight by $w_0,\ w_1,\ \ldots,\ w_k$ we get the approximation method

$$Y_{i+k} = Y_i + w_0 f(x_i,\ Y_i) + w_1 f(x_{i+1},\ Y_{i+1}) + \ \ldots\ + w_k f(x_{i+k},\ Y_{i+k}). \tag{10.13}$$

If we know $Y_i,\ Y_{i+1},\ \ldots,\ Y_{i+k-1}$, we can use (10.13) to get the next value Y_{i+k}. If $w_k \neq 0$ and f is a nonlinear function of y, some root-finding method has to be used to solve for Y_{i+k}. We call this kind of method *implicit*.

Example 10.6 Method (10.8) represents the case $k = 1$, with the trapezoidal quadrature method. If we use $k = 2$ and Simpson's rule, we get the method

$$Y_{i+2} = Y_i + \frac{h}{3}\left\{f(x_i,\ Y_i) + 4f(x_{i+1},\ Y_{i+1}) + f(x_{i+2},\ Y_{i+2})\right\}. \tag{10.14}$$

Both (10.8) and (10.14) are implicit methods.

Not all methods need to be implicit. By using open or semi-open quadrature methods that do not involve the right interval endpoint, we get methods in which the approximation to the next mesh point is given explicitly. Using the methods for constructing quadratures we discussed in Chapter 6, we can easily show that

$$\int_0^{4h} f(x)dx \cong \frac{h}{3}\left\{8f(h) - 4f(2h) + 8f(3h)\right\},$$

and that this approximation has order of convergence four. When applied to (10.12) we get the approximation method

$$Y_{i+4} = Y_i + \frac{h}{3}\left\{8f(x_{i+1},\ Y_{i+1}) - 4f(x_{i+2},\ Y_{i+2}) + 8f(x_{i+3},\ Y_{i+3})\right\}. \tag{10.15}$$

Here the new unknown Y_{i+4} is given directly and no root-finding is involved. Methods of this type are called *explicit.*

Notice that neither (10.14) nor (10.15) can be used for all points near $x = a$. In (10.14) we need Y_0 and Y_1 before we can compute Y_2. The first of these is given by the initial condition, but Y_1 has to be gotten in some other way. For (10.15) we must first get Y_1, Y_2, and Y_3 before we can proceed. The need for special starting values occurs with most formulas of this type. We will defer this issue until the next section.

We can also derive formulas for differential equation by replacing $y'(x)$ with a difference formula. For example, the centered difference

$$y'(x_{i+1}) \cong \frac{y(x_{i+2}) - y(x_i)}{2h}$$

suggests the method

$$Y_{i+2} = Y_i + 2hf(x_{i+1},\ Y_{i+1}). \tag{10.16}$$

One might expect that better approximations of $y'(x)$ give more accurate methods for the differential equations, but things do not always work as we expect them to.

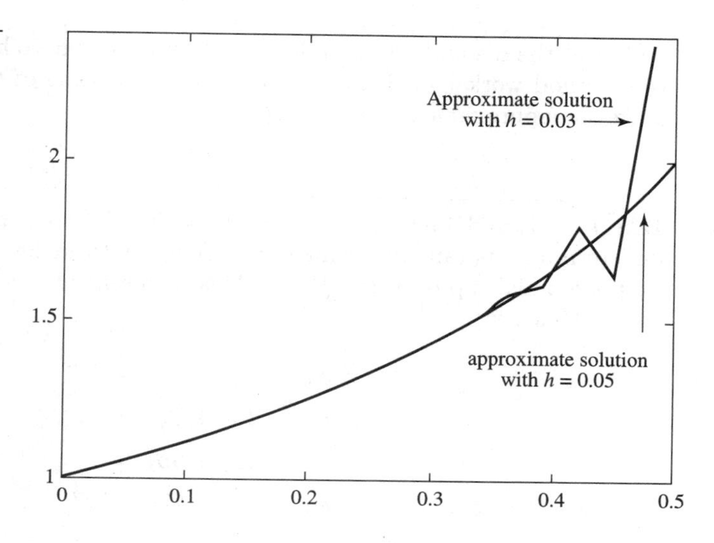

Figure 10.3
Solution of $y' = y^2$ by (10.17).

Example 10.7 The difference approximation

$$y'(x_{i+2}) \cong \frac{2y(x_{i+3}) + 3y(x_{i+2}) - 6y(x_{i+1}) + y(x_i)}{6h}$$

is exact for all polynomials up to degree three, and so has general accuracy $O(h^3)$. When we use this, we get the corresponding method for differential equations

$$Y_{i+3} = -\frac{1}{2}Y_i + 3Y_{i+1} - \frac{3}{2}Y_{i+2} + 3hf(x_{i+2},\ Y_{i+2}). \tag{10.17}$$

Although the method seems plausible, a few computations will quickly convince you of its unsuitability. Figure 10.3 is a graph of the results obtained on the equation in Example 10.3, with $h = 0.03$ and $h = 0.05$. Exact starting values were used. The method is clearly unstable. ■

The example illustrates an important point. When we deal with differential equations or even more complicated models, plausible methods do not always work. To predict the effectiveness of an algorithm and to understand such unexpected results, we need to carry out an error analysis. This is not always easy, as we will see in Section 10.5.

A general way of deriving formulas, which subsumes both methods based on the integration (10.12) as well as those based on approximating the derivative by a difference, is to start with the form

$$\begin{aligned} Y_{i+k} = \alpha_0 Y_i + \alpha_1 Y_{i+1} + \ \dots\ + \ \alpha_{k-1}Y_{i+k-1} + h\beta_0 f(x_i,\ Y_i) \\ + h\beta_1 f(x_{i+1},\ Y_{i+1}) + \ \dots\ + \ h\beta_k f(x_{i+k},\ Y_{i+k}) \end{aligned} \tag{10.18}$$

and treat the α's and β's as undetermined parameters to be chosen to make the method workable. Typically, one can select some of the coefficients to make the approximations exact for $y = 1,\ x,\ x^2,\ \ldots$.

Example 10.8 Take $k = 3$ and take $\beta_0 = \beta_1 = \beta_2 = 0$. We then are left with four undetermined parameters which we can select to make the method exact for polynomials up to degree three. This requirement gives the equation for the coefficients

$$\begin{aligned} \alpha_0 + \alpha_1 + \alpha_2 \qquad &= 1, \\ \alpha_1 + 2\alpha_2 + \beta_3 \ &= 3, \\ \alpha_1 + 4\alpha_2 + 6\beta_3 \ &= 9, \\ \alpha_1 + 8\alpha_2 + 27\beta_3 &= 27, \end{aligned} \tag{10.19}$$

which has the solution

$$\alpha_0 = \frac{2}{11},\ \alpha_1 = -\frac{9}{11},\ \alpha_2 = \frac{18}{11},\ \beta_3 = \frac{6}{11}$$

and the corresponding implicit method

$$Y_{i+3} = \frac{2}{11}Y_i - \frac{9}{11}Y_{i+1} + \frac{18}{11}Y_{i+2} + \frac{6h}{11}f(x_{i+3},\ Y_{i+3}). \tag{10.20}$$

■

Example 10.9 Any number of potential approximation methods can be derived this way. With $k = 4$ and $\alpha_0 = \alpha_1 = \alpha_2 = \beta_4 = 0$, we get the explicit method

$$\begin{aligned} Y_{i+4} = Y_{i+3} + \frac{h}{24}\{&55f(x_{i+3},\ Y_{i+3}) - 59f(x_{i+2},\ Y_{i+2}) \\ &+ 37f(x_{i+1},\ Y_{i+1}) - 9f(x_i,\ Y_i)\}. \end{aligned} \tag{10.21}$$

An implicit method is

$$\begin{aligned} Y_{i+4} = Y_{i+3} + \frac{h}{24}\{&9f(x_{i+4},\ Y_{i+4}) + 19f(x_{i+3},\ Y_{i+3}) \\ &- 5f(x_{i+2},\ Y_{i+2}) + f(x_{i+1},\ Y_{i+1})\} \end{aligned} \tag{10.22}$$

which can be obtained by starting with $\alpha_0 = \alpha_1 = \alpha_2 = \beta_0 = 0$. Whether any of these are suitable for computation remains to be seen.

■

Methods of the type (10.18) are called *multistep methods.* Multistep methods are easily constructed and are often quite effective, but they also have some undesirable features. For $k \geq 2$, the method requires some starting values that cannot be determined by the method. Thus, most multistep methods must be coupled with another method that is self-starting. Also, the methods are derived on the assumption of a uniform step size. In practice, one often takes a more adaptive approach that changes the step size based on the derived solution. Changing step size for multistep methods has to be done carefully and adds another complication. For these various reasons, other methods that are explicit and do not require special starting values are sometimes preferable.

A second difficulty is with implicit multistep methods. For reasons of accuracy and stability, implicit methods are often preferable to explicit ones. To reduce the need for many iterations we need a good guess for the unknown at each step. *Predictor-corrector* schemes use an explicit method to predict the value of the unknown, which is then used in an implicit method. If the explicit/implicit formulas are properly matched, no further iteration is needed.

Example 10.10 If we use (10.14) and (10.15) together, we get (with an index change in the first)

$$Y_{i+2}^P = Y_{i-2} + \frac{h}{3}\{8f(x_{i-1},\ Y_{i-1}) - 4f(x_i,\ Y_i) + 8f(x_{i+1},\ Y_{i+1})\} \tag{10.23}$$

which predicts the value of the new unknown Y_{i+2}, and

$$Y_{i+2} = Y_i + \frac{h}{3}\{f(x_i,\ Y_i) + 4f(x_{i+1},\ Y_{i+1}) + f(x_{i+2},\ Y_{i+2}^P)\} \tag{10.24}$$

which corrects it. This is one of several algorithms that go by the name of *Milne's method.* For this predictor/corrector pair we need two starting values Y_1 and Y_2.

Many other combinations of implicit and explicit methods are possible. For instance, (10.21) can be used as a predictor for (10.22). ■

EXERCISES

1. Consider a numerical quadrature of the form

$$\int_0^{3h} f(x)dx = w_0 f(0) + w_1 f(h) + w_2 f(2h).$$

Determine the weights w_0, w_1, w_2 and use them to get an explicit multistep method.

2. Develop an implicit method based on the three-eights rule.
3. Apply the method (10.16) to the solution of

$$y'(x) = xy(x), \quad 0 \le x \le 1, \quad y(0) = 1.$$

Examine the error and see how quickly it decreases as a function of h.

4. Derive the system (10.19).
5. Derive a method as in Example 10.8 with $k = 3$ and $\alpha_2 = \beta_2 = \beta_3 = 0$.
6. Show that the difference formula in Example 10.7 has $O(h^3)$ accuracy for numerical differentiation.
7. Use the methods of undetermined coefficients to get the weights in the formula

$$Y_{i+1} = Y_i + h\beta_0 f(x_{i-1}, Y_{i-1}) + h\beta_1 f(x_i, Y_i).$$

Is there a simple graphical interpretation of this method?

8. Derive the backward Euler method, described in Exercise 5, Section 10.2, via the general form (10.18).

10.4 Runge–Kutta Methods

Some of the difficulties associated with multistep methods are avoided by *single-step* methods in which the calculation of Y_{i+1} requires only a knowledge of Y_i. One class of single-step algorithms involves formulas known as the *Runge–Kutta* methods. They were first studied nearly 100 years ago and are named after their originators C. Runge and W. Kutta.

Runge–Kutta methods are constructed by starting with a general form containing undetermined coefficients, such as

$$Y_{i+1} = Y_i + ha_1k_1 + ha_2k_2, \tag{10.25}$$

where

$$\begin{aligned} k_1 &= f(x_i, Y_i), \\ k_2 &= f(x_i + hb_1, Y_i + hb_2k_1). \end{aligned} \tag{10.26}$$

The coefficients a_1, a_2, b_1, b_2 are then determined so that (10.25) approximates the differential equation in some sense. One way to do this is to match, as far as possible, the Taylor expansion of $y(x_{i+1})$ about x_i and the expansion of (10.25) and (10.26) after Y_i is replaced by $y(x_i)$. The reasoning behind this is that if $y(x_i)$ and Y_i were identical, the error in the expansion would be the difference between $y(x_{i+1})$ and Y_{i+1}; that is, the error would be committed in one step.

If we write the expansions out to three terms, we get

$$\begin{aligned} y(x_{i+1}) &= y(x_i) + hy'(x_i) + \frac{h^2}{2}y''(x_i) + O(h^3) \\ &= y(x_i) + hf(x_i,\ y(x_i)) + \frac{h^2}{2}(f_x(x_i,\ y(x_i)) \\ &\qquad + f_y(x_i,\ y(x_i))\, f(x_i,\ y(x_i)) + O(h^3), \end{aligned}$$

where f_x and f_y stand for the partial derivatives of f with respect to x and y, respectively. Next, we make a similar expansion for (10.25) and (10.26) with Y_i replaced by $y(x_i)$. This yields

$$\begin{aligned} y(x_{i+1}) &= y(x_i) + ha_1k_1 + ha_2k_2 \\ &= y + ha_1f + ha_2f + h^2a_2b_1f_x + h^2a_2b_2f_y\, f + O(h^3). \end{aligned} \tag{10.27}$$

We now neglect the $O(h^3)$ terms and choose the undetermined coefficients so that the two resulting expressions are equal. This gives the equations

$$\begin{aligned} a_1 + a_2 &= 1, \\ a_2b_1 &= \frac{1}{2}, \\ a_2b_2 &= \frac{1}{2}. \end{aligned} \tag{10.28}$$

Since there are four unknowns but only three equations, the nonlinear system has many solutions, each one leading to a different numerical scheme. If we pick

$$a_1 = a_2 = \frac{1}{2},$$

then

$$b_1 = b_2 = 1.$$

The associated method is

$$Y_{i+1} = Y_i + \frac{h}{2}\left\{f(x_i,\ Y_i) + f(x_{i+1},\ Y_i + hf(x_i,\ Y_i))\right\}, \tag{10.29}$$

which we previously derived in (10.10) by a different motivation.

A second solution is obtained by choosing $a_1 = 0$, giving

$$Y_{i+1} = Y_i + hf(x_i + \frac{h}{2},\ Y_i + \frac{h}{2}f(x_i,\ Y_i)). \tag{10.30}$$

This was also previously found in (10.11).

A third choice is

$$a_1 = \frac{1}{3}, \quad a_2 = \frac{2}{3},$$
$$b_1 = b_2 = \frac{3}{4}.$$

The resulting rule is

$$Y_{i+1} = Y_i + \frac{h}{3}\left\{ f(x_i,\ Y_i) + 2f(x_i + \frac{3h}{4},\ Y_i + \frac{3h}{4} f(x_i,\ Y_i)) \right\}. \tag{10.31}$$

It is clear that an unlimited number of similar formulas can be found in this manner.

The methods (10.29), (10.30), and (10.31) are comparable in their accuracy and performance. All three take two evaluations of f for each step and all are based on an expansion with accuracy $O(h^3)$, suggesting that the error in each step is proportional to h^3. Since there are $O(1/h)$ steps in the full interval, we expect that the overall error will be $O(h^2)$; that is, the method converges with second order. Such loose arguments may be suggestive, but they do not prove anything. Still, the conclusions are correct, as we will see in the next section.

Example 10.11 The equation

$$y'(x) = -y^2(x)$$

with

$$y(0) = 1$$

has an exact solution

$$y(x) = \frac{1}{1+x}.$$

Table 10.4 compares the approximate solution of this equation on $[0,\ 1]$, using (10.30) and (10.31) with several step sizes. There is little difference in the accuracy and both methods exhibit second order convergence. ■

Table 10.4 Error for the approximation of the equation in Example 10.11.

h	Max. error by (10.30)	Max. error by (10.31)
0.1	1.3×10^{-3}	1.0×10^{-3}
0.05	3.0×10^{-4}	2.5×10^{-4}
0.025	7.2×10^{-5}	6.0×10^{-5}

More accurate Runge–Kutta methods are developed in a similar way, starting from a more complicated form with additional free parameters. We can try, for instance,

$$
\begin{aligned}
Y_{i+1} &= Y_i + h(a_1 k_1 + a_2 k_2 + a_3 k_3 + a_4 k_4),\\
k_1 &= f(x_i,\ Y_i),\\
k_2 &= f(x_i + hb_1,\ Y_i + hb_2 k_1),\\
k_3 &= f(x_i + hb_3,\ Y_i + hb_4 k_2),\\
k_4 &= f(x_i + hb_5,\ Y_i + hb_6 k_3).
\end{aligned}
$$

Matching the Taylor expansion up to order h^4 gives many solutions, one of which is the *classical Runge–Kutta* method

$$
Y_{i+1} = Y_i + \frac{h}{6}(k_1 + 2k_2 + 2k_3 + k_4) \tag{10.32}
$$

with

$$
\begin{aligned}
k_1 &= f(x_i,\ Y_i),\\
k_2 &= f(x_i + \frac{h}{2},\ Y_i + \frac{h}{2}k_1),\\
k_3 &= f(x_i + \frac{h}{2},\ Y_i + \frac{h}{2}k_2),\\
k_4 &= f(x_i + h,\ Y_i + hk_3).
\end{aligned}
$$

The derivation of the constants is routine, but lengthy and tedious. The method takes four evaluations per step and has order of convergence four.

From these examples we see that Runge–Kutta methods are easy to implement, self-starting, and that changing step size presents no problem. On the other hand, each step involves several evaluations of f, so there is a question of their efficiency. This last point is not easily answered by just counting operations, since the step sizes for comparable accuracy may differ significantly for a multistep method and a Runge–Kutta method. It turns out that in practice Runge–Kutta methods are nearly as efficient as predictor-corrector methods, so that their advantage of simplicity is often a main factor in their use.

EXERCISES

1. Apply method (10.29) to the equation in Example 10.11 and compare the error with the errors in Table 10.4.
2. Derive the equations in (10.27).
3. Find the Runge–Kutta method one gets by using $a_1 = 1/4$ in (10.28).
4. Apply the rules (10.29), (10.30), and (10.31) to the equation

$$y'(x) = -2xy^2(x),$$

with $y(0) = 1$ in $[0,\ 1]$. Try different step sizes to see if there is any significant difference in their performance.

5. Apply the classical Runge–Kutta method, equation (10.32), to the equation in Example 10.11. How does its performance compare to that of (10.31)?

10.5 Systems and Equations of Higher Order

In practice, most of the applications of ordinary differential equations involve systems of equations. For simplicity, one usually describes and analyzes algorithms for single equations, and it is fortunate that much of what we know about single equations carries over quite easily to systems. Formally, we can convert the algorithms that we developed for single equations over to systems by writing everything in vector form. If we write (10.3) and (10.4) as

$$\mathbf{y}'(x) = \mathbf{f}(x, \mathbf{y}(x)), \tag{10.33}$$

with

$$\mathbf{y}(a) = \mathbf{y}_0,$$

then the methods we have described can be adapted for systems by simply replacing y, Y, and f by corresponding vectors.

Example 10.12 Euler's method for systems is

$$\mathbf{Y}_{i+1} = \mathbf{Y}_i + h\mathbf{f}\left(x,\ \mathbf{Y}_i\right),$$

which when written in component form is

$$\begin{aligned}
[\mathbf{Y}_{i+1}]_1 &= [\mathbf{Y}_i]_1 + h\mathbf{f}_1(x,\ [\mathbf{Y}_i]_1,\ [\mathbf{Y}_i]_2,\ \ldots,\ [\mathbf{Y}_i]_m),\\
[\mathbf{Y}_{i+1}]_2 &= [\mathbf{Y}_i]_2 + h\mathbf{f}_2(x,\ [\mathbf{Y}_i]_1,\ [\mathbf{Y}_i]_2,\ \ldots,\ [\mathbf{Y}_i]_m),\\
&\vdots\\
[\mathbf{Y}_{i+1}]_m &= [\mathbf{Y}_i]_m + h\mathbf{f}_m(x,\ [\mathbf{Y}_i]_1,\ [\mathbf{Y}_i]_2,\ \ldots,\ [\mathbf{Y}_i]_m).
\end{aligned}$$

Example 10.13 Linear systems can be written in matrix form as

$$\mathbf{y}'(x) = \mathbf{C}\mathbf{y}(x) + \mathbf{b}(x), \tag{10.34}$$

where $\mathbf{C}$ is an $m \times m$ matrix and $\mathbf{b}$ is an m-dimensional vector whose elements can be functions of x, but do not depend on $\mathbf{y}$. In the linear case, values for implicit methods are easily computed by solving a linear system. The method (10.9) gives the rule

$$\left(\mathbf{I} - \frac{h}{2}\mathbf{C}\right)\mathbf{Y}_{i+1} = \left(\mathbf{I} + \frac{h}{2}\mathbf{C}\right)\mathbf{Y}_i + \frac{h}{2}\left\{\mathbf{b}\left(x_i\right) + \mathbf{b}\left(x_{i+1}\right)\right\}.$$

Example 10.14 When one writes things out in component form, one has to be a little careful. Suppose that we apply method (10.10) to the equation

$$\begin{aligned}
y'(x) &= x + y(x) + z(x),\\
z'(x) &= \frac{1}{1 + y(x) + z(x)}.
\end{aligned}$$

If we let Y_i and Z_i denote the approximations to $y(x_i)$ and $z(x_i)$, respectively, the approximating equations can be written as

$$\begin{aligned}
Y_{i+1} &= Y_i + \frac{h}{2}\left\{x_i + Y_i + Z_i + x_{i+1} + Y_{i+1}^0 + Z_{i+1}^0\right\},\\
Z_{i+1} &= Z_i + \frac{h}{2}\left\{\frac{1}{1 + Y_i + Z_i} + \frac{1}{1 + Y_{i+1}^0 + Z_{i+1}^0}\right\},
\end{aligned}$$

where

$$Y_{i+1}^0 = Y_i + h\{x_i + Y_i + Z_i\},$$
$$Z_{i+1}^0 = Z_i + \frac{h}{1 + Y_i + Z_i}.$$

■

There are of course practical differences between single equations and systems, particularly if there are nonlinear equations in the system and we use an implicit method. At each step we then have to solve a nonlinear system, so if m is large and a small mesh size is used, there may be a great deal of computation involved. But from a conceptual viewpoint, there are few significant differences between single equations and systems, neither in the practical implementation nor in the theory.

Higher-order equations are usually solved by reduction to systems of first-order equations. While it is possible to devise direct methods for higher-order equations, the reduction is so simple and first-order systems are so much better understood that this is normally the most productive way.

Example 10.15 The second-order equation

$$y''(x) = f(x,\ y(x),\ y'(x)),$$

with initial conditions

$$y(0) = \alpha,$$
$$y'(0) = \beta,$$

can be reduced by the substitution

$$y_1(x) = y(x),$$
$$y_2(x) = y'(x).$$

The second-order equation then becomes the equivalent first-order system

$$y_1'(x) = y_2(x),$$
$$y_2'(x) = f(x,\ y_1(x),\ y_2(x)),$$

with the initial conditions

$$y_1(0) = \alpha,$$
$$y_2(0) = \beta.$$

■

EXERCISES

1. Reduce the third-order initial value problem

$$\begin{aligned} y'''(x) &= y''(x) - 3xy'(x) + y(x) + 1, \\ y(0) &= 0, \\ y'(0) - y''(0) &= y_1, \\ y'(0) + y''(0) &= y_2, \end{aligned}$$

to a system of first-order equations.

2. Reduce the system of second-order equations

$$\begin{aligned} y_1''(x) &= y_1'(x) + y_2(x) + 1, \\ y_2''(x) &= y_1(x) - y_2'(x), \\ y_1(0) &= 1, \\ y_1'(0) &= y_2(0) = y_2'(0) = 0, \end{aligned}$$

to a system of first-order equations.

3. How can standard methods for ordinary differential equations be used in an attempt to solve an equation in the nonstandard form

$$\frac{y'(x)}{y''(x) + y'(x)} = y(x).$$

Are there reasons to be concerned that such an approach may fail?

4. Reduce the equations in Example 10.1 to a system of first-order equations.

5. Use the classical Runge–Kutta method in (10.32) to compute an approximate solution for Exercise 1 in this section on the interval $[0, 0.5]$ with step sizes $h = 0.02$, 0.01, and 0.005. Examine and plot the three approximates to predict accuracy for the computed results.

10.6 Convergence of Approximations*

So far, we have concentrated on the constructive aspects of the methods, namely, on how to find plausible algorithms. Unfortunately, the plausibility of a method does not guarantee success, so we must now examine to what extent the methods are practically useful. In some of the examples we have already claimed convergence and orders of convergence, in analogy with similar terms in approximation theory. It still remains to define these terms precisely and demonstrate the claimed orders. To evaluate methods and to compare their effectiveness in a general sense, we need an error analysis. For simplicity we restrict the discussion to single equations and to methods that use a uniform mesh size h on the interval $[0,\ a]$, divided into n equal parts with $nh = a$.

Single and multistep methods give approximations at the mesh points. The vector

$$\mathbf{Y} = \begin{bmatrix} Y_0 \\ Y_1 \\ \vdots \\ Y_n \end{bmatrix}$$

is called the *mesh function* approximation to $y(x)$. Results can be compared only at the mesh points and we can only estimate the error

$$e_i = y(x_i) - Y_i, \quad i = 0,\ 1,\ \ldots,\ n. \tag{10.35}$$

We refer to this as the *global discretization error.*

Now it is rarely necessary to know the error at every mesh point; usually it is adequate to get a bound on its maximum magnitude over the mesh. Since the approximation $\mathbf{Y}$ and the true solution $y(x)$ at the mesh points are vectors, we can characterize the global discretization error by its maximum norm[2]

$$||\mathbf{e}||_\infty = ||\mathbf{y} - \mathbf{Y}||_\infty,$$

where $[\mathbf{y}]_i = y(x_i)$ and $[\mathbf{Y}]_i = Y_i$.

Definition 10.2

If $\mathbf{Y}$ is a mesh function approximation to the solution $y(x)$ of (10.1), computed by a single or multistep method, the method is said to be *convergent* if

$$\lim_{h \to 0,\, nh = a} ||\mathbf{y} - \mathbf{Y}||_\infty = 0.$$

If there exists some K such that

$$||\mathbf{y} - \mathbf{Y}||_\infty \le K h^p,$$

[2]Reminder: $||\mathbf{x}||_\infty = \max |[\mathbf{x}]_i|$.

for all h, then p is the *order of convergence* of the method. The parameter K can (and usually does) depend on y in some complicated way, but it is to be independent of h.

Error analysis for differential equations is a fairly complicated and technical matter, so we break it into pieces to make it easier to see what is going on. We first look at the relatively simple situation of one-step methods. This includes Euler's method (10.7), the methods given by equations (10.10), (10.11), and various Runge–Kutta methods. We incorporate all of these methods in a single result by introducing the general one-step method

$$Y_{i+1} = Y_i + h\Phi(Y_i). \tag{10.36}$$

The Φ is the *increment* function for the method. For Euler's method

$$\Phi(Y_i) = f(x_i, Y_i),$$

whereas (10.10) has the more complicated increment function

$$\Phi(Y_i) = \frac{1}{2}\left\{f(x_i,\ Y_i) + f(x_{i+1},\ Y_i + hf(x_i,\ Y_i))\right\}.$$

Definition 10.3

The quantity

$$\tau_i = \frac{y(x_{i+1}) - y(x_i)}{h} - \Phi(y(x_i)) \tag{10.37}$$

is the *local discretization error* at x_i of the method (10.36).[3]

Note that if we had an exact solution at x_i; that is, $Y_i = y(x_i)$, then

$$y(x_{i+1}) - Y_{i+1} = h\tau_i,$$

so that the term $h\tau_i$ is the error committed in going from x_i to x_{i+1}, not counting any previous errors.

[3] In some books this is called the local truncation error.

Example 10.16 In Euler's method

$$\begin{aligned} h\tau_i &= y(x_{i+1}) - y(x_i) - hf(x_i,\ y(x_i)) \\ &= hy'(x_i) + \frac{h^2}{2}y''(x_i) + O(h^3) - hf(x_i,\ y(x_i)) \\ &= \frac{h^2}{2}y''(x_i) + O(h^3). \end{aligned}$$

Therefore, if the solution of (10.1) is twice continuously differentiable, the local discretization error of Euler's method is $O(h)$.

For the method (10.10)

$$h\tau_i = y(x_{i+1}) - y(x_i) - \frac{h}{2}\left\{f(x_i,\ y(x_i)) + f(x_{i+1},\ y(x_i) + hf(x_i,\ y(x_i)))\right\}.$$

Making a Taylor series expansion about $(x_i,\ y(x_i))$ gives

$$\begin{aligned} h\tau_i &= hy' + \frac{h^2}{2}y'' + O(h^3) - hf - \frac{h^2}{2}f_x - \frac{h^2}{2}f_y f + O(h^3) \\ &= O(h^3). \end{aligned}$$

This shows that the local discretization error of (10.10) has second order. Similar arguments can be made for the methods (10.30) and (10.31). With a little more work, we can also show that the local discretization error for the classical Runge–Kutta method (10.32) is $O(h^4)$. ■

It should not be too surprising that a small local discretization error is essential to the accuracy of an approximating method. The local discretization error is simply a measure of how well the approximating method reflects the original equation.

Definition 10.4

A single-step method of type (10.36) is said to be a *consistent* approximation of (10.1) if

$$\lim_{h\to 0,\, nh=a} ||\mathbf{T}||_\infty = 0,$$

where $[\mathbf{T}]_i = \tau_i$. If there is a K such that

$$||\mathbf{T}||_\infty \le Kh^p$$

for all h, then the *order of consistency* of the method is p.

Bounding the local discretization error and establishing the order of consistency is usually a fairly straightforward process. But the local discretization error at each step combines with the error at the other steps and, depending on the nature of the differential equation, this combined effect grows as we integrate toward larger x. This makes connecting the individual local discretization errors with the global discretization error $||\,\mathbf{e}\,||_\infty$ a more difficult matter.

Theorem 10.2 Assume that the increment function in (10.36) satisfies a uniform Lipschitz condition

$$|\Phi(t_1) - \Phi(t_2)| \leq K|t_1 - t_2| \tag{10.38}$$

and that the method is consistent with

$$||\mathbf{T}||_\infty \leq M(h). \tag{10.39}$$

Then

$$||\mathbf{y} - \mathbf{Y}||_\infty \leq ae^{Ka}M(h). \tag{10.40}$$

Proof: We start by showing inductively that

$$|e_i| \leq (1 + hK)^i ih\,||\mathbf{T}||_\infty. \tag{10.41}$$

From (10.36) and (10.37)

$$e_{i+1} = e_i + h\,\{\Phi(y(x_i)) - \Phi(Y_i)\} + h\tau_i.$$

Using the Lipschitz condition (10.38)

$$|e_{i+1}| \leq (1 + hK)|e_i| + h\,||\mathbf{T}||_\infty.$$

With (10.41) as inductive assumption

$$\begin{aligned}|e_{i+1}| &\leq (1 + hK)\,\{(1 + hk)^i ih\,||\mathbf{T}||_\infty\} \\ &\quad + h\,||\mathbf{T}||_\infty \leq (1 + hK)^{i+1}(i + 1)\,h\,||\mathbf{T}||_\infty.\end{aligned}$$

Since (10.41) is satisfied for $i = 1$, it follows by induction that it is true for all i. Now

$$\begin{aligned}(1 + hK)^i &\leq (1 + hK)^n \\ &= (1 + hK)^{a/h} \\ &\leq e^{Ka}\end{aligned}$$

and (10.40) follows. ■

The immediate conclusion we can draw from this is that any consistent one-step method is convergent, with the order of convergence at least as large as the order of the local discretization error. For the methods (10.10), (10.30), and (10.31) this implies second-order convergence, while the classical Runge–Kutta scheme is a fourth-order method. Keep in mind, though, that some assumptions have been made. The order of consistency depends on the smoothness of the functions involved in (10.1), so all order of convergence statements should be qualified by "provided the solution and the function f are sufficiently smooth." If the smoothness conditions are violated, there may be a reduction in the order of convergence or the method may fail altogether.

The analysis of the general multistep method is more complicated. One reason is that in the solution of a nonlinear equation by an implicit method one cannot get an exact value of Y_{i+1}, but must settle for some approximate value. This additional error depends on exactly how the nonlinear equation is solved and so makes the analysis more tedious. But even if we can ignore this issue, the analysis is still harder than that in Theorem 10.2.

As in the analysis of one-step methods, we define the local discretization error τ_i by

$$\begin{aligned} y(x_{i+k}) = \alpha_0 y(x_i) + \alpha_1 y(x_{i+1}) + \ \dots \ + \alpha_{k-1} y(x_{i+k-1}) \\ + h\beta_0 f(x_i,\ y(x_i)) + h\beta_1 f(x_{i+1},\ y(x_{i+1})) + \ \dots \\ + h\beta_k f(x_{i+k},\ y(x_{i+k})) + h\tau_i. \end{aligned} \tag{10.42}$$

Consistency and its order are defined through the local discretization error in the same way as in Definition 10.3. As in the one-step methods, the local discretization error is easy to bound, but we need to connect it to the discretization error.

Subtracting (10.18) from (10.42) gives

$$\begin{aligned} & e_{i+k} - \alpha_{k-1} e_{i+k-1} - \ \dots \ - \alpha_0 e_i \\ & \quad = h\beta_0 f(x_i,\ y(x_i)) + h\beta_1 f(x_{i+1},\ y(x_{i+1})) + \ \dots \\ & \quad\quad + h\beta_k f(x_{i+k},\ y(x_{i+k})) - h\beta_0 f(x_i,\ Y_i) - h\beta_1 f(x_{i+1},\ Y_{i+1}) - \ \dots \\ & \quad\quad - h\beta_k f(x_{i+k},\ Y_{i+k}) + h\tau_i, \end{aligned} \tag{10.43}$$

where the values $e_0,\ e_1,\ \dots,\ e_{k-1}$ are given by the accuracy of the starting values.

To get an idea how the solution of such a recurrence behaves, look at the case $h = 0$. The errors are then determined by the solution of a homogeneous linear recurrence

$$e_{i+k} - \alpha_{k-1} e_{i+k-1} - \ \dots \ - \alpha_0 e_i = 0, \tag{10.44}$$

about which we know a great deal. The solutions of such an equation are governed by the roots of its characteristic polynomial, which here is

$$p(x) = x^k - \alpha_{k-1} x^{k-1} - \ \dots \ - \alpha_1 x - \alpha_0. \tag{10.45}$$

If we denote these roots by $\rho_1, \rho_2, \ldots, \rho_k$, the theory of linear recurrences tell us that[4] the solution of (10.44) can be written as

$$e_n = c_1\rho_1^n + c_2\rho_2^n \ldots c_k\rho_k^n$$

with the constants $c_1, c_2, \ldots, c_k$ determined by the starting values $e_1, e_2, \ldots, e_k$. This in turn tells us that the solutions of (10.44) will increase exponentially if any of the roots of the characteristic polynomial have a magnitude larger than one. Since we expect that for small h the solution of (10.43) will behave roughly like the solution of (10.44), the method can be expected to be unstable if $p(x)$ in (10.45) has any roots outside the unit circle in the complex plane.

Definition 10.5

A multistep method of the form (10.18) is said to satisfy the *root condition* if the roots $\rho_1, \rho_2, \ldots, \rho_k$ of its characteristic polynomial (10.45) have the properties that

(i) $|\rho_i| \leq 1$ for all i, and

(ii) any root that has modulus one must be simple.

It is the root property that is essential to the convergence of multistep methods, as expressed in the next theorem.

Theorem 10.3 A consistent multistep method of form (10.18) is convergent only if it satisfies the root condition. If the root condition is satisfied, and if the starting errors go to zero with h, then the method is convergent.

If the multistep method has order of consistency k_1, and if the starting errors go to zero with order k_2, then the order of convergence of the method is at least

$$k = \min(k_1, k_2).$$

Proof: The arguments for this theorem are tedious and of no importance to us here. Some motivation for it will be given in the next section. For a complete proof, see Isaacson and Keller [12]. ∎

[4] Assuming that all roots are distinct.

Example 10.17 Methods of type (10.13) based on the integration (10.12) all have characteristic polynomials

$$x^k = 1.$$

Since all the roots of unity are simple and lie on the circle $|z| = 1$ in the complex plane, the root condition is satisfied. Therefore methods of type (10.13) are always convergent, with the order depending on the quadrature used. ■

Example 10.18 The method (10.20) has the characteristic polynomial

$$p(x) = x^3 - \frac{18}{11}x^2 + \frac{9}{11}x - \frac{2}{11},$$

whose roots are 1 and $0.3182 \pm 0.2839i$. The root condition is therefore satisfied. We can also show that its order of consistency is four, so it converges with fourth order. ■

Example 10.19 The method (10.17) has characteristic polynomial

$$p(x) = x^3 + \frac{3}{2}x^2 - 3x + \frac{1}{2}.$$

The roots of this polynomial are 1.000, 0.1861, and −2.6861. The root condition is violated and the method is not convergent. This explains the results in Figure 10.3. It is also an example of how a perfectly plausible numerical method can fail to work. ■

EXERCISES

1. Show that, if the solution of (10.1) and the function f are sufficiently smooth, the local discretization error for both (10.30) and (10.31) is $O(h^2)$.
2. Show that the method (10.14) is convergent with order four.
3. Examine the orders of convergence of the methods described in Example 10.9. Why do the theorems given so far not apply to the predictor-corrector approach?

4. Examine the convergence of the method in (10.11).
5. Why is the second condition in Definition 10.5 necessary?
6. Find the characteristic polynomial for the method derived in the manner of Example 10.8, with $k = 3$ and $\alpha_2 = \beta_1 = \beta_3 = 0$. Is the root condition satisfied?

10.7 Stability*

Convergence is an essential property of any algorithm. Convergence results, such as Theorem 10.2, assure us that we can compute the solution to a differential equation to any desired accuracy by simply making h small enough. But Theorem 10.2 makes no reference to rounding and applies only for infinite-precision computations. To understand how an algorithm will work in practice, we need to bring in the other essential component of any error analysis, the stability of a method.

So far, we have used the term stability loosely as the sensitivity of a method to small perturbations. In this view stability is a matter of degree, with some algorithms more stable than others. For a rigorous analysis this is inadequate and we must replace it with a precise definition. In doing so, we find that we need several different concepts of stability.

Definition 10.6

A multistep method of type (10.18) is said to be stable if and only if it satisfies the root condition.

In this definition, stability plus consistency are essentially equivalent to convergence. It is the definition that is most frequently used in theoretical numerical analysis where it leads to comprehensive and elegant results. For the solution of differential equations it is only one of several stability definitions that have been found to be useful. In fact, there are a number of types of stability that are discussed in the literature and this variety can cause some of confusion. To understand why we need several definitions of stability, we first examine the different types of solutions that the differential equation (10.1) can have.

Equation (10.1) without an initial condition defines a family of curves in the xy-plane, each of which satisfies the equation. The initial condition serves to select one of these as the unique solution (Figure 10.4).

The initial condition gives us the starting point on one of the curves from which we can develop the solution by integration. When we perform

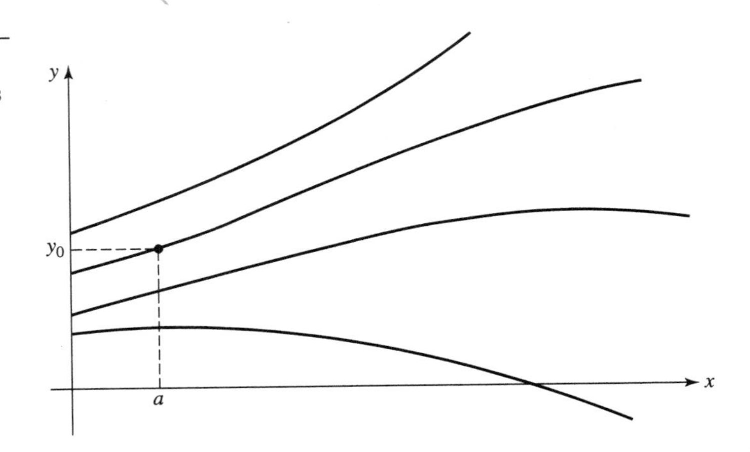

Figure 10.4 Family of solutions of a differential equation.

numerical computations, discretization and rounding will take us off the true curve, and as the computation proceeds we may deviate more and more from the correct solution. This is unavoidable and all we can hope is that the deviation does not become too large.

There are three qualitatively different situations that we have to take into account; they are shown in Figure 10.5. In Figure 10.5(a), the different curves of the solution family remain separated by roughly the same amount as x increases. We say that this is a *neutrally stable* solution set. In Figure 10.5(b), the different solutions diverge rapidly, a situation that we classify as *unstable*. Finally, in Figure 10.5(c), the solutions approach each other, something that we call *superstable*. Of course, these adjectives are intuitive and imprecise; there is not a clear separation between the three types, and a given equation may behave differently for different regions of x. Nevertheless, it is productive to think in these terms.

The unstable case is an ill-conditioned problem. Even a slight deviation from the true solution will eventually give a very large error. This difficulty is inherent in the problem and no numerical method can compensate for

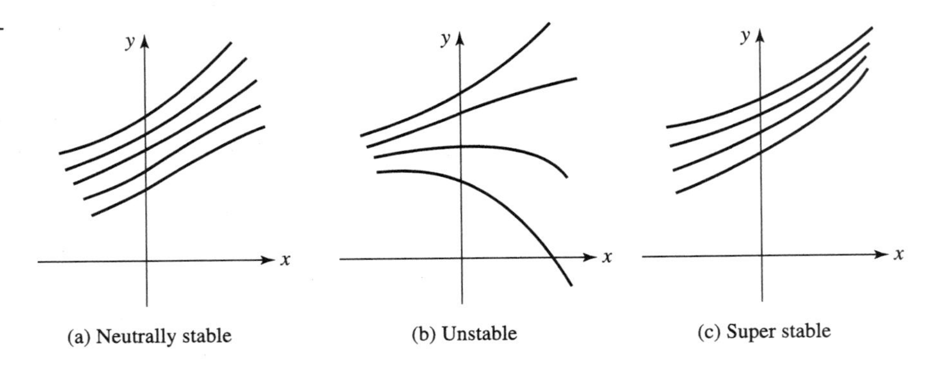

Figure 10.5 Different types of stability situations.

it. It is the other two cases that we have to look at carefully. If there are large errors, they are not mandated by the problem, but are instead caused by the numerical approximations and can therefore be avoided by a better method.

The easiest way of studying stability of numerical methods for ordinary differential equations is to look at the very simple equation

$$y'(x) = \lambda\, y(x), \tag{10.46}$$

with $y(0) = y_0$. This equation has the solution

$$y(x) = y_0 e^{\lambda x},$$

so that $\lambda = 0$, $\lambda > 0$, and $\lambda < 0$ represent the neutrally stable, unstable, and superstable cases, respectively. If we allow λ to be complex, there can also be an oscillating component. Even though the equation seems overly simple, the observations we make on it are mirrored by more complicated situations. In any case, the conclusions give us some necessary conditions for numerical algorithms, since anything that does not work on (10.46) is not likely to be successful in other instances.

If we set $f(x,\ y) = \lambda y$ in (10.18), we get the equation

$$\begin{aligned} Y_{i+k} = \alpha_{k-1} Y_{i+k-1} + \alpha_{k-2} Y_{i+k-2} + \ \dots \ + \alpha_0 Y_i \\ + h\lambda \left\{ \beta_k Y_{i+k} + \ \beta_{k-1} Y_{i+k-1} + \ \dots \ + \beta_0 Y_i \right\} \end{aligned} \tag{10.47}$$

for the approximation. This is a linear recurrence whose solution can be written as

$$Y_i = c_1 \rho_1^i(h\lambda) + c_2 \rho_2^i(h\lambda) + \ \dots \ + c_k \rho_k^i(h\lambda), \tag{10.48}$$

where the $\rho_i(h\lambda)$ are the roots of the characteristic polynomial

$$p(x) = (1 - h\lambda\beta_k)x^k - (\alpha_{k-1} + h\lambda\beta_{k-1})x^{k-1} \ - \ \dots \ - \ (\alpha_0 + h\lambda\beta_0), \tag{10.49}$$

and the c_i are determined by the starting values $Y_0,\ Y_1,\ \dots,\ Y_{k-1}$. If a method is consistent to any order it must give exact results for constant $y(x)$; that is, for $\lambda = 0$. This implies that

$$1 - \alpha_{k-1} - \alpha_{k-2} \ - \ \dots \ - \ \alpha_0 = 0,$$

so that for $h\lambda\ = 0$ there is always a root $\rho_1(0) = 1$. We know that the roots of a polynomial are continuous functions of the coefficients, so for any small positive $h\lambda$, the term $c_1\rho_1^i(h\lambda)$ is the part that approximates the true solution, while the other components $c_2\rho_2^i(h\,\lambda) + \ \dots \ + \ c_k\rho_k^i(h\lambda)$ represent spurious effects of the discretization. The numerical method will be effective for a given value of $h\lambda$ only if these spurious effects are small.

Consider now what happens when $h\lambda$ increases from zero and the roots of the characteristic polynomial move in the complex plane. To assess the effect of the spurious contribution we need to know what happens to the roots $\rho_i(h\lambda)$ for $i = 2, 3, \ldots, k$. If for some i, $|\rho_i(0)| > 1$, then even for small nonzero $h\lambda$ the term $c_i\rho_i(h\lambda)^i$ increases rapidly with i and causes instability that cannot be reduced by decreasing h. This is the reason for the necessity of the root condition for convergence. But even when the root condition is satisfied, not everything necessarily works smoothly. What we really need to be concerned with is the comparative rate at which the true solution and the spurious effects can grow.

Definition 10.7

For a given value of $h\lambda$, a multistep method is said to be *absolutely stable* if $|\rho_i(h\lambda)| \leq 1$ for all $i = 1, 2, \ldots, k$. Absolute stability implies that no components of the solution, including the one representing the actual solution, can grow at an exponential rate.

Definition 10.8

For a given value of $h\lambda$, a multistep method is said to be *relatively stable* if $|\rho_i(h\lambda)| \leq |\rho_1(h\lambda)|$ for all $i = 2, 3, \ldots, k$. Relative stability means that no spurious component can grow faster than the component representing the true solution.

Since for any $h\lambda$ a method is either stable or not, the complex $h\lambda$-plane is divided into two parts, a *region of stability* and a *region of instability*. These regions can have complicated shapes and it takes some work to find them for higher order multistep methods. For the simpler methods, the answers are not too hard to get.

Example 10.20 Find the region of absolute stability for Euler's method. The characteristic polynomial for Euler's method is

$$p(x) = x - (1 + h\lambda)$$

whose single root is

$$\rho_1(h\lambda) = 1 + h\lambda.$$

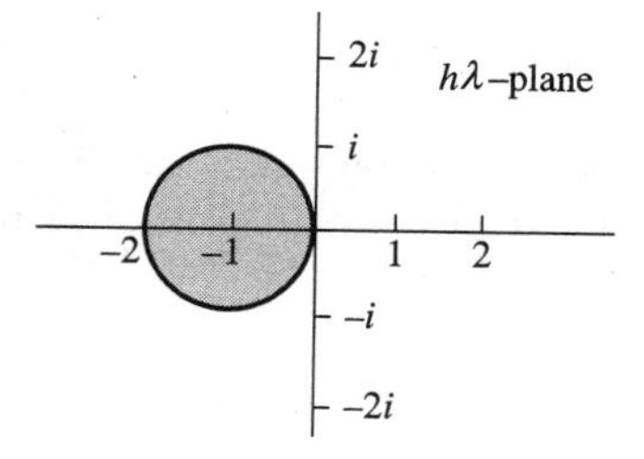

Figure 10.6 Region of absolute stability for Euler's method.

The condition $|\rho_1(h\lambda)| \leq 1$ then defines the absolute stability region shown in Figure 10.6. Since there is only one root, the method is always relatively stable.

It is not always necessary to have a complete description of the stability regions; sometimes it is sufficient to consider only the neighborhood of the origin. Since the roots near the origin vary continuously with λh, their behavior at the origin is crucial.

Definition 10.9

A multistep method is *strongly stable* if it satisfies the *strong root condition*

$$\rho_1(0) = 1,$$

and

$$|\rho_i(0)| < 1, \quad i = 2, 3, \ldots, k.$$

If a method is strongly stable, then there is always a neighborhood near $h\lambda = 0$ where the method is relatively stable. This follows from the continuity of the roots with respect to $h\lambda$. By a similar argument, a strongly stable method is also absolutely stable in a neighborhood of zero, but only in the left half of the complex plane. No method can be absolutely stable in the right half of the complex plane because if $\lambda > 0$ the problem itself is unstable.

Example 10.21 With $h = 0$, the multistep method (10.21) has the characteristic polynomial

$$p(x) = x^4 - x^3.$$

This has a single root at $x = 1$ and a triple root at $x = 0$. The method is therefore strongly stable.

Not all multistep methods are strongly stable. The method (10.14), for example, does not satisfy the strong root condition, since its characteristic polynomial is

$$p(x) = x^2 - 1,$$

with roots ± 1.

Definition 10.10

A multistep method which satisfies

$$|\rho_i(0)| \leq 1, \quad i = 2,\ 3,\ \dots,\ k,$$

and for which there is an i, with $2 \leq i \leq k$ such that

$$|\rho_i(0)| = 1$$

is said to be weakly stable.

Weakly stable methods allow the spurious components of the approximate solution to grow like e^{cx} with $c > 1$, so that if we integrate over large intervals this component may have a significant effect. This is the phenomenon of weak instability. Weak instability should not be confused with the instability caused by a failure to satisfy the root condition. In the absence of rounding and starting errors, a weakly stable method is still convergent and the error can be reduced by decreasing h and making the starting errors smaller. Practically, though, since rounding errors cannot be eliminated, weakly stable methods can cause problems.

Example 10.22 The equation

$$y'(x) = -y(x),$$
$$y(0) = 1,$$

was solved in $(0,\ 20)$, using the weakly stable method (10.14) with $h = 0.1$. As starting values we took the exact initial condition $Y_0 = 1$, but for Y_1 we took a value that differed from the exact answer by 10^{-4}. The

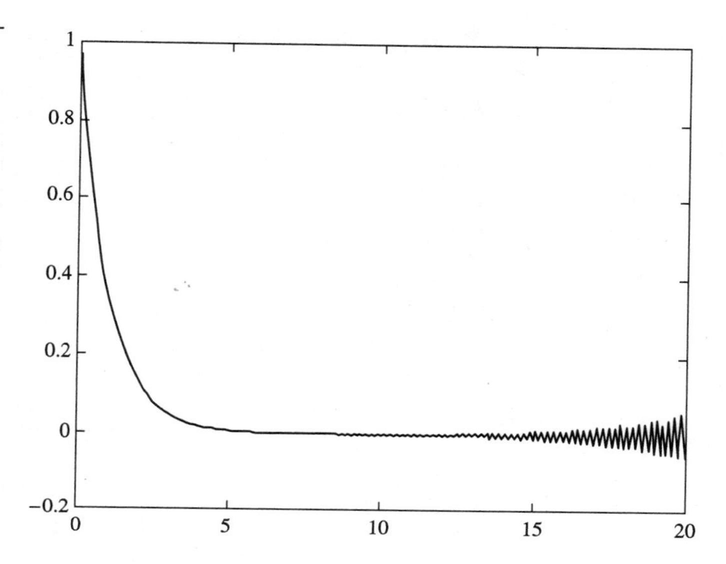

Figure 10.7 Integration of Example 10.22 with $h = 0.1$ and $e_1 = 10^{-4}$.

results, displayed in Figure 10.7, show the weak instability clearly. The same equation was then solved with $h = 0.05$ and a starting error $e_1 = 10^{-5}$. The instability now is barely noticeable to plotting accuracy (Figure 10.8). However, if we enlarge the interval, say to $(0,\ 25)$, the instability becomes again quite pronounced.

While weakly stable methods can give acceptable answers under certain conditions, they are not very reliable and as a result we prefer strongly stable methods. Because strongly stable methods are relatively stable for small h, and since small values of h are in any case necessary to keep the local discretization error under control, they perform quite well in most circumstances. Still, there are some exceptional cases where even strongly stable methods may not work well. This we consider in Section 10.8.

We have concentrated on the stability of multistep methods, but the same sort of analysis can be done for Runge–Kutta methods. If we put $f(x,\ y) = \lambda y$ in the formula for a Runge–Kutta method, we can perform a similar analysis. For example, (10.29) becomes

$$Y_{i+1} = (1 + h\lambda + \frac{h^2\lambda^2}{2})Y_i.$$

This clearly satisfies the root condition, so it is strongly stable. The entire region of absolute stability can easily be worked out and is plotted in Figure 10.9.

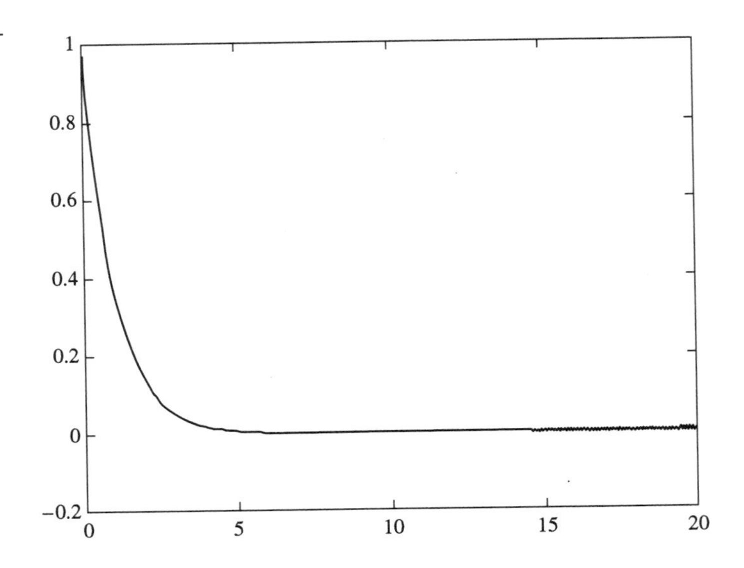

Figure 10.8
Integration of Example 10.22 with $h = 0.05$ and $e_1 = 10^{-5}$.

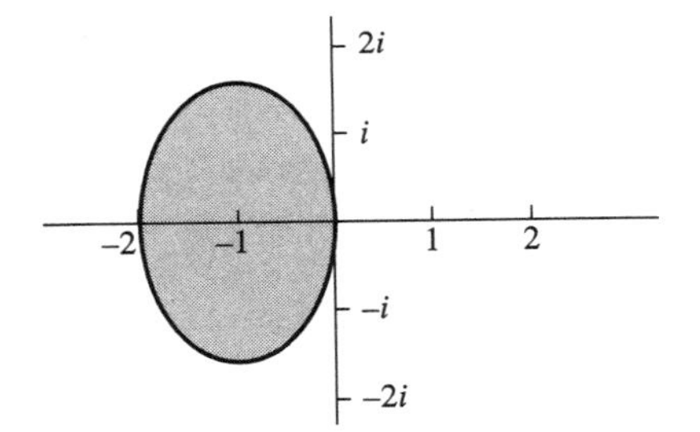

Figure 10.9
Region of absolute stability for the method (10.29).

For higher-order multistep and Runge–Kutta methods, stability regions can become quite complicated. For a thorough discussion of the stability issue and plots of the stability regions for many methods, see Lapidus and Seinfeld [17].

EXERCISES

1. Classify the solution set of

$$y'(x) = \frac{1}{1 + y^2(x)}$$

as neutrally stable, unstable, or superstable.

2. Find the region of absolute stability for the method (10.9).
3. Find the region of absolute stability for the method (10.30).

4. Is the method (10.16) strongly or weakly stable?
5. Show that any method based on (10.9) with $k = 1$ is strongly stable.
6. Show that any multistep method based on (10.9) with $k \geq 2$ cannot be strongly stable.
7. Determine if the multistep method

$$Y_{i+1} = Y_i + \frac{h}{12}\{5f(x_{i+1},\ Y_{i+1}) + 8f(x_i,\ Y_i) - f(x_{i-1},\ Y_{i-1})\}$$

is strongly stable.

8. Show that the method in the above exercise is consistent and determine its order of consistency.
9. Examine the stability properties of the multistep method

$$Y_{i+1} = Y_{i-1} + 2hf(x_i,\ Y_i).$$

10. Find the region of absolute stability for the classical fourth-order Runge–Kutta method.

10.8 Stiff Equations*

Strongly stable methods are suitable for most situations, but there are exceptions. There is a type of equation, called *stiff*, that needs special attention. Stiffness is a property that can only be loosely defined, but that occurs in many practical situations. Roughly speaking, we call an equation stiff if its solution has a component that decays rapidly, as well as components that are neutrally stable.

Example 10.23 The equation

$$y'(x) = -100y(x) + 100,$$
$$y(0) = 0,$$

has the solution

$$y(x) = 1 - e^{-100x}.$$

Since the solution contains a constant component as well as the rapidly changing term e^{-100x}, we consider it stiff.

When this equation is solved numerically, a small step size is needed near $x = 0$ because the solution changes rapidly there. As we move away from the origin, the solution becomes nearly constant, so very good accuracy is theoretically possible with quite large h. Unfortunately, there is also stability to be considered. If we use a method with a small stability region, we may need a small step size to keep the process stable. This will cause the method to perform inefficiently.

■

While one can construct simple examples like this to illustrate stiffness, most practical examples of stiffness involve systems of equations. Suppose we look at the simple linear system

$$\mathbf{y}' = \mathbf{A}\mathbf{y}, \tag{10.50}$$

where $\mathbf{A}$ is an $m \times m$ matrix with all constant entries. The theory of differential equations tells us that the components of the solution depend on the eigenvalues $\lambda_1, \lambda_2, \ldots, \lambda_m$ of $\mathbf{A}$ through

$$[\mathbf{y}]_i = c_{i1}e^{\lambda_1 x} + c_{i2}e^{\lambda_2 x} + \ldots + c_{im}e^{\lambda_m x}.$$

If all $\lambda_i \leq 0$, then the solutions are at least neutrally stable or superstable.

For more general and nonlinear systems,

$$\mathbf{y}'(x) = \mathbf{f}(x, \mathbf{y}). \tag{10.51}$$

This qualitative behavior can only be characterized locally, with the matrix $\mathbf{A}$ replaced by the Jacobian $\mathbf{J}$ with components

$$[\mathbf{J}]_{ij} = \frac{\partial\,[\mathbf{f}]_i}{\partial\, y_j}.$$

In either case, though, it is the eigenvalues of the local Jacobian that give the solution its essential characteristics.

If the spectrum of the Jacobian contains very large negative eigenvalues, the solution can exhibit rapid changes in some regions and the equation is considered stiff. For this reason, stiffness is often defined by the ratio of the smallest to the largest eigenvalue of the Jacobian.

Example 10.24 Take as the matrix $\mathbf{A}$ in (10.50)

$$\mathbf{A} = \begin{bmatrix} -55 & -2.25 & 2.25 \\ 90 & -55 & -45 \\ 99 & -49.5 & -50 \end{bmatrix}.$$

This matrix has eigenvalues $\{-4.812, -55.45, -99.74\}$. Since the ratio of the largest to smallest eigenvalue is about 20, the system is considered mildly stiff. In practice, systems with much larger eigenvalue ratios are common.

When we attempted to solve this system with the classical Runge–Kutta method, we obtained reasonable values with step sizes less than 0.02, but serious instabilities appeared quickly for larger step sizes.

■

The stability of the numeric process of any method is controlled by $h\lambda$ where

$$\lambda = \min(\lambda_1,\ \lambda_2,\ \ldots,\ \lambda_m).$$

The difficulty with stiff problems comes from the fact that the large negative eigenvalues make $|h\,\lambda|$ so large as to fall outside the stability region of the method used. We can of course keep the method stable by reducing h, but this will increase the work well beyond what is required by the discretization. In other words, the step size is no longer controlled by accuracy considerations but by excessive stability requirements so the method can become quite inefficient. A way around this is to use methods with large regions of absolute stability; ideally, we would like the whole negative $h\lambda$ axis to be included.

Definition 10.11

A method is *A-stable* if its stability region includes the entire negative real axis.

If a method is A-stable it does not matter how large $|h\lambda|$ is because even for large negative λ there will be no instability. The only problem with this suggestion is that it is hard to find good A-stable methods.

Example 10.25 The implicit method

$$Y_{i+1} = Y_i + hf(x_{i+1},\ Y_{i+1})$$

is known as the *backward Euler* method. Its characteristic polynomial has one root

$$\rho_1 = \frac{1}{1 - h\,\lambda},$$

so the method is A-stable.

■

Example 10.26 The method (10.8) is also A-stable. The root of its characteristic polynomial is

$$\rho_1 = \frac{2 + h\,\lambda}{2 - h\,\lambda}$$

so that $|\rho_1| \leq 1$ for all negative λ. ■

Unfortunately, the success with these low-order methods does not transfer to higher-order schemes. It is known that there are no A-stable multistep methods of order higher than two, so the trapezoidal method is about the best we can do. Also, the Runge–Kutta methods that we have presented here are not A-stable. There is another problem with the implicit methods, such as the trapezoidal rule, when they are applied to nonlinear systems: At each step we have to solve a nonlinear system, which requires some root-finder. We have so far only suggested successive substitution, but this will work only if

$$h\lambda << 1.$$

For stiff problems, this again calls for small h. To make these methods work we have to use other root-finding methods, adding considerably to the complexity of the algorithms.

Attempts to overcome these difficulties have led to much research on methods for stiff problems. A number of suitable approaches have been found, but they all require a considerable amount of modification of the methods discussed here, so they are beyond our scope. Solving stiff problems is still not as easy as solving nonstiff problems. For a brief discussion of how to solve stiff problems, see Kahaner, Moler, and Nash [14]. A more thorough treatment can be found in Lapidus and Seinfeld [17].

EXERCISES

1. Show that Euler's method is not A-stable.
2. Show that the predictor-corrector method (10.10) is not A-stable.
3. Show that the method in (10.20) has a large stability region, but is not A-stable.
4. In the equation

$$y'(x) = -100y(x) + 100,$$

is the observation that the classical Runge–Kutta method is stable for $h < 0.02$ but unstable for larger step sizes consistent with what we know about the stability of the method?

10.9 Error Estimation and Step Size Selection*

In software for initial value problems it is important that the step size h be selected effectively. For a complicated system, a user may have a very poor idea of what a good step size is and, when forced to make this decision, the user may be overly conservative and choose an h much smaller than necessary. The situation is even less transparent if the nature of the equation calls for changing step size in the interval. In such a situation, the step size selection will have to be done by the program. Good software should at least give the user the option of specifying the desired global accuracy and have the program find a good strategy for achieving this goal.

Normally, step size is selected adaptively. In each step, we get an estimate of the local discretization error, then use this estimate to adjust the step size. In this way the fineness of the discretization is based on the accuracy requirements at each step. We have already seen this work effectively in adaptive quadrature. For the integration of differential equations there are a few new wrinkles.

Estimating the local discretization error is not too hard. The most obvious way to go about this is to carry out the computations from x_i to x_{i+1} twice, first with a full step size, then with half of this, using the kinds of estimates discussed in Section 6.5. For one-step methods this presents no problems, but it is not quite as easy for multistep methods. In the latter, the lack of approximations at the right points makes it harder to work with the half-step approximation.

Example 10.27 Compute the solution of

$$y'(x) = 3e^{-3x}y(x),$$
$$y(0) = 1,$$

in $[0,\ 2]$ to an accuracy of 10^{-6}, using the trapezoidal method (10.8).

Because of the linearity of the equation, (10.8) can be arranged to compute Y_{i+1} without a predictor or iteration by

$$Y_{i+1} = \frac{1 + \dfrac{3h}{2}e^{-3x_i}}{1 - \dfrac{3h}{2}e^{-3x_{i+1}}} Y_i. \tag{10.52}$$

The method can therefore be viewed as one-step. It is obviously of order two.

If we approximate $y(0.2)$ with a single step with step size 0.2, we get

$$\hat{Y} = 1.5562.$$

If we next compute an approximation with $h = 0.1$, the approximation is

$$\tilde{Y} = 1.5665.$$

Since the method is of second order, we assume that the error at any point is roughly[5]

$$e \cong cdh^2, \tag{10.53}$$

where d is the interval size and c an unknown constant. Using the two observed values $\hat{Y}$ and $\hat{y}$, we can arrive at the approximate value of

$$c \cong 1.71. \tag{10.54}$$

If we extend (10.53) to the whole interval $[0,\ 2]$, we get an estimate for the error in the computed value Y at $x = 2$,

$$|y(2) - Y| \leq 3.44h^2. \tag{10.55}$$

For the required accuracy we then want a step size

$$h \cong 5.4 \times 10^{-4}.$$

We conclude that if we use such a step size we should get the required accuracy. In reality, this is overly conservative. When the computations were carried out with $h = 1 \times 10^{-3}$, the result was

$$y(2) \cong 2.711552.$$

This is correct to the last digit. ■

This example is a little oversimplified and we may want to use a more sophisticated scheme that is less likely to give an unnecessarily small h. In addition, we may want to make the computations in each step as efficient as possible. For this, the error estimation is often done in a way that reduces the amount of work necessary to compute the solution twice. Commonly, the new value Y_{i+1} is computed by two different formulas and the error estimated from the two answers. If one of the methods is of higher order than the other, the difference between the two approximations is a good estimate for the accuracy of the result of the lower-order method. A number of methods have been developed which share computed values so that the two answers can be obtained with about the same effort as one of them.

[5] This assumption implies that the error depends linearly on the size of the interval over which we integrate. This is not easy to justify but, in the absence of more specific information, not unreasonable.

Example 10.28 The *Runge–Kutta–Fehlberg* method of order 4 and 5 is quite popular. It starts by computing the values

$$\begin{aligned}
k_1 &= hf(x_i,\ Y_i),\\
k_2 &= hf(x_i + \frac{h}{4},\ Y_i + \frac{1}{4}k_1),\\
k_3 &= hf(x_i + \frac{3h}{8},\ Y_i + \frac{3}{32}k_1 + \frac{9}{32}k_2),\\
k_4 &= hf(x_i + \frac{12h}{13},\ Y_i + \frac{1932}{2197}k_1 - \frac{7200}{2197}k_2 + \frac{7296}{2197}k_3),\\
k_5 &= hf(x_i + h,\ Y_i + \frac{439}{216}k_1 - 8k_2 + \frac{3680}{513}k_3 - \frac{845}{4104}k_4),\\
k_6 &= hf(x_i + \frac{h}{2},\ Y_i - \frac{8}{27}k_1 + 2k_2 - \frac{3544}{2565}k_3 + \frac{1859}{4104}k_4 - \frac{11}{40}k_5).
\end{aligned}$$

Then

$$Y_{i+1} = Y_i + \frac{25}{216}k_1 + \frac{1408}{2565}k_3 + \frac{2197}{4104}k_4 - \frac{1}{5}k_5$$

is a fourth-order approximation. For estimating the error in this we compute the locally fifth-order approximation

$$\hat{Y}_{i+1} = Y_i + \frac{16}{135}k_1 + \frac{6656}{12825}k_3 + \frac{28561}{56430}k_4 - \frac{9}{50}k_5 + \frac{2}{55}k_6.$$

This is a complicated but very effective method. ■

The local error estimate is used to determine the step size, decreasing it if the estimated error is too large and increasing it when it appears that we have more accuracy than is needed. A real difficulty arises because there is no obvious connection between the local error estimate and a bound on the global error normally requested by the user. If the equation is neutrally stable or superstable, the local error estimate should reflect the global error at least in an order of magnitude sense. However, if the problem shows some instability, then we may need to make the local error several orders of magnitude smaller than the desired global error. In general, there is no practical way in which the program can determine locally what the situation is globally. Most of the initial value solvers assume some sort of neutral stability, so it is not hard to find examples where their error estimations fail.

EXERCISES

1. Give an intuitive explanation that in Example 10.28 the local error estimate

$$\begin{aligned} r_{i+1}(h) &= \frac{\hat{Y}_{i+1} - Y_{i+1}}{h} \\ &= \frac{1}{h}\left\{\frac{1}{360}k_1 - \frac{128}{4275}k_2 - \frac{2197}{75240}k_4 + \frac{1}{50}k_5 + \frac{2}{55}k_6\right\} \end{aligned}$$

is $O(h^4)$.

2. Suppose that in Exercise 1 $r_{i+1}(h) \approx Kh^4$ and $|r_{i+1}(h)| > \varepsilon > 0$. Show that the scaling factor

$$s = \sqrt[4]{\frac{\varepsilon}{2|r_{i+1}(h)|}}$$

is a good choice so that

$$|r_{i+1}(sh)| < \varepsilon \ .$$

Chapter 11

Boundary and Eigenvalue Problems for Ordinary Differential Equations

In initial value problems, the subsidiary conditions needed for a unique solution are all given at the same point $x = a$. This allows us to find the solution for any x by forward integration from the starting point. There are other types of problems where conditions are specified at several different points, most commonly at the endpoints or the boundaries of the region of interest. These are known as *boundary value problems.*

Example 11.1 Boundary value problems occur in many areas of engineering, for example, in the simple case of the bending of an elastic beam shown in Figure 11.1.

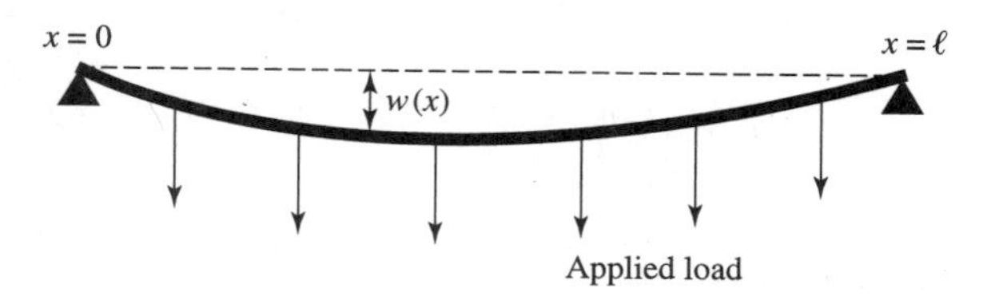

Figure 11.1 The deflection of a beam with uniform loading.

The physical situation in Figure 11.1 is modeled by the equation

$$\frac{d^2w}{dx^2} = a(x)w + b(x)x(x-l),$$

where $w(x)$ is the deflection at distance x from the left end of the beam, and $a(x)$ and $b(x)$ are functions that depend on the applied load, the geometry of the beam, and its elastic properties. With the assumption that no deflection occurs at the ends of the beam, the differential equation has the two boundary conditions; i.e., $w(0) = w(l) = 0$.

If the functions $a(x)$ and $b(x)$ are not simple, the equation does not have a closed form solution and we must rely on numerical methods. But now the subsidiary conditions are specified at two distinct points, so that the initial value methods of the last chapter are not immediately applicable. ■

We begin by discussing the general *two-point boundary value problem* for ordinary differential equations, which we can write as

$$y'' = f(x, y, y'), \tag{11.1}$$

with boundary conditions at two distinct points a and b of the form

$$\begin{aligned} a_0 y(a) + a_1 y'(a) &= \alpha, \\ b_0 y(b) + b_1 y'(b) &= \beta. \end{aligned} \tag{11.2}$$

For boundary value problems, the questions of existence and uniqueness of a solution are more difficult and require more assumptions than are necessary for initial value problems. The next theorem gives conditions that guarantee a unique solution; these conditions are sufficient, but not necessary. There could still be a unique solution even if they are not satisfied, but we may not be able to guarantee this a priori. A lack of assurance that a problem has a solution often creates serious practical difficulties, an issue that plagues many nonlinear equations.

Theorem 11.1 Suppose that

(a) $f(x, u, v)$ is continuous for all u, v, and $a \le x \le b$,

(b) $f(x, u, v)$ satisfies a Lipschitz condition in both u and v; that is, there exists a constant c such that

$$\begin{aligned} |f(x, u_1, v) - f(x, u_2, v)| &\le c|u_1 - u_2|, \\ |f(x, u, v_1) - f(x, u, v_2)| &\le c|v_1 - v_2|, \end{aligned}$$

for all x, u, v,

(c) $\dfrac{\partial f(x, u, v)}{\partial u} > 0$ for all x, u, v,

(d) $\left|\dfrac{\partial f(x, u, v)}{\partial v}\right|$ is bounded for all x, u, v,

(e) $a_0 a_1 \leq 0$, $b_0 b_1 \leq 0$, and $|a_0| + |b_0| > 0$.

Then equation (11.1) with boundary conditions (11.2) has a unique solution.

Proof: See Keller [15], p. 9. ■

An important special case of (11.1) is the *linear* two-point boundary value problem

$$y''(x) = p(x)y'(x) + q(x)y(x) + r(x). \tag{11.3}$$

For existence and uniqueness of the solution of the linear two-point boundary value problem, it suffices that p, q, and r are continuous and that $q(x) > 0$.

Example 11.2 The equation

$$y''(x) = y(x) + r(x),$$
$$y(0) = y(\pi) = 0,$$

is guaranteed by Theorem 11.1 to have a unique solution for all suitable $r(x)$. On the other hand, the seemingly similar equation

$$y''(x) = -y(x) + r(x),$$
$$y(0) = y(\pi) = 0,$$

cannot have a unique solution for any $r(x)$. This is because the homogeneous case with $r(x) = 0$ has the nontrivial solution

$$y(x) = \sin(x).$$

■

11.1 Shooting Methods

Because we have many effective algorithms for solving initial value problems, it is natural to ask if these algorithms can be utilized to solve boundary value problems. As we will see, this can be done and two-point boundary value problems can be recast in terms of initial value problems. The resulting methods are called *shooting methods.*

For simplicity, we will look at a special case of the boundary conditions

$$\begin{aligned} y(a) &= \alpha, \\ y(b) &= \beta. \end{aligned} \tag{11.4}$$

These conditions make the discussion easier, but most of what we will say can be extended without too much difficulty to the general case.

To employ initial value techniques, we first convert (11.1) into a system of two first-order equations

$$\begin{aligned} z'(x) &= f(x, y, z), \\ y'(x) &= z(x). \end{aligned} \tag{11.5}$$

If we were given the correct value $z(a) = y'(a)$, this would be an initial value problem. To get around the difficulty of not knowing $y'(a)$, we guess the value $z(a) = \gamma$. We then use the techniques of Chapter 10 to integrate the system and obtain a value of $y(b)$. Unless we are very lucky, this will not have the required value β, so we need to take another guess and solve another initial value problem. If we repeat this, choosing $z(a)$ carefully, we will eventually get $y(b)$ as close as we want to β. Since this is an iterative process, it will work best if we have a good initial guess for γ. Two-point boundary value problems often have a physical origin, and this origin may suggest a reasonable initial value. Otherwise we might choose a fairly large positive and a fairly large negative value first, then go from there.

For linear problems, getting the right initial slope is easy and requires no iteration. Let y_1 and y_2 be the respective solutions of

$$\begin{aligned} y_1'' &= p(x)y_1' + q(x)y_1(x) + r(x), \\ y_1(a) &= \alpha, \\ y_1'(a) &= 0, \end{aligned} \tag{11.6}$$

and

$$\begin{aligned} y_2'' &= p(x)y_2' + q(x)y_2(x), \\ y_2(a) &= 0, \\ y_2'(a) &= \gamma. \end{aligned} \tag{11.7}$$

Assuming that both (11.6) and (11.7) are uniquely solvable, we take the linear combination

$$y(x) = y_1(x) + sy_2(x), \tag{11.8}$$

with

$$s = \frac{\beta - y_1(b)}{y_2(b)}. \tag{11.9}$$

A simple check shows that y satisfies (11.3) and the boundary conditions (11.4), so that the solution to the linear boundary value problem can be found from the solutions of two linear initial value problems.

When we compute numerically we do not get the correct values of $y_1(x)$ and $y_2(x)$, so we need to use the approximate ones. Suppose that the approximate solutions to (11.6) and (11.7) are computed by a method that gives the approximate solutions on the same mesh. Let U_i and V_i denote the mesh function approximations to $y_1(x)$ and $y_2(x)$, with U_n and V_n the respective approximations to $y_1(b)$ and $y_2(b)$. The mesh function approximation to y is then computed by

$$Y_i = U_i + \hat{s}V_i, \tag{11.10}$$

where

$$\hat{s} = \frac{\beta - U_n}{V_n}. \tag{11.11}$$

It is not hard to analyze how the errors in the solution of the initial value problems affect the accuracy of the solution of the boundary value problem.

Theorem 11.2 Suppose the approximations U_i and V_i are computed by an initial value method that has order of convergence p. Then

$$\max_{0 \leq i \leq n} |Y_i - y(x_i)| = O(h^p).$$

Proof: We first show that $|s - \hat{s}| = O(h^p)$. This follows easily from (11.9) and (11.10), provided that $y_2(b) \neq 0$ and $V_n \neq 0$. The first of these conditions is guaranteed by the assumption of the unique solvability of (11.7); the second can also be expected to hold since V_n approximates $y_2(b)$. Then

$$y(x_i) - Y_i = y_1(x_i) - U_i + (s - \hat{s})y_2(x_i) + \hat{s}\{y_2(x_i) - V_i\}$$

and the desired result follows directly. ∎

If the two solutions are not computed on the same mesh, some adjustment has to be made. We can, for example, use spline interpolation to produce two continuous approximations. From these, we get two mesh function approximations on the same mesh, then use (11.1). Alternatively, we can always treat (11.5) iteratively with different values of γ until the right boundary condition is satisfied to the accuracy we need.

The iterative approach always has to be taken in the nonlinear case, where the process of guessing the correct slope may have to be repeated many times. We do know from the many-dimensional version of Theorem

10.1 that, under appropriate conditions, the solution to an initial value problem is a continuous function of the initial values, so that $y(b)$ is a continuous function of $y'(a)$. If the boundary value problem has a solution, there is some choice of $y'(a)$ that gives the right boundary condition at b. The problem is then a root-finding problem in the single unknown $y'(a)$. We can use the root-finding methods discussed in Chapter 7, but not all of them are easy to apply. In particular, it is hard to use Newton's method since we do not have an explicit expression for the function whose root we want to find. Derivative-free methods such as the secant method will work fine, but there may be a considerable amount of work involved in iterating.

Example 11.3 To compute an approximate solution to the equation

$$y''(x) = \frac{1}{1 + y^2(x)},$$

with boundary conditions

$$y(0) = 0,$$
$$y(1) = 2,$$

we used the MATLAB function `ode45`, calling for an accuracy of 10^{-6}, and solving the resulting system with various values of $y'(0) = \gamma$. Denoting the resulting approximation of $y(1)$ by $Y_n(\gamma)$, we found that $Y_n(1) = 1.4169$ and $Y_n(2) = 2.3389$. Using linear interpolation with these values (which is an application of the secant method) we found the new trial value

$$\gamma = 1.6325$$

and the corresponding $Y_n(1.6325) = 1.9977$. The relation between the various solutions is shown in Figure 11.2.

Even though the last solution shows the qualitative features of the true solution, it is unlikely to be highly accurate since the right boundary condition is still not satisfied very closely. To get a better solution we have to repeat the root-finding step several more times. ■

The analysis for the shooting method is complicated by the fact that there are two kinds of errors involved: the discretization error for the integration of the differential equation, and the error due to not satisfying the right boundary condition exactly. Particularly for nonlinear problems, these two errors interact in a complicated way and make it difficult to talk about the order of convergence of the method, or to get any useful error bounds. In practice, one mostly solves the problem with several different error criteria and compares results.

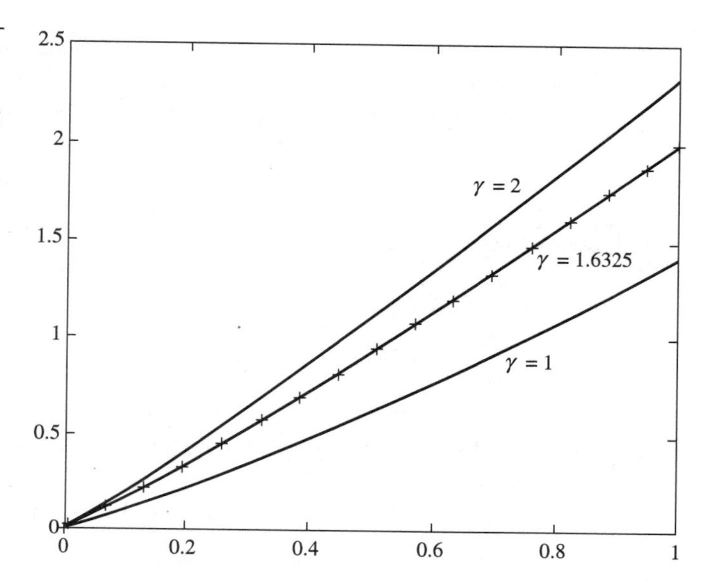

Figure 11.2 Results for Example 11.3 using the shooting method. The curve in the center is the solution obtained with $\gamma = 1.6325$. The points marked with + represent the true solution.

The main advantage of the shooting method is its convenience; with a program for solving initial-value systems, the implementation of the shooting method is a simple matter. This is particularly true for nonlinear equations. Since the nonlinearity is only a modest complication for initial-value problems we avoid large, hard-to-solve nonlinear systems that arise in more direct methods. However, the difficulties inherent in nonlinear boundary value problems cannot be overcome by any numerical technique. Often nonlinear two-point boundary value problems have a smooth solution only for a limited range of initial slopes $y'(a) = \gamma$; guesses outside these limited ranges lead to solutions that have singularities. In such cases many trial values of γ yield no significant information at all, and a lengthy search may have to be conducted before we find a starting slope that allows us to compute usable results.

Example 11.4 When the equation

$$y''(x) = y(x)y'(x) + 1,$$

with

$$y(0) = y(4) = 0$$

is solved with various values of γ, we find that values of $\gamma > -1.2$ lead to solutions that go to infinity somewhere in [0, 4]. Only when we use $\gamma < -1.2$ can we get any useable answers.

■

Many variants have been proposed to extend the usefulness of the shooting methods. Some of these are discussed in detail in Keller's book. With a certain amount of care the shooting method is the easiest and most efficient approach for many two-point boundary value problems.

EXERCISES

1. Prove that the two-point boundary value problem

$$y''(x) = y'(x) + y(x) + 1,$$

with $y(0) = 1$ and $y(1) - y'(1) = 0$ has a unique solution.

2. Show how the shooting method can be adapted to the boundary conditions $y(a) = \alpha$ and $y'(b) = \gamma$.

3. Discuss how one could solve the three-point boundary value problem

$$y'''(x) = f(x,\, y(x),\, y'(x),\, y''(x)),$$

with $y(a) = \alpha$, $y'(a) = \gamma$, and $y(b) = \beta$.

4. Find a way to apply the linear combination method of equations (11.6) to (11.9) for the more general boundary conditions (11.2).

5. What do you think might happen with the shooting method when (11.1) does not have a unique solution?

6. Show how a shooting method can be devised to solve systems of two-point boundary value problems, such as

$$y'' = f_1(x\,, y,\, z,\, y',\, z'),$$
$$z'' = f_2(x,\, y,\, z,\, y',\, z'),$$

with

$$y(a) = \alpha_1, \quad y(b) = \beta_1,$$
$$z(a) = \alpha_2, \quad z(b) = \beta_2.$$

7. Equation (11.9) cannot be used when $y_2(b) = 0$ and will lead to inaccuracy if this value is very small. Is this a difficulty with the shooting method or does the exception arise from the equation?

8. Discuss how the shooting method could be used to solve two-point boundary value problems for equations of higher order, such as

$$y^{(iv)}(x) = a(x)y(x) + b(x)y''(x) + c(x),$$

with

$$y(a) = \alpha_1, \quad y'(a) = \alpha_2,$$
$$y(b) = \beta_1, \quad y'(b) = \beta_2.$$

9. A certain game is modeled by the differential equation

$$2y'''(x) + y(x)y''(x) = 0,$$

with

$$\begin{aligned} y(0) = y'(0) &= 0, \\ y'(10) &= 1. \end{aligned}$$

Discuss how one could solve this two-point boundary value problem.

11.2 Finite Difference Methods*

Direct methods for solving boundary value problems discretize the equation in its original form to produce a finite system. One of the most obvious ways of doing this is to use finite differences instead of derivatives. We first look at the linear case (11.3).

Chapter 6 showed how to approximate derivatives by differences. Here we use centered difference approximations on a uniform mesh of width h, specifically the formulas

$$y''(x) \cong \frac{y(x+h) - 2y(x) + y(x-h)}{h^2}, \tag{11.12}$$

and

$$y'(x) \cong \frac{y(x+h) - y(x-h)}{2h}. \tag{11.13}$$

For sufficiently smooth y, both approximations have $O(h^2)$ accuracy. If we apply this to (11.3), satisfying it at the mesh points, we arrive at the system

$$\begin{aligned} \frac{Y_{i+1} - 2Y_i + Y_{i-1}}{h^2} - p(x_i)\frac{Y_{i+1} - Y_{i-1}}{2h} - q(x_i)Y_i = r(x_i), \\ i = 1, 2, \dots, n-1. \end{aligned} \tag{11.14}$$

These $n - 1$ equations, together with the boundary conditions

$$\begin{aligned} Y_0 &= \alpha, \\ Y_n &= \beta, \end{aligned}$$

are sufficient to determine the unknowns Y_i, $i = 0, 1, \dots, n$. The method seems plausible, so we can hope that Y_i will be a good approximation of $y(x_i)$. But before we tackle the accuracy question, let us see what is involved in solving the system (11.14).

We can write (11.14) in matrix form

$$\mathbf{A}_n \mathbf{y} = \mathbf{r}, \tag{11.15}$$

where

$$\mathbf{A}_n = \frac{2}{h^2}\begin{bmatrix} b_1 & c_1 & & & & \\ a_2 & b_2 & c_2 & & & \\ & \ddots & \ddots & \ddots & & \\ & & \ddots & \ddots & \ddots & \\ & & & a_{n-2} & b_{n-2} & c_{n-2} \\ & & & & a_{n-1} & b_{n-1} \end{bmatrix} \tag{11.16}$$

with

$$a_i = \frac{1}{2} + \frac{h}{4}p(x_i),$$
$$b_i = -1 - \frac{h^2}{2}q(x_i),$$
$$c_i = \frac{1}{2} - \frac{h}{4}p(x_i),$$

and

$$\mathbf{r} = \begin{bmatrix} r(x_1) - \dfrac{2\alpha\, a_1}{h^2} \\ r(x_2) \\ \vdots \\ r(x_{n-2}) \\ r(x_{n-1}) - \dfrac{2\beta\, c_{n-1}}{h^2} \end{bmatrix}.$$

The unknown vector $\mathbf{y}$ then has components $Y_1, Y_2, \ldots, Y_{n-1}$.

The matrix $\mathbf{A}_n$ is tridiagonal, so the efficient algorithm described in Chapter 8 can be used to solve the system in $O(n)$ operations. The solution of the finite difference discretization of the two-point linear boundary value problem can therefore be found easily even for very small mesh sizes.

The construction of finite difference methods is very intuitive and, at least in the linear case, the resulting methods are very efficient. These characteristics make finite difference methods very popular, but there are some limitations. One of these is the extension to nonlinear systems. While discretization is usually not hard, the result is a nonlinear system of equations that, as we know, can be quite difficult to solve. For nonlinear systems coming from the discretization of the boundary value problem, some savings are possible.

Example 11.5 The equation

$$y''(x) = y^2(x)$$

has the discretization

$$\frac{Y_{i+1} - 2Y_i + Y_{i-1}}{h^2} = Y_i^2.$$

If we apply Newton's method to the resulting nonlinear system, the Jacobian is

$$\mathbf{J}(\mathbf{y}) = \begin{bmatrix} 2h^2Y_1 + 2 & -1 & & \\ -1 & 2h^2Y_2 + 2 & -1 & \\ & & \ddots & \\ & & -1 & 2h^2Y_{n-1} + 2 \end{bmatrix}.$$

Since this is a tridiagonal matrix, each iteration of the Newton method can be performed quickly.

■

A serious difficulty with nonlinear discretizations is finding good starting values. Sometimes the physical origin of the problem will suggest a rough solution, or perhaps a similar problem was solved before, but unless there is some such extraneous help available this is a hard question. In such cases, the shooting method may be preferable since the nonlinearity affects it much less (although, as we have seen, the shooting method is not always immune to nonlinearities).

The error analysis also presents a problem, although the terminology is much the same as for initial value problems. For simplicity, we will restrict ourselves to the linear case (11.3).

Definition 11.1

The local discretization error of (11.14) is defined as

$$\mathbf{t} = \begin{bmatrix} \tau_1 \\ \tau_2 \\ \vdots \\ \tau_{n-1} \end{bmatrix}, \tag{11.17}$$

with

$$\tau_i = y''(x_i) - \frac{y(x_i+h) - 2y(x_i) + y(x_i-h)}{h^2}$$
$$- p(x_i)y'(x_i) + p(x_i)\frac{y(x_i+h) - y(x_i-h)}{2h}.$$

As with initial value problems, the local discretization error measures how well the difference formulas approximate the true derivatives. For sufficiently smooth y we know that τ_i will be of order h^2, so that $||\mathbf{t}||_\infty$ goes to zero as $h \to 0$. We expect that for any viable method the local discretization should go to zero as h becomes small.

Definition 11.2

Any finite difference method of the type similar to (11.14) for which

$$\lim_{h\to 0} ||\mathbf{t}||_\infty = 0$$

is called *consistent.* If

$$||\mathbf{t}||_\infty = O(h^p)$$

we say that the method is consistent of order p.

In general, the order of consistency depends on the coefficients in the equation and the smoothness of the solution. One can expect that the higher the order of consistency the better the numerical results, but to prove this we have to look at the connection between the local discretization error and the actual errors in the approximation

$$e_i = y(x_i) - Y_i. \tag{11.18}$$

If we subtract (11.14) from (11.3) we find that

$$\frac{e_{i+1} - 2e_i + e_{i-1}}{h^2} - p(x_i)\frac{e_{i+1} - e_{i-1}}{2h} - q(x_i)e_i = \tau_i. \tag{11.19}$$

We can write this in matrix form as

$$\mathbf{A}_n\mathbf{e} = \mathbf{t},$$

where $\mathbf{e}$ is the vector of the errors $(e_1,\ e_2,\ \ldots,\ e_{n-1})^T$ and $\mathbf{A}_n$ is the matrix in (11.16). If $\mathbf{A}_n$ is invertible, we then get directly that

$$||\mathbf{e}||_\infty \le ||\mathbf{A}_n^{-1}||_\infty\ ||\mathbf{t}||_\infty. \tag{11.20}$$

Therefore, if $||\mathbf{A}_n^{-1}||_\infty$ is of order unity, the approximation error is the same order of magnitude as the local discretization error.

Definition 11.3

The finite difference method of type (11.14) is said to be *stable* if there is a constant c such that

$$||\mathbf{A}_n^{-1}|| \leq c \tag{11.21}$$

for all n.

Theorem 11.3 If a method of type (11.14) is stable and consistent, then the approximate solution converges to the true solution at the mesh points. The order of convergence is at least as large as the order of consistency.

Proof: This follows immediately from (11.20) and (11.21). ■

Most of the time consistency is readily verified, but stability tends to be a challenge whose resolution is not obvious and that requires considerable technical skills, utilizing the special nature of $\mathbf{A}_n$. This leads to arguments that are too difficult for an elementary exposition. For the method under consideration the analysis has been done and it has been shown that the method does indeed satisfy the stability condition (11.21). This implies that the accuracy of the computed solution is at least $O(h^2)$, provided that y is sufficiently smooth. Stability has also been demonstrated for many other equations and methods, but there are cases where stability is apparent but cannot be rigorously defended.

Example 11.6 The problem

$$y''(x) = xy'(x) + y(x) - xe^x,$$

with boundary conditions $y(0) = 1$ and $y(1) = e$, was solved using the finite difference method (11.14) with several mesh sizes. Since in this simple case we know that the exact solution is

$$y(x) = e^x,$$

Table 11.1
Errors and Condition Numbers for Example 11.6.

h	$\|\|\mathbf{e}\|\|_\infty$	k	$\|\|\mathbf{A}_n^{-1}\|\|_2$
$\frac{1}{8}$	5.1×10^{-5}	24.0	0.0969
$\frac{1}{16}$	1.3×10^{-5}	97.8	0.0962
$\frac{1}{32}$	3.3×10^{-6}	392.7	0.0961

we can check the performance of the method. In Table 11.1 we show the observed errors. The results illustrate clearly the stability and the expected second order convergence of the method. We also give the condition number

$$k = ||\mathbf{A}_n||_2 \, ||\mathbf{A}_n^{-1}||_2$$

and values of $||\mathbf{A}_n^{-1}||_2$. Stability tells us that $||\mathbf{A}_n^{-1}||_2$ is bounded; this is confirmed by the numerical results. But the condition number increases, essentially by a factor of four with each halving of the mesh size. The reason for this is that the matrix $\mathbf{A}_n$ is proportional to h^{-2}. The practical implication of the increase in the condition number is that rounding and errors in the elements of $\mathbf{A}_n$ may be magnified somewhat. But with the usual values of h this is a minor concern.

■

Inequality (11.20) tells us that, if the method is stable, the error is always less than a bound that decreases with the order of consistency. For a second order method, halving the mesh size decreases the bound by a factor of four. In practice, numerical results often indicate a stronger pattern in which the actual error (not just the bound) shows such a systematic decrease. If this is the case, repeated computations are very effective in estimating the error and possibly in improving the result.

Suppose that in the method (11.14) we do two sets of computations, first with mesh size h, then with mesh size $h/2$. Because of the choice of the two meshes, the mesh points in the second computation will include all the mesh points of the first. Let x denote a mesh point common to both meshes, and let $Y_{1,x}$, $Y_{2,x}$, $e_{1,x}$, and $e_{2,x}$ stand for the approximate values and the errors of the two computations at x, respectively. If the error behaves so systematically that

$$e_{1,x} = ch^2 \tag{11.22}$$

implies

$$e_{2,x} \cong \frac{1}{4}ch^2, \tag{11.23}$$

then

$$y(x) - Y_{1,x} \cong ch^2$$

and

$$y(x) - Y_{2,x} \cong \frac{1}{4}ch^2.$$

From these two estimates we eliminate the unknown $y(x)$ to get

$$\frac{3}{4}ch^2 \cong Y_{2,x} - Y_{1,x}$$

or

$$e_{1,x} \cong \frac{4}{3}(Y_{2,x} - Y_{1,x}). \tag{11.24}$$

The errors in $Y_{1,x}$ and $Y_{2,x}$ can therefore be estimated from the two solutions.

Note that this is closely related to the method for estimating the error in adaptive quadrature. As in this quadrature, we can use the estimate to extrapolate and possibly improve the result. We find that

$$Y_{e,x} = Y_{1,x} + \frac{4}{3}(Y_{2,x} - Y_{1,x}) \tag{11.25}$$

is often considerably more accurate than either $Y_{2,x}$ or $Y_{1,x}$. The process is an example of Richardson's extrapolation, encountered before in connection with numerical integration.

Example 11.7 The computations in Example 11.6 were extrapolated to give the results in Table 11.2. The improvement is considerable, showing more accuracy and a higher convergence rate.

Table 11.2 Effect of extrapolation in Example 11.7.

h	$\|\|\mathbf{e}\|\|_\infty$	$\|\|\mathbf{e}\|\|_\infty$ in extrapolated results
$\frac{1}{8}$	5.1×10^{-5}	
$\frac{1}{16}$	1.3×10^{-5}	2.3×10^{-7}
$\frac{1}{32}$	3.3×10^{-6}	1.5×10^{-8}

We next consider equation (11.1) with the more complicated boundary conditions (11.2). Now it is no longer possible to satisfy the boundary conditions exactly, but the derivatives must also be approximated. If we use a second order scheme such as (11.14), we should make this approximation also of second order. Since we cannot use centered differences we need to use suitable backward or forward schemes. Typically, the first condition at $x = a$ is replaced by

$$a_0 Y_0 + \frac{a_1}{2h}(-3Y_0 + 4Y_1 - Y_2) = \alpha, \tag{11.26}$$

and the condition at $x = b$ is approximated by

$$b_0 Y_n + \frac{b_1}{2h}(-3Y_{n-2} + 4Y_{n-1} - Y_n) = \beta. \tag{11.27}$$

This changes the matrix in (11.16) but does not seriously affect the practical solution of the problem. What does become more difficult is the analysis. The stability of the matrix in (11.16) does not directly guarantee the stability of the new matrix. Also, while we may be able to justify Richardson's extrapolation for the simple boundary conditions, the discretization of the boundary conditions introduces errors that are less smooth and often invalidate the extrapolation. Similar issues arise if we use other difference schemes to approximate the equation, or when we have third- or fourth-order boundary value problems. While finite difference schemes are easily constructed, the analysis is generally hard. However, these theoretical difficulties do not generally stop people from using finite difference methods, which are very popular for the solution of many differential equations.

EXERCISES

1. In Example 11.6, find an explicit expression for the local discretization error.
2. Describe a finite difference method that one might use for solving the three-point boundary value problem

   $$y'''(x) = a(x)y''(x) + b(x)y'(x) + c(x)y(x) + d(x),$$

   with

   $$y(0) = \alpha, \;\; y(0.5) = \beta, \;\; y(1) = \gamma.$$

3. Design a finite difference method for the fourth-order two-point boundary value problem

   $$y^{(iv)}(x) = a(x)y(x) + b(x)y''(x) + c(x)$$

   with boundary conditions

   $$y(a) = \alpha_1, \;\; y'(a) = \alpha_2,$$
   $$y(b) = \beta_1, \;\; y'(b) = \beta_2.$$

4. Look at the numerical results in Example 11.7. What do they suggest about the order of convergence of the method after extrapolation?
5. What does the matrix in (11.16) look like when the boundary conditions are

$$y(a) = \alpha, \quad y'(b) = \gamma?$$

6. What do you expect will happen when the scheme (11.14) is used with a derivative boundary condition approximated by the simple forward difference

$$y'(x) \cong \frac{y(x+h) - y(x)}{h}?$$

7. (a) Discretize the nonlinear equation

$$y''(x) = \frac{y(x)}{1 + 0.1y^2(x)}$$

using centered differences. Is the Jacobian still tridiagonal? If the boundary conditions are $y(0) = 1$ and $y(1) = 0$, how would you get a good initial approximation for y?

(b) Use Theorem 11.1 to show that this problem has a unique solution.

11.3 The Collocation, Least Squares, and Galerkin's Methods*

Another type of approximation is constructed by an expansion of the form

$$y_n(x) = \sum_{j=0}^{n} c_j \phi_j(x), \tag{11.28}$$

selecting the coefficients c_j so that (11.28) satisfies equation (11.1) approximately. The basis functions ϕ_j must be selected so that they can effectively approximate the unknown solution. Since the solution of the two-point boundary value needs to be twice continuously differentiable, we ask the same thing of the approximation. The approximation (11.28) is then substituted into (11.1) and the undetermined coefficients c_j are selected in one of several ways. The linear case is easier than the nonlinear one, so we will restrict our discussion to equation (11.3).

When we substitute (11.28) into (11.3) we normally cannot get equality, no matter how we select the expansion coefficients. The best we can do is to satisfy the equation approximately in some sense, for example, by asking that the equation be satisfied at a finite number of points. This idea is called *collocation*. There are $n+1$ undetermined coefficients in (11.28); two can be chosen so that the approximation satisfies the boundary conditions.

For the rest, we pick $n-1$ collocation points $x_1,\ x_2,\ \ldots,\ x_{n-1}$ where the equation is satisfied, so that

$$\sum_{j=0}^{n} c_j \phi_j''(x_i) = p(x_i) \sum_{j=0}^{n} c_j \phi_j'(x_i) + q(x_i) \sum_{j=0}^{n} c_j \phi_j(x_i) + r(x_i), \tag{11.29}$$

for $i = 1,\ 2,\ \ldots,\ n-1$. The two additional equations depend on the boundary conditions. If, for example, we have the simple conditions (11.4), we also need

$$\sum_{j=0}^{n} c_j \phi_j(a) = \alpha, \tag{11.30}$$

and

$$\sum_{j=0}^{n} c_j \phi_j(b) = \beta. \tag{11.31}$$

Equations (11.29), (11.30), and (11.31) constitute an $(n+1) \times (n+1)$ linear system

$$\mathbf{Ac} = \mathbf{y}, \tag{11.32}$$

with

$$\begin{aligned} [\mathbf{A}]_{ij} &= \phi_{j-1}''(x_i) - p(x_i)\phi_{j-1}'(x_i) - q(x_i)\phi_{j-1}(x_i), \quad i = 1, 2, \ldots, n-1, \\ &= \phi_{j-1}(a), \quad i = n, \\ &= \phi_{j-1}(b), \quad i = n+1, \end{aligned}$$

$$\begin{aligned} [\mathbf{y}]_i &= r(x_i), & i &= 1, 2, \ldots, n-1, \\ &= \alpha, & i &= n, \\ &= \beta, & i &= n+1. \end{aligned}$$

The solution of (11.32) gives the vector of coefficients $\mathbf{c} = (c_0,\ c_1,\ \ldots,\ c_n)^T$ which, through (11.28), defines the approximate solution.

Polynomial approximations are very simple, but as we know, the use of the monomials x^j can lead to poor conditioning. Usually we prefer orthogonal polynomials and often Chebyshev polynomials are effective.

Example 11.8 The equation in Example 11.6 over the interval $[-1,\ 1]$ with boundary conditions

$$y(-1) = \frac{1}{e},$$

Table 11.3
Results for Example 11.8.

x	Approximate solution	True solution
−1.0	0.3679	0.3679
−0.8	0.4592	0.4493
−0.6	0.5606	0.5488
−0.4	0.6801	0.6703
−0.2	0.8262	0.8187
0.0	1.0069	1.0000
0.2	1.2305	1.2214
0.4	1.5052	1.4918
0.6	1.8392	1.8221
0.8	2.2409	2.2255
1.0	2.7183	2.7183

and

$$y(1) = e$$

was approximated by

$$y_n(x) = \sum_{j=0}^{n} c_j T_j(x).$$

The computed solution with $n = 3$ and collocation points $(-\frac{1}{3}, \frac{1}{3})$ is compared with the true solution in Table 11.3. The computed solution differs from the true solution $y(x) = e^x$ by less than 0.018. With $n = 4$ and collocation points $(-\frac{1}{2}, 0, \frac{1}{2})$ the error is reduced to less than 0.003.

■

We see from this example that polynomial collocation can be simple and effective, especially when only low accuracy is needed. Similar observations can be made about trigonometric approximations, where the basis functions are sines and cosines. Often, just a few terms will give the result within plotting accuracy and, when all functions involved are very smooth, high accuracy can be obtained with relatively small n. Nevertheless, the approach has limitations. Apart from the problematic convergence properties of polynomials, there is the question of selecting the collocation points. There is empirical evidence that some problems are quite sensitive to the collocation points, but it is not known how to choose them optimally. Finally, with a nonlocal basis the matrix in (11.32) will be full, rather than

Table 11.4 Error in the approximate solution of Example 11.6 by cubic spline collocation.

h	Max. error
0.2	6.7×10^{-4}
0.1	1.7×10^{-4}
0.05	4.1×10^{-5}

sparse as it is in the finite difference method. Although this is often stated as a disadvantage, it is an issue only if n is large. But approximation by polynomials of high degree has other drawbacks and is rarely used, so perhaps this is not quite so serious. In any case, some of these objections can be overcome by taking a local basis, typically by a piecewise polynomial approximation. Here we must be concerned with the differentiability of the basis elements. With a second-order problem, we normally require approximations that are twice continuously differentiable. For this purpose, the cubic B-splines form an appropriate basis.

Example 11.9 If we use B-splines approximations on an interval with the knot set $\{x_0, x_1, \ldots, x_n\}$, then, as remarked in Chapter 5, to avoid nonuniqueness we should use the basis $\{B_{-1}, B_0, \ldots, B_{n+1}\}$. For collocation, we have $n+3$ basis functions. Using the $n+1$ knots as collocation points and two boundary conditions gives us the required $n+3$ conditions.

The equation in Example 11.6 on $[0, 1]$ was solved by this cubic spline collocation on a uniform mesh of width h. The results for several values of h are summarized in Table 11.4. The numbers suggest second-order convergence.

■

Other methods can be developed with expansion (11.28) by considering various conditions on the residual

$$\rho(y_n, x) = y_n''(x) - p(x)y_n'(x) - q(x)y_n(x) - r(x). \tag{11.33}$$

The coefficients in the collocation method are chosen by satisfying the boundary conditions and by requiring that this residual vanish at all the collocation points; that is,

$$\rho(y_n, x_i) = 0, \quad \text{for } i = 1, 2, \ldots, n-1,$$

but this is not the only possibility. We can avoid picking special collocation points by making the residual small throughout the interval, so that we might choose the expansion coefficients to minimize the integral of the square of the residual

$$||\rho||_2^2 = \int_a^b \rho^2(y_n, x)dx.$$

Substituting (11.28) into this and carrying out the appropriate minimization, we find that now the expansion coefficients are given by the system

$$\mathbf{Ac} = \mathbf{y}, \tag{11.34}$$

with

$$[\mathbf{A}]_{ij} = \int_a^b \chi_{i-1}(x)\chi_{j-1}(x)dx, \tag{11.35}$$

where

$$\chi_i(x) = \phi_i''(x) - p(x)\phi_i'(x) - q(x)\phi_i(x), \tag{11.36}$$

and

$$[\mathbf{y}]_i = \int_a^b r(x)\chi_{i-1}(x)dx. \tag{11.37}$$

This is the least squares method for the two-point boundary value problem.

An immediate problem is that the linear system (11.34) does not take care of the boundary conditions. One way to account for this necessary information is to transform the original differential equation so that it has homogeneous boundary conditions; that is, so that $\alpha = \beta = 0$ in (11.2). The next step is to pick the basis so that each element ϕ_j satisfies the same homogeneous boundary conditions. In this case, the linear combination (11.28) will also satisfy the boundary values and (11.34) will give the least squares solution with the right conditions. Often, these requirements are not too hard to meet.

Example 11.10 To compute an approximate solution to

$$y''(x) = xy'(x) + y(x) + 1,$$

with conditions $y(0) = 0$ and $y(1) = 1$, we first make a transformation by introducing

$$z(x) = y(x) - x.$$

This new function now satisfies the equation

$$z''(x) = xz'(x) + z(x) + 2x + 1,$$

with the homogeneous boundary conditions $z(0) = z(1) = 0$.

To use this, we need to find expansion functions that satisfy these boundary conditions. For polynomial approximations, we might try

$$\phi_j(x) = x(1-x)T_j(x).$$

For B-splines, most of the interior splines satisfy the zero boundary conditions, but the end ones do not. For these we can use a linear combination of several B-splines. One suitable set is

$$\begin{aligned}
\phi_0(x) &= \frac{1}{4}B_0(x) - B_{-1}(x),\\
\phi_1(x) &= \frac{1}{4}B_0(x) - B_1(x),\\
\phi_i(x) &= B_i(x), \quad i = 2, 3, \ldots, n-2, \\
\phi_{n-1}(x) &= \frac{1}{4}B_n(x) - B_{n-1}(x),\\
\phi_n(x) &= \frac{1}{4}B_n(x) - B_{n+1}(x).
\end{aligned} \tag{11.38}$$

With the transformation and the correct choice of basis functions we can then use (11.35) and (11.37).

■

In addition to all the manipulations required to set up the least squares method, there is a major complication in the integrals (11.35) and (11.37). If these have to be done numerically a lot of extra effort is needed, making the least squares method very expensive. Often a better alternative is to use collocation with $m > n - 1$ collocation points. Then (11.32) is an overdetermined system and cannot be expected to have a solution. But we can solve it in the least squares sense, giving us a method that falls somewhere between collocation and least squares, a *least squares collocation* method. The least squares collocation approach tends to be easier to implement and more efficient than the full least squares method. Note, however, that this does not eliminate the need for a special treatment of the boundary conditions. As in the full least squares method, the boundary conditions can be handled by transforming the problem and using appropriate expansion functions. But it is also possible to impose the boundary values as collocation conditions. Now the boundary conditions are satisfied only approximately. In this case, the boundary conditions have to be weighted more and more heavily as the number of collocation points increases. If this is not done, eventually the boundary conditions will be poorly satisfied, leading to an inaccurate result.

Finally, there is also the *Galerkin method* in which the expansion coefficients are chosen so that the residual is orthogonal to the expansion functions; that is,

$$\int_a^b \rho(y_n, x)\phi_j(x)dx = 0$$

for all j. This leads to a system of the form (11.34) with

$$[\mathbf{A}]_{ij} = \int_a^b \chi_{i-1}(x)\phi_{j-1}(x)dx, \tag{11.39}$$

and

$$[\mathbf{y}]_i = \int_a^b r(x)\phi_{i-1}(x)dx. \tag{11.40}$$

Again, for this to work, the problem has to be recast with homogeneous boundary conditions that are satisfied by all the expansion functions.

The analysis of collocation, least squares, and Galerkin's method is not an easy matter, but a good bit of work has been done on it. In fact, it is a very attractive topic for theoreticians as the analysis is challenging yet doable. From a practical viewpoint, though, the least squares and the Galerkin methods are competitive with finite differences only under special circumstances. Normally, the finite difference discretization is easy to set up and to solve, so there is little reason not to use it. But when the coefficient functions p, q, or r in (11.3) have abrupt changes or are unbounded, the finite difference method may fail and the methods involving the integrals of these functions may do better.

Quite a few practical problems still remain, one of which is the effective choice of the mesh. As we did for numerical quadrature and for solving initial value problems, we would like to find a way to make methods for boundary value problems more automatic, having the algorithm choose the discretization most suitable for the problem at hand. For this, we now have a well-established pattern:

- Find a way to estimate the local discretization error.
- Determine how the local and global errors are connected.
- Select the discretization locally so that it reduces the global error effectively.

For adaptive quadrature, the implementation of this scheme is relatively straightforward. For the initial value problem, this proves to be more difficult because the connection between the local and the global error is hard to determine. For boundary value problems it is even harder, and each of the three steps presents us with obstacles. In fact, the situation is so complicated that it still is an active research area.

EXERCISES

1. Explain why the results in Table 11.4 suggest second-order convergence.
2. Discuss why the expansion functions in (11.38) satisfy the stated homogeneous boundary conditions.
3. What is a set of expansion functions, based on B-splines that satisfy the boundary conditions $y(a) = 0$ and $y'(b) = 0$?
4. Discuss how one could apply collocation to nonlinear equations. Are the resulting systems more tractable than those in the finite difference case?
5. What is an appropriate transformation for Example 11.10 if the boundary conditions are $y(0) = 1$ and $y'(1) = 1$?
6. Use collocation with trigonometric functions $\sin(x)$, $\cos(x)$, $\sin(2x)$, and $\cos(2x)$ to get an approximate solution to the equation in Example 11.6.
7. In least squares collocation with inexact boundary approximation, why is it important to weigh the boundary conditions more heavily as the number of interior collocation points increases?
8. Examine the sparsity pattern in the collocation method with cubic splines.

11.4 Eigenvalue Problems for Differential Equations*

Methods for approximating two-point boundary values can also be used for the solution of eigenvalue problems. The major difficulties lie in the analysis, with both the theory and the numerical aspects quite complicated. We will consider only the classical Sturm–Liouville eigenvalue problem that has been studied thoroughly and is well understood.

The Sturm–Liouville problem for second order differential equations is

$$(p(x)y'(x))' - q(x)y(x) = \lambda r(x)y(x), \qquad (11.41)$$

with

$$y(a) = y(b) = 0,$$

where y and λ are both to be determined. Such a homogeneous equation always has the trivial solution $y(x) = 0$, but for certain parameters λ there are also nonzero solutions. Any nontrivial y for which this equation has a solution is an eigenvector, with λ the corresponding eigenvalue of the equation. It is known that if the given functions $p(x)$, $q(x)$, and $r(x)$ are continuously differentiable and positive on $[a, b]$, then (11.41) has a discrete set of real eigenvalues. In many ways, the Sturm–Liouville problem acts like a symmetric matrix, which greatly facilitates its numerical solution.

Adapting the shooting method to the eigenvalue problem is straightforward. We first convert the equation into a system of two simultaneous first-order equations, with $y(a) = 0$ and $y'(a) = 1$. We then pick a value for λ and integrate the system. If the approximate solution at b is not zero, we modify our guess for λ and repeat the process until the right boundary condition is closely satisfied. In this procedure the unknown is λ and not $y'(a)$. The latter can be kept at its original value since its only effect is to scale the eigenvector. Note also that even though equation (11.41) is linear in the unknowns, the right boundary value is not a linear function of λ. The correct value can therefore not be gotten by a closed form formula of trial values, but must be obtained by iteration.

Example 11.11 The eigenvalue problem

$$((1+x)y'(x))' + \frac{\lambda}{1+x}y(x) = 0$$

on $0 \leq x \leq 1$ was converted into a system

$$y'(x) = \frac{z(x)}{1+x},$$

$$z'(x) = -\frac{\lambda y(x)}{1+x},$$

with $y(0) = 0$ and $z(0) = 1$. The system was then integrated using the MATLAB function `ode45` with various values of λ. After several trials, we obtained the approximate values $y(1) \cong 0.0093$ for $\lambda = 20$ and $y(1) \cong -0.0076$ for $\lambda = 21$, indicating that the correct eigenvalue is between 20 and 21. After a few more iterations, we found that $\lambda = 20.54$ gave zero for $y(1)$ to four significant digits.

■

Working with examples like this shows the advantages and disadvantages of the shooting method for eigenvalues. Simplicity of implementation is a major advantage. Using the existing programs for initial value problems, eigenvalues can be found quickly with very little programming. On the other hand, if more than one eigenvalue is needed, a good bit of searching is involved and it is important to have a good starting guess because there are many roots. Some of these difficulties can be avoided by using a finite difference approach.

In the finite difference method, we apply the centered difference approximation for the first derivative (11.13) twice, so that, on the usual uniform mesh of width h,

$$
\begin{aligned}
(p(x)y'(x))' &\cong \frac{p(x_i + \frac{h}{2})y'(x_i + \frac{h}{2}) - p(x_i - \frac{h}{2})y'(x_i - \frac{h}{2})}{h} \\
&\cong \frac{p(x_i + \frac{h}{2})\frac{y(x_i + h) - y(x_i)}{h} - p(x_i - \frac{h}{2})\frac{y(x_i) - y(x_i - h)}{h}}{h} \qquad (11.42) \\
&= \frac{p(x_i + \frac{h}{2})y(x_i + h) - \left\{p(x_i + \frac{h}{2}) + p(x_i - \frac{h}{2})\right\} y(x_i) + p(x_i - \frac{h}{2})y(x_i - h)}{h^2}.
\end{aligned}
$$

When we use this in (11.41) and denote the approximate eigenvalue by Λ and the corresponding approximate eigenvector by $\mathbf{Y}$, we get the matrix eigenvalue problem

$$\mathbf{AY} = \Lambda \mathbf{BY}. \qquad (11.43)$$

Here

$$
\mathbf{A} = \frac{1}{h^2}\begin{bmatrix} b_1 & c_1 & & & & \\ a_2 & b_2 & c_2 & & & \\ & \ddots & \ddots & \ddots & & \\ & & \ddots & \ddots & \ddots & \\ & & & a_{n-2} & b_{n-2} & c_{n-2} \\ & & & & a_{n-1} & b_{n-1} \end{bmatrix}, \qquad (11.44)
$$

with

$$
\begin{aligned}
a_i &= p(x_i - \tfrac{h}{2}), \\
c_i &= p(x_i + \tfrac{h}{2}), \\
b_i &= -(a_i + c_i + h^2 q(x_i)),
\end{aligned}
$$

and $\mathbf{B}$ a diagonal matrix with elements

$$[\mathbf{B}]_{ii} = r(x_i).$$

For a uniform mesh size $p(x_{i+1} - \frac{h}{2}) = p(x_i + \frac{h}{2})$, so that the matrix $\mathbf{A}$ is symmetric and tridiagonal. While (11.43) is not a standard eigenvalue problem, the assumption that $r(x) > 0$ guarantees that $\mathbf{B}$ is invertible, so that the values of Λ are the eigenvalues of the tridiagonal matrix $\mathbf{B}^{-1}\mathbf{A}$.

Example 11.12 Approximations to the three smallest eigenvalues of the equation in Example 11.11 were computed by the finite difference method (11.43). Results for

Table 11.5 Computed eigenvalues for Example 11.12.

λ_i	$h = 0.25$	$h = 0.125$	$h = 0.0625$
1	19.56	20.28	20.48
2	67.19	77.67	81.02
3	133.2	162.2	179.0

several h are shown in Table 11.5. The numbers suggest convergence, but some results appear to be quite inaccurate. In this example, we know the true results

$$\lambda_i = \left(\frac{i\pi}{\log_e 2}\right)^2$$

The errors, shown in Table 11.6, clarify the situation. All three results appear to converge with order two, but the approximations for the higher eigenvalues become inaccurate very quickly.

■

We can understand the observed results by looking at the following theorem, that tells us what to expect with the finite difference method. We discuss only the special case $r(x) = 1$, but similar results can be shown to hold for the more general equation (11.41).

Consider the solution of

$$(p(x)y'(x))' - q(x)y(x) = \lambda y(x), \tag{11.45}$$

by the finite difference method

$$\mathbf{AY} = \Lambda \mathbf{Y}, \tag{11.46}$$

where $\mathbf{A}$ is the matrix in (11.44). In analogy with the finite difference approximation to the two point boundary value problem, we introduce the local discretization error $\mathbf{t}(h,\ y) = (\tau_1,\ \tau_2,\ \dots,\ \tau_{n-1})^T$, where

$$\tau_i = (p(x)y(x)')'_{x=x_i} - \frac{1}{h^2}\left\{a_i y(x_{i-1}) - (a_i + c_i)y(x_i) + c_i y(x_{i+1})\right\}. \tag{11.47}$$

As in the boundary value problem, we expect that the size of the local discretization error will be a measure of the accuracy of the eigenvalue approximation.

Table 11.6 Error in the computed eigenvalues for Example 11.12.

λ_i	$h = 0.25$	$h = 0.125$	$h = 0.0625$
1	0.98	0.26	0.07
2	15.0	4.5	1.2
3	51.6	22.6	5.9

Theorem 11.4 Assume that λ is an eigenvalue of (11.45) with the corresponding eigenvector y. Then (11.46) has at least one eigenvalue Λ such that

$$|\lambda - \Lambda| \leq \frac{||\mathbf{t}(h, y)||_2}{||\bar{\mathbf{y}}||_2}, \tag{11.48}$$

where

$$\bar{\mathbf{y}} = (y(x_1), y(x_2), \ldots, y(x_{n-1}))^T.$$

Proof: Using (11.45) we get

$$[\mathbf{A}\bar{\mathbf{y}} - \lambda\bar{\mathbf{y}}]_i = [\mathbf{A}\bar{\mathbf{y}}]_i - (p(x)y'(x))'_{x=x_i} + q(x_i)y(x_i),$$

so that, from (11.47)

$$[\mathbf{A}\bar{\mathbf{y}} - \lambda\bar{\mathbf{y}}]_i = \tau_i,$$

or

$$\mathbf{A}\bar{\mathbf{y}} - \lambda\bar{\mathbf{y}} = -\mathbf{t}(h, y).$$

If $\lambda \in \sigma(\mathbf{A})$ there is nothing to prove; otherwise $\mathbf{A} - \lambda\mathbf{I}$ is invertible so that

$$\bar{\mathbf{y}} = -(\mathbf{A} - \lambda\mathbf{I})^{-1}\mathbf{t}(h, y),$$

and

$$||\bar{\mathbf{y}}||_2 \leq ||(\mathbf{A} - \lambda\mathbf{I})^{-1}||_2 \, ||\mathbf{t}(h, y)||_2.$$

Since $\mathbf{A}$ is a symmetric matrix, we can show that

$$\begin{aligned} \left|\left|(\mathbf{A} - \lambda\mathbf{I})^{-1}\right|\right|_2 &= \max_{\Lambda\in\sigma(\mathbf{A})} \frac{1}{|\lambda - \Lambda|} \\ &= \frac{1}{\min\limits_{\Lambda\in\sigma(\mathbf{A})} |\lambda - \Lambda|}. \end{aligned} \tag{11.49}$$

From this, the result

$$\min_{\Lambda\in\sigma(\mathbf{A})} |\lambda - \Lambda| \leq \frac{||\mathbf{t}\,(h, y)||_2}{||\bar{\mathbf{y}}||_2}$$

follows. ■

Since $||\bar{\mathbf{y}}||_2/\sqrt{n}$ is essentially independent of h, the accuracy of the eigenvalue approximation can be expected to be roughly proportional to

$||\mathbf{t}(h, y)||_2/\sqrt{n}$. In the method under discussion, with the assumption of sufficient smoothness, this quantity can be shown to be $O(h^2)$, so the observed second-order convergence is to be expected. Also note that $\mathbf{t}(h, y)$ depends on the eigenvector y; if y oscillates rapidly $||\mathbf{t}(h, y)||_2$ can be quite large. This is the case for the higher eigenvalues in Example 11.12, explaining the poor results. The theorem suggests that getting eigenvalues associated with rapidly oscillating eigenvectors is difficult.

The conclusions of Theorem 11.4 are quite general and depend only on our being able to put a bound on the local discretization error. No stability condition is needed, so the theorem can be extended to other methods without much difficulty. Note, however, that the theorem is incomplete. It tells us that, for any fixed eigenvalue of the original problem, there is a converging sequence of eigenvalues of the discretized problem. It does not rule out the possibility of spurious eigenvalues; that is, a converging approximation sequence that does not correspond to any true eigenvalue. Answering this and related issues is a much more difficult undertaking and does indeed bring in a stability concept.

EXERCISES

1. Show that, for sufficiently smooth p and y, the approximation (11.42) has accuracy $O(h^2)$.
2. Derive the relation in equation (11.49).
3. What is meant by saying that $||\bar{\mathbf{y}}||_2/\sqrt{n}$ is essentially independent of h? Justify the claim.
4. In Example 11.12, what values would you expect for the first three eigenvalues with $h = 0.03125$?
5. Describe how one could use collocation to approximate eigenvalues and eigenvectors.

Chapter 12

Initial Value Problems for Partial Differential Equations*

We now come to one of the most important topics in applied mathematics, and a focus of much of numerical analysis: the solution of partial differential equations. Many aspects of partial differential equations, both theoretical and practical, are difficult. The subject of the existence, uniqueness, and properties of solutions is an extensive and advanced topic in applied mathematics. Numerically, the construction of algorithms for partial differential equations involves many complications and much tedium. Finally, the error analysis is often difficult, if not impossible, and requires a great deal of mathematical sophistication. Much of this material is too complicated for any extensive discussion in an elementary text, and anyone who becomes seriously involved in the use and solution of partial differential equations needs to go well beyond what we present here. Still, solving partial differential equations is one of the main concerns of scientists and engineers, so it is important to understand at least the main principles of the approximate solution of partial differential equations. For this purpose we will examine several prototype problems, each of which, while simpler than equations that are normally encountered in practice, still exhibit the major features of the more complicated cases. We first look at initial value problems. These problems, although they may involve some boundary conditions, have a time dimension, with the solution evolving from some known initial state. Even though the presence of several independent variables complicates matters significantly, the numerical methods have a great deal in common with the initial value methods for ordinary differential equations.

The first prototype problem is the partial differential equation

$$\frac{\partial u(x,\,t)}{\partial t} = \frac{\partial^2 u(x,\,t)}{\partial x^2}, \qquad 0 \le x \le 1, \qquad 0 \le t \le T, \tag{12.1}$$

subject to the initial conditions

$$u(x,\,0) = g(x), \quad 0 \le x \le 1, \tag{12.2}$$

and the boundary conditions

$$u(0,\,t) = u(1,\,t) = 0, \quad 0 \le t \le T. \tag{12.3}$$

For continuity reasons we assume that $g(0) = g(1) = 0$. Equations (12.1) to (12.3) describe the temperature distribution in a thin conducting wire of uniform thermal conductivity, extending in one space dimension over the interval $[0,\,1]$, so it is generally called the *heat equation.* At the initial time $t = 0$ the temperature of the wire is prescribed as $g(x)$. Throughout the experiment, the ends of the wire at $x = 0$ and $x = 1$ are kept at temperature $u = 0$. Given these initial and boundary conditions, we want to determine the temperature distribution in the wire at successive time $t \in [0,\,T]$. There are several other physical phenomena, such as diffusion, that lead to this equation, so (12.1) is sometimes referred to as the *diffusion equation.* The heat conduction or diffusion equation is a prototype for the class of *parabolic* partial differential equations.

The second prototype equation we will study is

$$\frac{\partial^2 u(x,\,t)}{\partial t^2} = \frac{\partial^2 u(x,\,t)}{\partial x^2}, \qquad 0 \le x \le 1, \qquad 0 \le t \le T, \tag{12.4}$$

with

$$u(x,\,0) = g(x), \tag{12.5}$$

$$\frac{\partial u(x,\,0)}{\partial t} = h(x), \quad 0 \le x \le 1, \tag{12.6}$$

and

$$u(0,\,t) = u(1,\,t) = 0, \quad 0 \le t \le T. \tag{12.7}$$

This equation models various wave phenomena in one-dimensional media, and we refer to it as the *wave equation.* The wave equation is the simplest example of the class of *hyperbolic* partial differential equations.

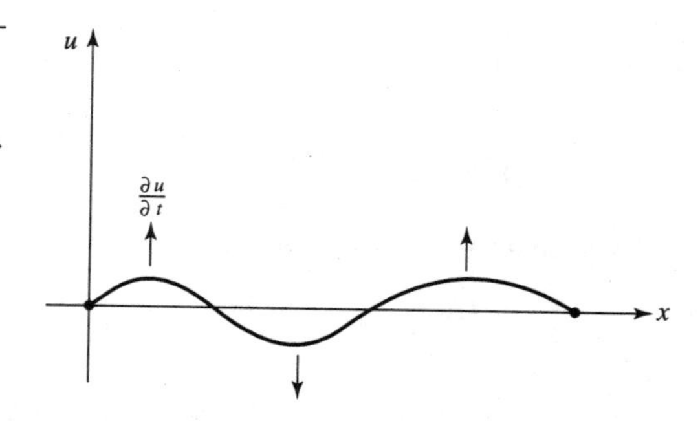

Figure 12.1
A string stretched between two points.

Example 12.1 The equation (12.4) describes the motion of a string, with uniform and unit density, stretched between two points, given an initial displacement $g(x)$ and an initial velocity $h(x)$. (See Figure 12.1.)

12.1 Finite Difference Methods for the Heat Equation

The application of finite difference methods to (12.1) is straightforward. A two-dimensional mesh, or *grid*, is imposed on the domain $\{0 \leq x \leq 1, 0 \leq t \leq T\}$, as shown in Figure 12.2.

The grid spacing will be assumed uniform in each dimension, with the interval sizes in the spatial and time dimensions denoted by Δx and Δt, respectively. We define n_1 such that $n_1 \Delta x = 1$ and n_2 such that $n_2 \Delta t = T$. The coordinates of the grid points are $(x_i = i\Delta x,\ t_j = j\Delta t)$. The approximate values of the temperature at the grid points will be denoted by $U_{i,j}$, which is to be an approximation to $u(x_i,\ t_j)$. To discretize (12.1) we replace the derivatives by differences and the solution u by its approximation U. For the second derivative in the space-dimension we take the centered

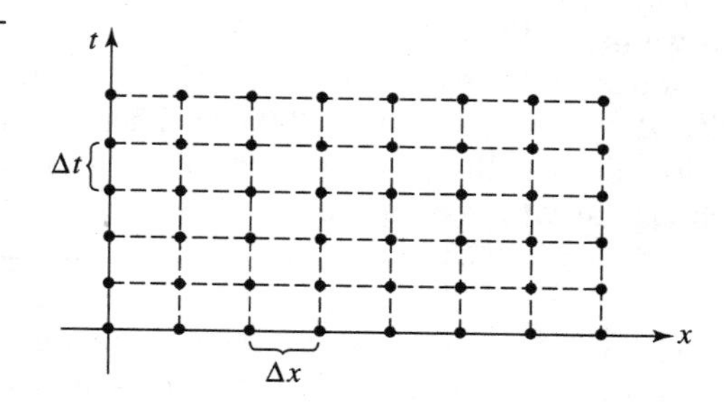

Figure 12.2
A two-dimensional space-time grid.

difference approximation

$$\frac{\partial^2 u(x_i,\, t_j)}{\partial x^2} \cong \frac{U_{i-1,j} - 2U_{i,j} + U_{i+1,j}}{(\Delta x)^2}. \tag{12.8}$$

For the time derivative we can use a forward difference

$$\frac{\partial u(x_i,\, t_j)}{\partial t} \cong \frac{U_{i,j+1} - U_{i,j}}{\Delta t}. \tag{12.9}$$

When these approximations are inserted into (12.1), we find that successive values of $U_{i,\,j}$ can be computed by

$$U_{i,j+1} = U_{i,j} + \frac{\Delta t}{(\Delta x)^2}(U_{i-1,j} - 2U_{i,j} + U_{i+1,j}), \tag{12.10}$$

with

$$U_{i,0} = g(x_i), \quad i = 0,\, 1,\, \ldots,\, n_1 \tag{12.11}$$

and

$$U_{0,j} = U_{n_1,j} = 0, \quad j = 0,\, 1,\, \ldots,\, n_2. \tag{12.12}$$

This is an example of an *explicit method*, because the solution at each point $(x_i,\; t_{j+1})$ can be calculated directly from (12.10), using only values at the previous time level j. It is also an example of a two-level scheme, as only two time levels, t_j and t_{j+1}, are involved at any stage. Since the values of $U_{i,\,0}$ are given by the initial conditions

$$U_{i,0} = g(x_i),$$

the solution can be carried forward from $t = 0$ in a trivial manner. The method is simple to implement and inexpensive to execute, but it suffers from stability problems.

Example 12.2 Equation (12.1) with $g(x) = x(1-x)e^{-x}$ and $T = 0.1$ was solved by (12.10). In Figure 12.3 we have the solution with $\Delta x = 0.05$ and $\Delta t = 0.0012$. The approximation looks quite reasonable. For Figure 12.4, the same problem was solved with $\Delta x = 0.05$ and $\Delta t = 0.0014$. These results are completely meaningless and show oscillations that indicate the presence of instability. Just a small change in Δt makes a great difference in the method's stability. ■

Clearly, there are some difficulties with the explicit method (12.10), so we need to look for alternatives. Instead of using the spatial difference

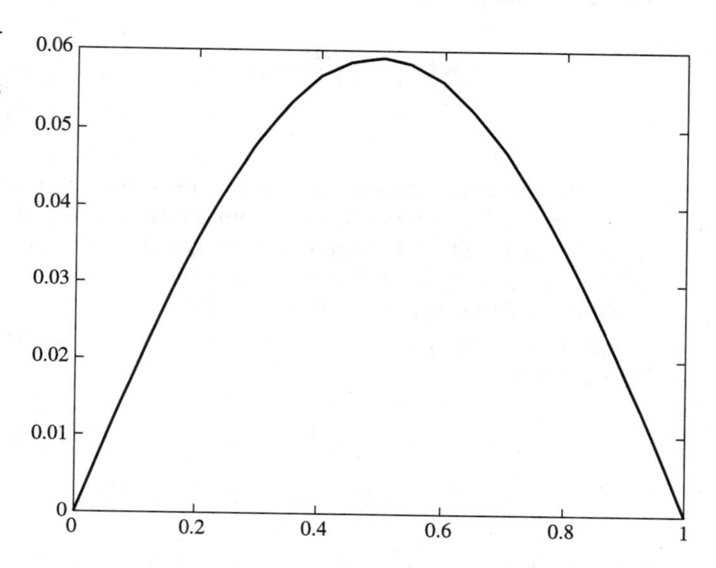

Figure 12.3
Solution of the heat equation with $\Delta t = 0.0012$.

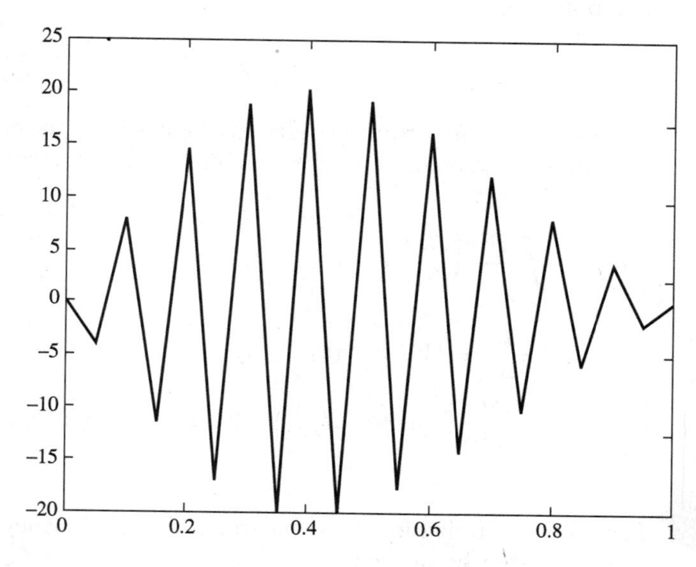

Figure 12.4
Solution of the heat equation with $\Delta t = 0.0014$.

approximation at time level j, we can use an average at levels j and $j+1$. This gives the popular *Crank–Nicholson* method

$$U_{i,j+1} = U_{i,j} + \frac{\Delta t}{2(\Delta x)^2}\left(U_{i-1,j} - 2U_{i,j} + U_{i+1,j} + U_{i-1,j+1} - 2U_{i,j+1} + U_{i+1,j+1}\right). \tag{12.13}$$

Now the computations at each time level are more complicated, as (12.13) defines the values of $U_{i,\,j+1}$ implicitly through a linear system. Fortunately, the matrix involved is tridiagonal so that it can be solved efficiently. Our analysis will show that the implicit method (12.13) is often preferable to the explicit method (12.10).

To generalize, we will carry out the analysis for the more general *two-level* scheme

$$U_{i,j+1} = U_{i,j} + \frac{\Delta t}{(\Delta x)^2}\left\{(1-r)(U_{i-1,j} - 2U_{i,j} + U_{i+1,j}) + r(U_{i-1,j+1} - 2U_{i,j+1} + U_{i+1,j+1})\right\}, \tag{12.14}$$

where $0 \leq r \leq 1$. Method (12.10) is a special case of this with $r = 0$, whereas $r = \frac{1}{2}$ gives the Crank–Nicholson scheme.

We are interested in the error of the approximate solution and, as in all finite difference methods, the most immediate relevant quantities are the errors at the grid points. We will denote the approximation error at the grid points by

$$e_{i,j} = u(x_i,\, t_j) - U_{i,j}.$$

As was the case for ordinary differential equations, the challenge is to relate this approximation error to the local discretization error made in each step

$$\begin{aligned} d_{ij} = & \int_{t_j}^{t_{j+1}} \frac{\partial^2 u(x_i,\, t)}{\partial x^2}\, dt \\ & - \frac{\Delta t}{(\Delta x)^2}\{(1-r)[u(x_{i-1},\, t_j) - 2u(x_i,\, t_j) + u(x_{i+1},\, t_j)]\} \\ & - \frac{\Delta t}{(\Delta x)^2}\{r[u(x_{i-1},\, t_{j+1}) - 2u(x_i,\, t_{j+1}) + u(x_{i+1},\, t_{j+1})]\}. \end{aligned} \tag{12.15}$$

The local discretization error measures the discrepancy that is created at each time-step by approximating the spatial derivative by differences and replacing the integration from one time-step to the next by a sum.

Since each time-step produces a vector of approximations for the x direction, we use vector notation to denote the quantities of interest at each value of t. Thus, $\mathbf{U}_j$, $\mathbf{e}_j$, and $\mathbf{d}_j$ will be defined by

$$[\mathbf{U}_j]_i = U_{i,j},$$
$$[\mathbf{e}_j]_i = e_{i,j},$$
$$[\mathbf{d}_j]_i = d_{i,j},$$

where $i = 1, 2, \ldots, n_1 - 1$. We assume that the 2-norm is used in the following discussion. In (12.14), values at successive time levels are computed by

$$\mathbf{U}_{j+1} = \mathbf{C}\mathbf{U}_j, \qquad (12.16)$$

where the specific form of the matrix $\mathbf{C}$ depends on the value of r. A simple argument shows that the discretization error then satisfies

$$\mathbf{e}_{j+1} = \mathbf{C}\mathbf{e}_j + \mathbf{d}_j, \qquad (12.17)$$

with $\mathbf{e}_0 = 0$. For methods of type (12.14) we can show that, under typical conditions,

$$\max_{0 \le j \le n_2} ||\mathbf{d}_j|| \to 0$$

as $\Delta x \to 0$ and $\Delta t \to 0$. In analogy with our previous results, we call this the consistency condition. We are interested in proving convergence, namely that

$$\max_{0 \le j \le n_2} ||\mathbf{e}_j|| \to 0$$

as $\Delta x \to 0$ and $\Delta t \to 0$.

From (12.17) it follows that

$$||\mathbf{e}_{j+1}|| \le (1 + ||\mathbf{C}|| + ||\mathbf{C}||^2 + \cdots + ||\mathbf{C}||^j) \max_{0 \le i \le j} ||\mathbf{d}_i||, \qquad (12.18)$$

and we can get a bound on the discretization error in terms of the local consistency error if we can bound the norms of the powers of the matrix $\mathbf{C}$.

Theorem 12.1 Suppose that

$$\sup_{0 \le j < \infty} ||\mathbf{C}||^j \le c < \infty \qquad (12.19)$$

and

$$\sup_{0 \le j < \infty} ||\mathbf{d}_j|| = O(|\Delta t|^p) + O(|\Delta x|^q). \qquad (12.20)$$

Then

$$\sup_{0 \le j < \infty} ||\mathbf{e}_j|| = O(|\Delta t|^{p-1}) + O(|\Delta x|^q). \qquad (12.21)$$

Proof: From (12.18) it follows that

$$||\mathbf{e}_{j+1}|| \leq c(j+1) \sup_{0 \leq i < \infty} ||\mathbf{d}_i||.$$

Since $j|\Delta t| \leq T$, (12.21) follows immediately. ■

Part of the local discretization error is created by replacing the second derivative with respect to x by a centered difference. As we know, this error is of order $|\Delta x|^2$ if $u(x,\ t)$ is sufficiently smooth. The second part of the error comes from approximation of the integration from t_j to t_{j+1} by a numerical quadrature. Method (12.10), which uses $r = 0$, is derived by using a rectangular quadrature to approximate the integral; therefore the local discretization error is

$$||\mathbf{d}_j|| = O(|\Delta t|^2) + O(|\Delta x|^2),$$

and

$$||\mathbf{e}_j|| = O(|\Delta t|) + O(|\Delta x|^2).$$

The method (12.13), on the other hand, is based on an application of the trapezoidal method, so that

$$||\mathbf{d}_j|| = O(|\Delta t|^3) + O(|\Delta x|^2),$$

and

$$||\mathbf{e}_j|| = O(|\Delta t|^2) + O(|\Delta x|^2).$$

For the same time-step we therefore can expect considerably more accuracy from the Crank–Nicholson method than from (12.10). However, all of this assumes that (12.19) holds. It takes a good bit of work to show when this is so.

Definition 12.1

Assume that $\mathbf{C}$ is a symmetric matrix that depends on Δt. Then $\mathbf{C}$ is said to satisfy the *von Neumann stability condition* if its spectral radius[1] satisfies the condition

[1] The spectral radius of a matrix $\mathbf{C}$ is $\rho(\mathbf{C}) = \sup_{\lambda \in \sigma(\mathbf{C})} |\lambda|$.

$$\rho(\mathbf{C}) \le 1 + c\Delta t, \tag{12.22}$$

where c is some constant.

Theorem 12.2 If $\mathbf{C}$ satisfies the von Neumann condition, then (12.19) holds.

Proof: If the spectral radius of $\mathbf{C}$ satisfies (12.22), then the spectral radius of $\mathbf{C}^j$ is bounded by

$$\rho(\mathbf{C}^j) \le (1 + c\Delta t)^j.$$

But $j\Delta t \le T$ so that

$$\sup(\rho(\mathbf{C}^j)) \le e^{cT},$$

completing the argument. ■

The von Neumann stability condition not only guarantees the convergence of a consistent method, but it also makes the method stable under small perturbations. If we can show that a method satisfies the von Neumann stability condition, we consider it viable numerically. The stability analysis then becomes a matter of bounding the spectral radius of a matrix. In the present case this can be done without too much difficulty. There are a good many details involved and we will only sketch the argument. First notice that (12.14) can be written as

$$\mathbf{A}\mathbf{U}_{j+1} = \mathbf{B}\mathbf{U}_j, \tag{12.23}$$

where

$$\mathbf{A} = r\kappa\mathbf{D} + \mathbf{I}, \tag{12.24}$$

$$\mathbf{B} = -\kappa(1-r)\mathbf{D} + \mathbf{I}, \tag{12.25}$$

with $\kappa = \dfrac{\Delta t}{(\Delta s)^2}$. Here $\mathbf{D}$ is the tridiagonal matrix

$$\mathbf{D} = \begin{pmatrix} 2 & -1 & 0 & \cdots & 0 \\ -1 & 2 & -1 & \cdots & 0 \\ \vdots & & & & \\ 0 & \cdots & -1 & 2 & -1 \\ 0 & \cdots & 0 & -1 & 2 \end{pmatrix}. \tag{12.26}$$

It can easily be verified that the eigenvectors of $\mathbf{D}$ are

$$\mathbf{v}_i = \begin{pmatrix} \sin\ i\pi/n_1 \\ \sin\ 2i\pi/n_1 \\ \vdots \\ \sin\ i(n_1-1)\pi/n_1 \end{pmatrix},$$

with corresponding eigenvalues

$$\lambda_i = 4\ \sin^2\left(\frac{i\pi}{2n_1}\right), \qquad i = 1,\ 2,\ \ldots,\ n_1 - 1.$$

Since $\mathbf{C} = \mathbf{A}^{-1}\mathbf{B}$ it follows by an easy calculation that the eigenvectors of $\mathbf{C}$ are the $\mathbf{v}_i$ and the corresponding eigenvalues are

$$\mu_i = \frac{1 - 4\kappa(1-r)\sin^2(i\pi/2n_1)}{1 + 4\kappa r \sin^2(i\pi/2n_1)}. \tag{12.27}$$

From this we see that $\max|\mu_i| \leq 1$ for all values of κ as long as $r \geq \frac{1}{2}$. The Crank–Nicholson method (12.13) is therefore always stable. On the other hand, when $r < \frac{1}{2}$ we need

$$\kappa \leq \frac{1}{2 - 4r},$$

which in the case of (12.10) means that we must have

$$\frac{\Delta t}{(\Delta x)^2} \leq \frac{1}{2} \tag{12.28}$$

to ensure stability. The numerical results in Example 12.2 show this clearly. For

$$\kappa = \frac{0.0012}{(0.05)^2} = 0.48$$

the method works, but for

$$\kappa = \frac{0.0014}{(0.05)^2} = 0.56$$

the instability is quite pronounced. For the explicit method, stability places a restriction on the relation between the time and space discretizations that may force us to use an unnecessarily small time-step. Such a restriction is not present in the implicit method.

The general scheme (12.14) is not the only way we can develop formulas for parabolic equations. More than two time levels can be involved, along with a variety of difference approximations.

Example 12.3 If we use three time levels t_{j-1}, t_j, and t_{j+1} we can approximate the time derivative by

$$\frac{\partial u(x_i, t_j)}{\partial t} \cong \frac{U_{i,j+1} - U_{i,j-1}}{2\Delta t},$$

and replace the spatial derivative at t_j by

$$\frac{\partial^2 (x_i, t_j)}{\partial x^2} \cong \frac{U_{i+1,j} - 2U_{i,j} + U_{i-1,j}}{(\Delta x)^2}.$$

This leads to the explicit scheme

$$U_{i,j+1} = U_{i,j-1} + \frac{2\Delta t}{(\Delta x)^2} \{U_{i+1,j} - 2U_{i,j} + U_{i-1,j}\}.$$

Although this looks like a plausible method, it is unstable for all values of $\Delta t/(\Delta x)^2$. However, if we replace $U_{i,j}$ by the average of $U_{i,j+1}$ and $U_{i,j-1}$, the resulting method

$$\frac{U_{i,j+1} - U_{i,j-1}}{2\Delta t} = \frac{U_{i+1,j} - U_{i,j+1} - U_{i,j-1} + U_{i-1,j}}{(\Delta x)^2},$$

or equivalently,

$$U_{i,j+1} = \frac{(\Delta x)^2 - 2\Delta t}{(\Delta x)^2 + 2\Delta t} U_{i,j-1} + \frac{2\Delta t}{(\Delta x)^2 + 2\Delta t} \{U_{i+1,j} + U_{i-1,j}\}, \quad (12.29)$$

is still explicit, but now stable for all Δx and Δt. It is also known that the scheme is consistent and hence convergent as long as $\Delta t/\Delta x \to 0$. This is the well-known *Dufort–Frankel* method.

The Dufort–Frankel method has the advantage of being explicit, so there are very few computations needed to advance from one time level to the next. Its explicitness also makes it suitable for more complicated, nonlinear parabolic equations. A disadvantage is that it involves three time levels, so the approximation at t_1 has to be obtained by a special starting method.

■

The heat equation serves as a model for a variety of more complicated equations and the finite difference method for the heat equation can be adapted without too much trouble to more complicated parabolic equations. Deriving the difference equations is nearly trivial, but the error analysis is quite another matter.

Example 12.4 Heat conduction in a wire of variable thermal conductivity is modeled by the equation

$$\frac{\partial u(x,t)}{\partial t} = \frac{\partial}{\partial x}\left(p(x)\frac{\partial u(x,t)}{\partial x}\right),$$

where $p(x)$ describes the thermal conductivity along the wire. We leave it as an exercise to develop a suitable finite difference method for this equation. ■

Example 12.5 Consider the *diffusion-convection* equation

$$\frac{\partial(x,t)}{\partial t} = \frac{\partial^2(x,t)}{\partial x^2} - b\frac{\partial u(x,t)}{\partial x}, \qquad (12.30)$$

with subsidiary conditions similar to (12.2). The second part on the right of the equation is the convection term and the sizes of b measure the relative importance of convection. To construct a finite difference equation for this case, all we need to do is add a term that represents the convection. When we use a centered difference, we can take care of this by changing the matrix $\mathbf{D}$ in (12.24) and (12.25) to

$$\mathbf{D} = \begin{pmatrix} 2 & -1+b\Delta x & 0 & \cdots & 0 \\ -1+b\Delta x & 2 & -1+b\Delta x & \cdots & 0 \\ \vdots & & & & \\ 0 & \cdots & -1+b\Delta x & 2 & -1+b\Delta x \\ 0 & \cdots & 0 & -1+b\Delta x & 2 \end{pmatrix}. \qquad (12.31)$$

The solution of the resulting system is straightforward, but it is hard to predict how well this is likely to work. The stability analysis for the heat equation was very much dependent on the specific matrices that arose; when we change the equation, this breaks down. The new analysis is much more difficult and if we complicate our equations even more, it eventually will become impossible. This often forces the practitioners of partial differential equation solutions to reason by analogy and hope that what has been worked out for the simple prototype problems will also hold for the more complicated situations. ■

These examples show the difficulties that we encounter in the analysis of numerical methods for partial differential equations. Even in this very simple case, the stability argument is nontrivial and any changes in the equation require a new analysis. Sometimes, as in Example 12.2, an apparently minor change can make an unstable method stable or vice versa. As equations become more complicated this difficulty escalates, and for many problems that occur in practice it is simply not known how to do the stability analysis. For more details on methods and their error analysis, see Vemuri and Karplus [26] or the exhaustive treatment of Richtmyer and Morton [23].

EXERCISES

1. Use ?? to derive the formula (12.13) and find an expression for the local discretization error d_{ij}.
2. Show how (12.18) follows from (12.17).
3. Verify the claim made about the eigenvalues and eigenvectors of the matrix $\mathbf{D}$ in (12.26).
4. Verify that the expression in (12.27) gives the eigenvalues of the matrix $\mathbf{C}$.
5. What is the stability requirement for method (12.14) if $r = 1/3$?
6. Show how the explicit method (12.10) to (12.12) can be modified for the boundary conditions $u'(0) = 1$ and $u(1) = 0$.
7. Does the general method (12.14) have to be modified significantly for nonzero boundary conditions $u(0,\ t) = \alpha(t)$ and $u(1,\ t) = \beta(t)$?
8. Devise an explicit finite difference method for the more general heat equation

$$\frac{\partial u(x,\ t)}{\partial t} = \frac{\partial}{\partial x}\left(p(x)\frac{\partial u(x,\ t)}{\partial x}\right).$$

9. Adapt the Crank–Nicholson method to the equation in Exercise 8.
10. What do you think would be a reasonable stability criterion for the general heat equation in Exercise 8?
11. Show what modifications have to be made to the Crank–Nicholson method for the solution of the parabolic partial-differential equation

$$\alpha\frac{\partial u(x,\ t)}{\partial t} = \frac{\partial^2 u(x,\ t)}{\partial x^2} + F(x), \qquad 0 \le x \le 1, \qquad 0 \le t \le T,$$

with initial and boundary conditions given by (12.2) and (12.3). Assume that α is a nonzero constant and $F(x)$ is a smooth function of x.

12. The temperature $u(x,\ t)$ in a long, thin rod of constant cross section and homogeneous conducting material is governed by the one-dimensional heat equation. If heat is generated in the material at a constant rate, for example, by resistance to a current or a nuclear reaction, the heat equation becomes

$$K\frac{\partial u(x,\ t)}{\partial t} = \frac{\partial^2 u(x,\ t)}{\partial x^2} + \frac{Kr}{\rho q}, \qquad 0 \le x \le d, \qquad 0 \le t \le T,$$

where r represents the heat generated per unit volume. Here d is the length of the rod, ρ its density, q the specific heat, and K the thermal conductivity. If the ends of the rod are kept at 0°, then $u(0,\ t) = u(d,\ t) = 0$ for $t > 0$.

Suppose that $d = 1.5$, $K = 1.04$, and $\rho = 10.6$, $r(x,\ t,\ u) = 5$ and the initial temperature distribution is given by

$$u(x, 0) = \sin\frac{x}{d},\ 0 \le x \le d.$$

Use the modified method in Exercise 11 to approximate the temperature distribution at $t = 15$ with $\Delta t = r = 0.5$ and $\Delta x = 0.15$.

13. An equation describing a particular one-dimensional, single phase, slightly compressible flow in a petroleum reservoir is given, for $0 \le x \le 1000$ and $0 \le t$, by

$$4 \times 10^{-4} \frac{\partial p}{\partial t}(x,\ t) = \frac{\partial^2 p}{\partial x^2}(x,\ t),$$

with the initial and boundary conditions

$$p(x,\ 0) = 2.5 \times 10^7,$$
$$p(0,\ t) = p(1000, t) = 0,$$

where t is the time (in days), x the distance (in ft), and p the pressure (in lb/in^2). Find the pressure p at $t = 5$, using the modified Crank–Nicholson method in Exercise 11 with $\Delta t = r = 0.5$ and $\Delta x = 100$.

12.2 Finite Difference Methods for the Wave Equation

We next look at the wave equation

$$\frac{\partial^2 u(x,\ t)}{\partial t^2} = \frac{\partial^2 u(x, t)}{\partial x^2},$$

with initial and boundary conditions given by (12.5) to (12.7). In discretizing this equation, we follow the pattern established for the heat equation, taking into account that the equation now has a second derivative in time. This makes it possible, for instance, to use centered differences in both x and t. There are several choices for how this can be done. The space derivative can be approximated at time level j, using values at $(x_{i-1},\ t_j)$, $(x_i,\ t_j)$, and $(x_{i+1},\ t_j)$. Similarly, the time derivative can be approximated using

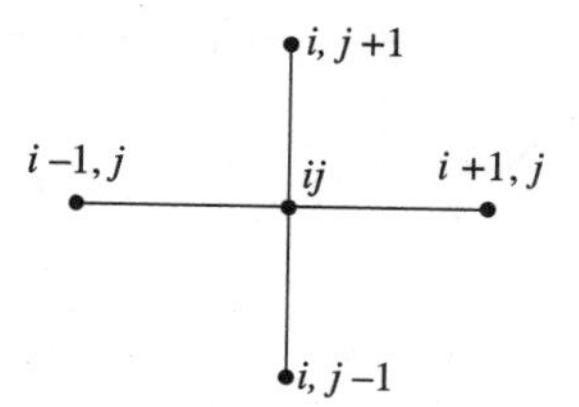

Figure 12.5 Points for a centered difference approximation to the wave equation.

values entirely at space level i. The method is represented schematically in Figure 12.5. The corresponding formula is

$$U_{i,j+1} = m^2(U_{i-1,j} + U_{i+1,j}) + 2(1 - m^2)U_{ij} - U_{i,j-1}, \tag{12.32}$$

where $m = \Delta t/\Delta x$. The method is explicit, so forward integration from one time level to the next can be carried out without the need to solve any systems of equations.

The need for a special starting process is apparent. Because of the second-order time derivative, we need two time levels to start. For time level $t = 0$ we use (12.5). To get approximations at the second time level, we use (12.6), replacing the derivative by a forward difference, so that

$$U_{i,1} = g(x_i) + \Delta t\, h(x_i). \tag{12.33}$$

After this, the forward integration can use (12.32) without any further difficulty.

Stability is an issue that needs to be considered, but this is not an easy task. In Section 12.4 we will provide some motivation for a necessary stability requirement

$$\frac{\Delta t}{\Delta x} \leq 1,$$

but for the moment we illustrate only this claim with an example. Note that this is a much less stringent requirement on the time-step than the condition (12.28). Consequently, explicit methods are more useful for hyperbolic than for parabolic equations.

Example 12.6 The wave equation with $g(x) = x(1-x)e^{-x}$ and $h(x) = 0$ was solved twice; the first time with $\Delta x = 1/20$, $\Delta t = 1/20.1$, and the second time with $\Delta x = 1/20$, $\Delta t = 1/19.9$. The results, after 200 time-steps, are plotted in Figures 12.6 and 12.7. The results strongly suggest that $\Delta t/\Delta x = 1$ is the dividing line for stability.

■

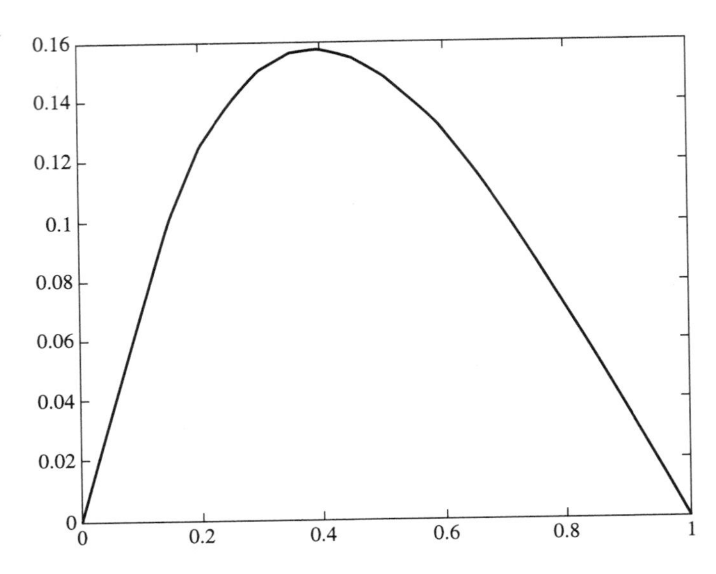

Figure 12.6 Integrating the wave equation with $m = 0.995$.

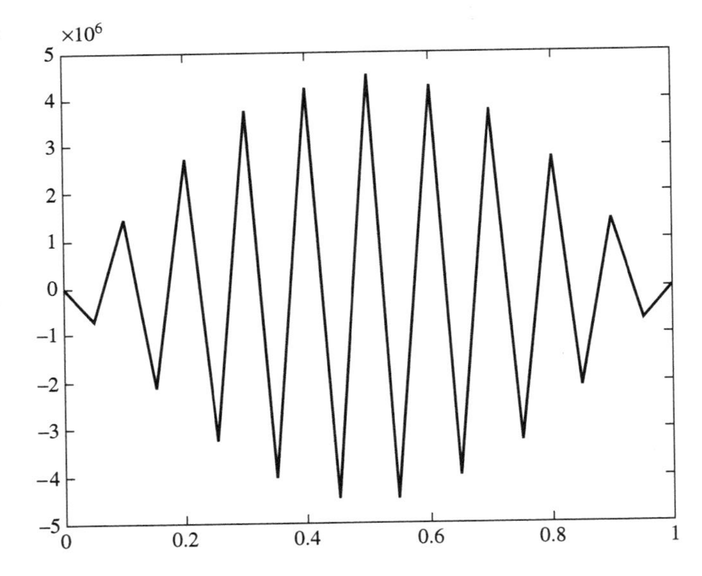

Figure 12.7 Integrating the wave equation with $m = 1.005$.

EXERCISES

1. Show how (12.32) is derived.
2. Find the order of the local discretization error for the explicit scheme (12.32).
3. Construct a scheme for solving the wave equation, using a three-point backward difference formula at t_{j+1}, t_j, and t_{j-1} for the time derivative, and a

centered difference at t_{j+1} for the spatial derivative. Apart from any stability issues, would such a scheme have any advantages over the method (12.32)?

4. Modify (12.32) to approximate the solution of

$$\frac{\partial^2 u(x, t)}{\partial t^2} = \alpha \frac{\partial^2 u(x, t)}{\partial x^2}, \quad 0 \le x \le 1, \quad 0 \le t \le T,$$

with initial and boundary conditions given by (12.5) to (12.7). What is a reasonable conjecture for the stability condition in this case?

5. The air pressure $p(x, t)$ in an organ pipe is governed by the wave equation

$$\frac{\partial^2 p(x, t)}{\partial t^2} = \alpha \frac{\partial^2 p(x, t)}{\partial x^2}, \quad 0 \le x \le l, \quad 0 \le t \le 1,$$

where l is the length of the pipe and α is a physical constant. When the pipe is open, the boundary conditions are given by

$$p(0, t) = p(l, t) = p_0,$$

with the initial conditions

$$p(x, 0) = p_0 \cos 2\pi x, \text{ and } \frac{\partial p(x, 0)}{\partial t} = 0, \quad 0 \le x \le l.$$

Approximate the pressure for an open pipe with $\alpha = 2$, $l = 1$, and $p_0 = 0.9$, at $t = 0.5$ and $t = 1$, using the algorithm developed in Exercise 4 with $\Delta t = 0.05$ and $\Delta t = 0.1$.

12.3 Semi-Discretization and the Method of Lines

There are other ways of treating initial value problems for partial differential equations. We will describe some of them briefly in connection with the heat equation. The extension to the wave equation and other equations of this type is straightforward. We can, for example, approximate the solution $u(x, t)$ in (12.1) by

$$u_n(x, t) = \sum_{i=1}^{n} c_i(t)\phi_i(x), \tag{12.34}$$

where the ϕ_i form a properly chosen basis for approximation in x, with each expansion function satisfying the homogeneous boundary conditions (12.3). The $c_i(t)$ are time-dependent coefficients to be chosen so that $u_n(x, t)$ approximates $u(x, t)$. Generally, there is of course no way we can select these coefficients so that $u_n(x, t)$ satisfies the equation exactly, but if we impose a suitably chosen set of conditions we will get a convergent approximation. As before, the easiest conditions come from collocation: we choose n collocation points $(x_1, x_2, \ldots, x_n)$ and ask that equation (12.1) be satisfied exactly at these points. This gives

$$\frac{\partial u_n(x_j, t)}{\partial t} = \frac{\partial^2 u_n(x_j, t)}{\partial x^2},$$

for $j = 1, 2, \ldots, n$, or

$$\sum_{i=1}^{n} c_i'(t)\phi_i(x_j) = \sum_{i=1}^{n} c_i(t)\phi_i''(x_j), \qquad j = 1, 2, \ldots, n. \tag{12.35}$$

This is a set of n initial-value ordinary differential equations that can be integrated by standard numerical methods to give approximate values for the expansion coefficients $c_i(t)$. The necessary initial conditions for this can be obtained from another interpolation

$$\sum_{i=1}^{n} c_i(0)\phi_i(x_j) = g(x_j). \tag{12.36}$$

Alternatively, we could use least squares or Galerkin conditions instead of collocation, but the process is then more complicated. All of these methods are classified as *semi-discrete* because initially we discretize only in the x direction, retaining t as a continuous variable.

A somewhat simpler way to replace the partial differential equation by a system of ordinary differential equations is the *method of lines.* Here we use a difference approximation in the space direction and treat the approximation at a set of uniformly spaced grid points $x_0, x_1, \ldots, x_n$ as a function of time. We use $U_i(t)$ to denote the approximation to $u(x_i, t)$. Using centered differences, a plausible scheme is

$$\frac{dU_i(t)}{dt} = \frac{U_{i+1}(t) - 2U_i(t) + U_{i-1}(t)}{(\Delta x)^2}, \quad i = 1, 2, \ldots, n-1, \tag{12.37}$$

with

$$U_0(t) = 0,$$
$$U_n(t) = 0,$$
$$U_i(0) = g(x_i).$$

Solving this system of ordinary differential equations gives $U_i(t)$ as an approximation to $u(x_i, t)$.

Semi-discretization has the attractive feature of replacing a partial differential equation by a system of ordinary differential equations. With readily available and highly efficient software for ordinary differential equations, this is a simple way of attacking various kinds of partial differential equations. The analysis of these semi-discretization methods is not easy, but it is known that the systems of differential equations is often stiff. We therefore have to select our software carefully.

When we eventually solve the system of ordinary differential equations, we must discretize in time. Depending on how this is done, we may end up with an algorithm that we know from other sources.

Example 12.7 Because the method given by (10.9) is A-stable, it is an attractive choice for systems that are potentially stiff. But if we use this in (12.37), we are back to the Crank–Nicholson method. If we are to get any benefit from the method of lines we should use a higher order method suitable for stiff systems. ■

It is easy to see that the particular form of the heat equation is not essential to the discussion, so semi-discretization also works for other problems, including the wave equation and various other linear or even nonlinear partial differential equations.

Exercises

1. Show how semi-discretization can be used for the wave equation. How do the initial conditions come into play here?
2. Apply the method of lines to

$$\frac{\partial u(x,\, t)}{\partial t} = \frac{\partial^2 u(x,\, t)}{\partial x^2} - u(x,\, t)\frac{\partial\, u(x,\, t)}{\partial\, x}.$$

3. Apply the method of lines to the wave equation.
4. What do you get if you use Euler's method to integrate the system (12.37)?

12.4 Distinction between Parabolic and Hyperbolic Equations

So far, we have stressed the common aspects of parabolic and hyperbolic equations: they are both initial value problems and they both can be solved by forward integration. There are some differences in the formulas and in the stability requirements, but these seem to be relatively unimportant. In practice, though, the differences are much more fundamental and justify separate classification.

As we can easily see from some numerical calculations, in the heat equation the starting solution gradually gets smaller and peaks are smoothed out.

In the wave equation, on the other hand, the solution changes and oscillates, but its magnitude remains roughly constant over long periods of time. This of course reflects the underlying physical phenomena, the dissipative nature of heat conduction and diffusion versus the periodic nature of vibration. In hyperbolic equations, sharp peaks and discontinuities do not dissipate but travel through the medium. It is this particular aspect that gives hyperbolic equations their special character.

To understand some of these issues, we look at a special case of the wave equation by removing the boundary conditions at $x = 0$ and $x = 1$ and assume that g and h are given on the whole x-axis. The equation then has a nice closed-form solution which can be easily verified:

$$u(x,\, t) = \frac{1}{2}\left\{ g(x-t) + g(x+t) + \int_{x-t}^{x+t} h(\tau)d\tau \right\}. \tag{12.38}$$

We see from this that the conditions at $(x_0,\, 0)$ propagate to later times only within the region defined by $x - t = x_0$ and $x + t = x_0$. This is the *region of influence* of the point $(x_0,\, 0)$. Conversely, the values at any point (x, t) are influenced only by the initial values in the segment $(x - t,\; x + t)$. This is the *domain of dependence* of a point. The lines that bound the region of influence and the domain of dependence are called the *characteristics* of the equation (Figure 12.8).

The domain of dependence expresses the fact that, in a hyperbolic equation, effects do not travel with infinite speed. Instead they are confined to a physical region that grows with time, but at any particular time is bounded. Whenever we discretize, this effect must be preserved. Any numerical method must be such that the domain of dependence of the approximating scheme must contain the domain of dependence of the original equation; otherwise we could make changes in the domain of dependence of the equation that could not possibly be reflected in the approximation. This would make convergence impossible. For the method (12.32) this implies that we must have $\frac{\Delta t}{\Delta x} \leq 1$ as we previously stated. This is the *domain of dependence condition* necessary for the stability of the method, but the argument does not establish its sufficiency. In fact, it is known that it

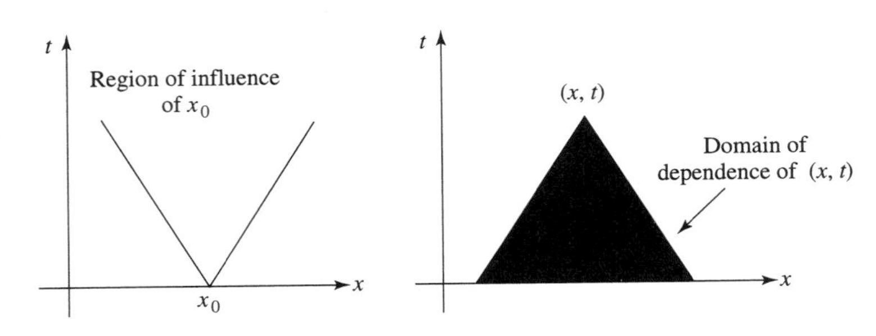

Figure 12.8 Propagation of values in hyperbolic equations.

is not sufficient, so a more thorough stability analysis is necessary. For a discussion of these issues, see Isaacson and Keller [12].

The behavior we see in the simple wave equation is typical for other hyperbolic equations. A hyperbolic equation defines two families of mutually intersecting characteristics and these define the appropriate regions of influence and domains of dependence. Any numerical method must reflect this behavior so knowing about the characteristics helps us analyze the method. Actually, there is even more significance to the characteristics. As we can see from (12.38), any discontinuity in the initial values $g(x)$ will be propagated without change along the characteristics. Since in many hyperbolic problems such discontinuities exist and may in fact be the most interesting aspect of the problem (e.g., shock waves), explicit computations of the characteristics becomes a feature of the numerical algorithms. We can, for example, think of a characteristic grid along which we integrate. Now we must not only find the approximate solution, but also the approximate grid of characteristics. The resulting algorithms are quite a bit more complicated than the finite difference methods, but are often used. For details, see Richtmyer and Morton [23].

Chapter 13

Boundary Value Problems for Partial Differential Equations*

lliptic equations constitute the third category of partial differential equations. As a prototype, we take the *Poisson equation*

$$\nabla^2 u(x,\ y) = \frac{\partial^2 u(x,\ y)}{\partial x^2} + \frac{\partial^2 u(x,\ y)}{\partial y^2} = g(x,\ y), \tag{13.1}$$

which is to be satisfied by the function $u(x,\ y)$ in some region Ω of the xy plane, with the right side $g(x,\ y)$ given on all of Ω. The special case $g(x,\ y) = 0$ is *Laplace's* equation and the differential operator ∇^2 is called the *Laplacian*. Elliptic partial differential equations occur frequently as models for static situations in potential theory, elasticity, and thermal conduction. There is no time variable involved, so the approximation has to be determined for all points at once.

The subsidiary conditions for elliptic equations are entirely boundary conditions. We assume that Ω is a finite two-dimensional region whose boundary is denoted by Γ. Boundary conditions will be specified on Γ. We assume that the boundary is a single closed curve and that it is smooth in the sense that ν, the outward normal to the boundary curve, is a continuously differentiable function along Γ (Figure 13.1).

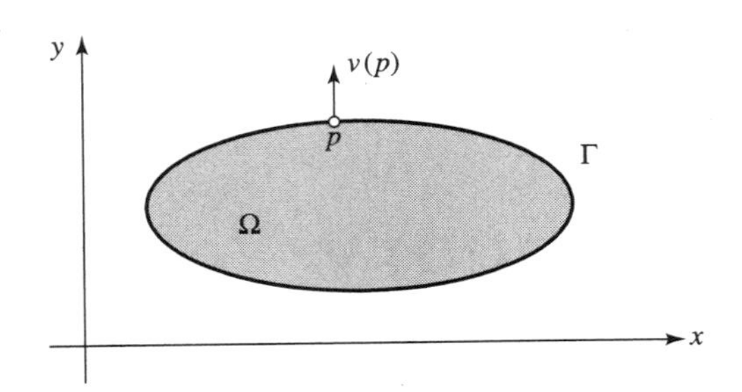

Figure 13.1 Region Ω and its boundary Γ.

We will limit our discussion to a particularly simple type of boundary condition in which we specify that the value of u on the boundary by

$$u(x,\, y) = \alpha(x,\, y)\,, \quad \text{for all } (x,\, y) \in \Gamma. \tag{13.2}$$

In partial differential equations literature these are known as *Dirichlet* boundary conditions.

13.1 Finite Difference Methods

The construction of finite difference methods for the Poisson equation is quite intuitive. We first construct a rectangular grid on Ω. Here we will use a grid with uniform spacing h in both the x and y directions, but this can be easily changed. We then use finite differences to approximate the Laplacian. When a grid point has all of its four neighbors in Ω, we say that it is a *regular* point and use the *five-point star* arrangement shown in Figure 13.2, which leads to the approximation

$$\nabla^2 u(x_i, y_j) \cong \frac{u(x_{i-1}, y_j) + u(x_{i+1}, y_j) + u(x_i, y_{j-1}) + u(x_i, y_{j+1}) - 4u(x_i, y_j)}{h^2}. \tag{13.3}$$

At most interior points we can use this standard approximation for the Laplacian, but some changes may have to be made near the boundaries. When the boundary is curved the grid lines will intersect the boundary at some nongrid points and we get *irregular* interior points. The distance of

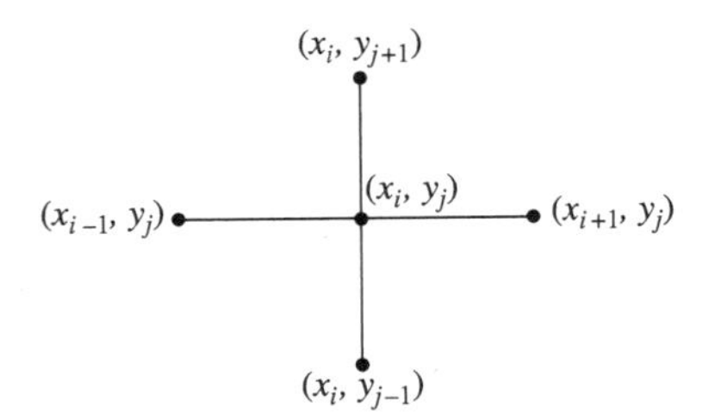

Figure 13.2 Regular five-point star pattern.

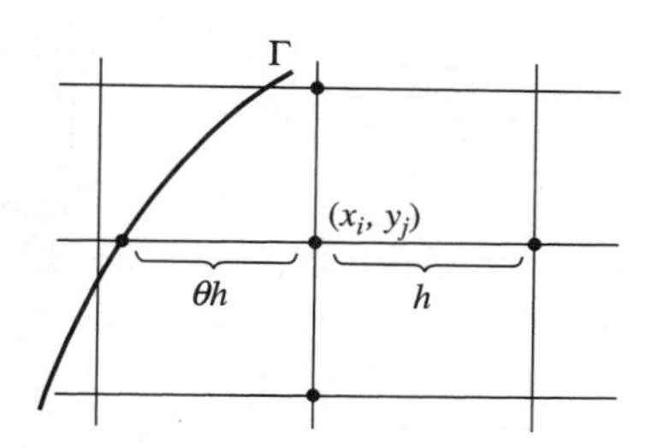

Figure 13.3
Grid spacing at an irregular point.

these irregular interior points from the boundary points is less than the grid spacing, so the formulas for approximating the derivatives have to be adjusted. For the situation shown in Figure 13.3, we find from the discussion of finite difference approximations that

$$\frac{\partial^2 u(x_i, y_j)}{\partial x^2} \cong \frac{2\theta u(x_{i+1}, y_j) + 2\alpha(x_i - \theta h, y_j) - 2(1+\theta)u(x_i, y_j)}{\theta(1+\theta)h^2} \tag{13.4}$$

is an approximation of the second partial with respect to x at the grid point (x_i, y_j). Using this or similar formulas for the derivative in y, we can derive a suitable approximation for the Laplacian at other irregular points.

The finite difference approximations are then used to replace the derivatives in equation (13.1), and the equation is satisfied at all the interior points. If we use $U_{i,j}$ to denote the approximation to $u(x_i, y_j)$, we get equations of the form

$$\frac{U_{i-1,j} + U_{i+1,j} + U_{i,j-1} + U_{i,j+1} - 4U_{i,j}}{h^2} = g(x_i, y_j) \tag{13.5}$$

at all regular points, with suitably modified formulas for the irregular points. In any case, if there are m interior points we will have exactly m equations and m unknowns, so in principle the system can be solved by standard matrix methods.

The double index scheme in the grid labeling is convenient for visualization, but to make the equations (13.5) into a standard matrix problem we have to translate this into a single index scheme. We do this by sequentially labeling interior points as $P_1, P_2, \ldots, P_m$ and the boundary points as $B_1, B_2, \ldots, B_k$. The unknowns in (13.5) are converted into a vector $\mathbf{U}$, such that $[\mathbf{U}]_i$ represents the approximation corresponding to the grid point P_i; that is,

$$[\mathbf{U}]_i = u(P_i) = U_{k,l},$$

where (x_k, y_l) are the coordinates of the point P_i. The set of equations can then be written in matrix form as

$$\mathbf{AU} = \mathbf{y}, \tag{13.6}$$

where the elements of the matrix $\mathbf{A}$ and the right side $\mathbf{y}$ depend on the particular arrangement of the grid and the boundary.

The specific form of the matrix $\mathbf{A}$ depends on the way we go from the double index to the single index notation. Often one uses a systematic numbering scheme in which we proceed along one column of grid points, say bottom to top. When we reach the top of one row, we begin again at the bottom of the adjacent column. This *natural* ordering, illustrated in the next example, is often used.

Example 13.1

Suppose that we want to solve Poisson's equation in the interior of an ellipse Ω with

$$\Gamma = \{(x, y)\colon -1 \le x \le 1,\ x^2 + 1.44y^2 = 1\}.$$

Choosing $h = 0.4$, we get the grid shown in Figure 13.4. The labeling of the points is column-wise, as suggested. There are a total of 16 interior points, so that $\mathbf{A}$ will be a 16×16 matrix. The point P_6 is a regular point, with neighbors P_2, P_5, P_7, and P_{10}. The sixth row of $\mathbf{A}$ will therefore have the elements

$$[\mathbf{A}]_{66} = -\frac{4}{h^2},$$

$$[\mathbf{A}]_{62} = [\mathbf{A}]_{65} = [\mathbf{A}]_{67} = [\mathbf{A}]_{6,10} = \frac{1}{h^2}.$$

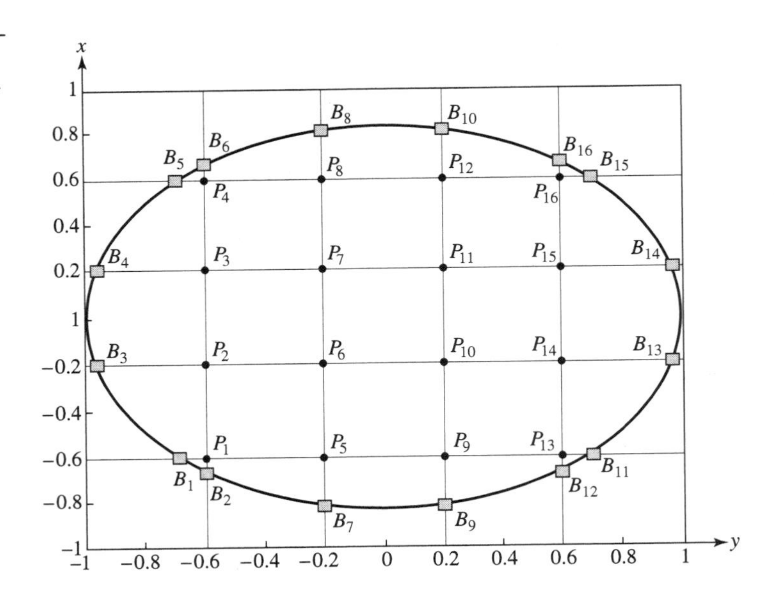

Figure 13.4
Grid and boundary points ordering for Example 13.1.

All other entries in this row will be zero. The right side has the component

$$[\mathbf{y}]_6 = g(P_6).$$

The point P_8 is irregular, so the boundary comes into play. For differentiation in the y direction we need to use the appropriately modified formula (13.4),

$$\frac{\partial^2 u(P_8)}{\partial y^2} \cong \frac{2\theta\alpha(B_8) + 2u(P_7) - 2(1+\theta)u(P_8)}{\theta(1+\theta)h^2}.$$

The y coordinate of B_8 is 0.8165, so that $\theta = 0.5414$. Replacing u by its approximation $\mathbf{U}$ and using the boundary value, this gives

$$\frac{\partial^2 u(P_8)}{\partial y^2} \cong \frac{1.083\alpha(B_8) + 2[\mathbf{U}]_7 - 3.083[\mathbf{U}]_8}{0.817h^2}.$$

Putting this together with the centered difference approximation in the x direction, we find that the nonzero elements in the eighth row of $\mathbf{A}$ are

$$[\mathbf{A}]_{8,8} = -\frac{5.695}{h^2},$$

$$[\mathbf{A}]_{8,4} = [\mathbf{A}]_{8,12} = \frac{1}{h^2},$$

$$[\mathbf{A}]_{8,7} = \frac{1.300}{h^2}.$$

The right side has

$$[\mathbf{y}]_8 = g(P_8) - \frac{2.398\alpha(B_8)}{h^2}.$$

All the other elements of $\mathbf{A}$ and $\mathbf{y}$ can be found in a similar way. ■

This example illustrates an aspect that complicates the task of solving partial differential equations. The curved boundaries affect the construction of the matrix $\mathbf{A}$ and make it a nontrivial problem. For more complicated equations with difficult boundaries, constructing the approximating equations is a major source of concern and writing a computer program that does it is a lengthy undertaking. Creating software for realistic partial differential equations usually involves serious data management problems and rarely is an easy task.

There are two other major issues that we have to consider. We need to produce an error analysis that tells us how the method can be expected to work, and we need to find efficient methods for solving the large systems that we get.

EXERCISES

1. Derive the approximation (13.4).
2. Show how (13.3) needs to be modified when the grid spacing in the x and y directions differs.
3. In Example 13.1, find the elements of the second row of $\mathbf{A}$.
4. In Example 13.1, find the pattern of the nonzero elements of $\mathbf{A}$.
5. Suppose that Ω is the unit square $0 \leq x \leq 1,\ 0 \leq y \leq 1$. Describe what the coefficient matrix $\mathbf{A}$ looks like in this case if the natural ordering illustrated in Example 13.1 is used.
6. How does the discussion in this section have to be changed for the solution of the equation

$$\frac{\partial^2 u(x,\, y)}{\partial\, x^2} + \frac{\partial^2 u(x,\, y)}{\partial\, y^2} + u(x,\, y) \ = g(x,\, y)?$$

13.2 Error Analysis for the Finite Difference Method

To analyze the error in this finite difference method, we proceed as we did in the analysis of the two-point boundary value problem. We use

$$\mathbf{u} = \begin{bmatrix} u(P_1) \\ u(P_2) \\ \vdots \\ u(P_m) \end{bmatrix}$$

to denote the vector of true solutions at the grid points, and consider the error vector $\mathbf{e} = (e_1,\ e_2,\ \ldots,\ e_m)^T$ with

$$e_i = u(P_i) - [\mathbf{U}]_i.$$

We are interested in the behavior of $||\mathbf{e}||$ as h decreases.

The local discretization error $\mathbf{t}(u,\ h) = (\tau_1,\ \tau_2,\ \ldots,\ \tau_m)^T$ is defined as

$$\tau_i = \nabla^2 u(P_i) - [\mathbf{Au}]_i \tag{13.7}$$

and our goal is to establish the connection between the local discretization error and the error in the approximate solution. Noting that

$$\begin{aligned}[\mathbf{AU}]_i &= g(P_i) \\ &= \nabla^2 u(P_i),\end{aligned}$$

we get

$$[\mathbf{A}(\mathbf{U} - \mathbf{u})]_i = \tau_i,$$

or

$$\mathbf{Ae} = -\mathbf{t}(u, h).$$

If $\mathbf{A}$ is invertible, then

$$||\mathbf{e}|| \leq ||\mathbf{A}^{-1}|| \; ||\mathbf{t}(u, h)\,||. \tag{13.8}$$

The local discretization error is usually quite tractable. If assumptions on the smoothness of u can be made, it is only a matter of bounding the error in numerical differentiation. The question of the convergence of the finite difference then becomes a matter of bounding $||\mathbf{A}^{-1}||$ as a function of h. This is the stability question which is, as usual, a difficult issue. For finite difference methods for partial differential equations this often means appealing to special results from matrix theory. For the case under discussion, the arguments can be found in various places (see for example Isaacson and Keller [5], p.447) and it is known that $||\mathbf{A}^{-1}||_\infty$ is bounded as $h \to 0$. Accepting this, we can conclude that the finite difference method described converges and that the error is proportional to the local discretization error. We can also conclude that the condition number of (13.5),

$$\begin{aligned} k &= ||\mathbf{A}||_\infty \; ||\mathbf{A}^{-1}||_\infty \\ &= O(\frac{1}{h^2}), \end{aligned}$$

so there will be some limited accumulation of rounding error.

EXERCISES

1. Suppose that Ω is the unit square and that u is four times continuously differentiable in both variables. Show that then $||\mathbf{t}(u,\ h)||_\infty = O(h^2)$.
2. Show that the discretized Laplacian on the unit square satisfies a *maximum* principle that guarantees that the maximum grid value must be located on the boundary. Use this to demonstrate that the numerical method is stable. (Hint: Assume that the maximum value occurs at an interior point and show that this leads to a contradiction.)
3. How is the bound in Exercise 1 affected by curved boundaries?
4. How is the stability argument in Exercise 2 affected by curved boundaries?
5. Construct the matrix $\mathbf{A}$ for a unit square and empirically examine $||\,\mathbf{A}^{-1}\,||$ as a function of h.

13.3 Solving Finite Difference Systems

Since the number of unknowns in the discretized system $m = O(1/h^2)$, even a moderately small h leads to a large number of equations. For complicated regions, or when high accuracy is required, the system (13.5) may contain in excess of 10,000 equations. This makes it impractical to use a normal GEM algorithm. Fortunately, the matrix $\mathbf{A}$ in (13.6) has special properties that we can take advantage of.

As we know, direct methods for sparse matrices are potentially effective as long as we can control the fill-in. For banded matrices we can do this to some extent. For rectangular regions, or even for simple convex regions such as that in Example 13.1, the structure of the matrix is predictable. For a rectangle and a natural ordering, the matrix will be banded with bandwidth roughly

$$\beta \cong 2\sqrt{m},$$

as shown in Figure 13.5. If we apply the Gauss elimination method restricted to the band, we find that the reduction of each column requires about β^2 multiplications and additions so that, if no row interchanges are necessary, the entire triangularization takes

$$N \cong 4m^2$$

operations. For $m = 10^4$, this is within the range of present-day computers, but still takes a great deal of time.

The band reduction methods are not as efficient as possible because they do not take advantage of the large number of zeros within the band. In recent years there has been a great deal of interest in faster direct methods. Many algorithms have been developed that can solve large problems quite well, but this is a special topic that is not easily treated in an elementary fashion. Details can be found in books devoted to this subject, for example, Duff, Erisman, and Reid [8].

Although direct methods have many advantages, the simplicity of iterative methods often makes them attractive alternatives. Historically, it-

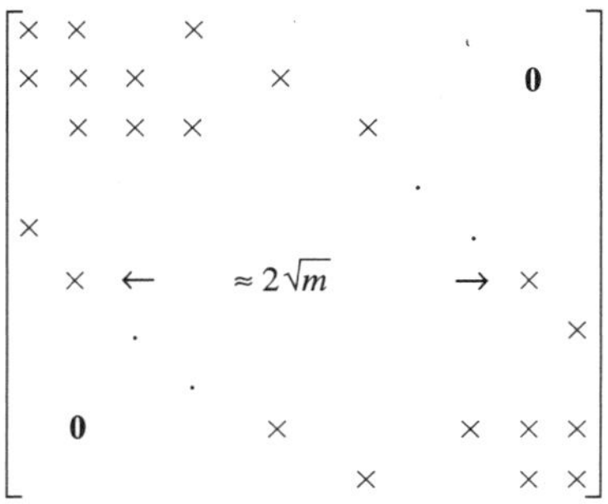

Figure 13.5 Band structure for the finite difference method on a rectangle.

erative methods for finite difference systems were the first to be used, and their performance has been studied for many years. A large number of useful results are known, of which we will present only the simplest. To get an intuitive grasp for how the iterations are performed, it is helpful to switch back to the two-dimensional indexing scheme for the unknowns, in which $U_{i,\,j}$ is an approximation to $u(x_i,\ y_j)$. With this convention the equation at a regular grid point can be written as

$$U_{i,j} = \frac{1}{4}\Big((U_{i-1,j} + U_{i+1,j} + U_{i,j-1} + U_{i,j+1}) + h^2 g(x_i, y_j)\Big), \tag{13.9}$$

with suitable modifications for the irregular grid points. If we let $U_{i,\,j}^{[k]}$ denote the kth iterate for $U_{i,\,j}$, then (13.9) suggests the simple iteration

$$U_{i,j}^{[k+1]} = \frac{1}{4}\left(U_{i-1,j}^{[k]} + U_{i+1,j}^{[k]} + U_{i,j-1}^{[k]} + U_{i,j+1}^{[k]} + h^2 g(x_i, y_j)\right). \tag{13.10}$$

This is known as the *Jacobi iteration.*

If we think of a cycle of the Jacobi method as one iteration on all unknowns, we see that in one cycle only the values of the previous cycle are used. The entire array is updated when the cycle has been completed, hence the result of one cycle does not depend on the order in which the new iterates are computed. We can change this and use the new iterates as soon as they are available, but in this case, the order in which the points are taken does matter. For the ordering scheme in Figure 13.4, the iteration can be defined by

$$U_{i,j}^{[k+1]} = \frac{1}{4}\left(U_{i-1,j}^{[k+1]} + U_{i+1,j}^{[k]} + U_{i,j-1}^{[k+1]} + U_{i,j+1}^{[k]} + h^2 g(x_i, y_j)\right). \tag{13.11}$$

If we compute the unknowns at the points P_i in the order $i = 1,\ 2,\ \dots$, then the quantities on the right of (13.11) are known whenever they are needed, and so the formula is a straightforward successive substitution. The algorithm is called the *Gauss-Seidel* method.

To analyze the convergence of these schemes, we return to vector notation and write

$$\mathbf{A} = \mathbf{D} + \mathbf{L} + \mathbf{R},$$

where $\mathbf{D}$ is a diagonal matrix, $\mathbf{L}$ is lower triangular, and $\mathbf{R}$ is upper triangular, both of the latter with zero diagonals. The Jacobi method can then be written as

$$\mathbf{U}^{[k+1]} = -\mathbf{D}^{-1}(\mathbf{L} + \mathbf{R})\mathbf{U}^{[k]} + \mathbf{D}^{-1}\mathbf{y}, \tag{13.12}$$

while the Gauss-Seidel method is

$$\mathbf{U}^{[k+1]} = -(\mathbf{D} + \mathbf{L})^{-1}\mathbf{R}\mathbf{U}^{[k]} + (\mathbf{D} + \mathbf{L})^{-1}\mathbf{y}. \tag{13.13}$$

We know from the study of iterative methods that a sufficient condition for the convergence of successive substitutions

$$\mathbf{U}^{[k+1]} = \mathbf{B}\mathbf{U}^{[k]} + \mathbf{y}$$

is that

$$||\mathbf{B}|| < 1.$$

For symmetric matrices and the 2-norm, this is equivalent to the bound on the spectral radius,

$$\rho(\mathbf{B}) = r < 1. \tag{13.14}$$

While $\mathbf{B}$ is not always symmetric, it can be shown that this condition is also sufficient for nonsymmetric matrices. Therefore, the speed of convergence is governed by the magnitude of r, with the error in the kth iteration proportional to r^k. The convergence question is then reduced to finding the dominant eigenvalues of the matrices involved in the iteration.

For special cases, such as rectangular Ω where all grid points are regular, one can find the eigenvalues explicitly, but generally this is a difficult task. We want to therefore appeal to general results in spectral theory that allow us to establish (13.14). We cannot use Gershgorin's theorem directly because the necessary conditions are not satisfied. For example, in Jacobi's method, where

$$\mathbf{B} = -\mathbf{D}^{-1}(\mathbf{L} + \mathbf{R}),$$

we find that for all i

$$\mathbf{B}_{ii} = 0,$$

and

$$\sum_{j \neq i} |\, [\mathbf{B}]_{i,j}| \leq 1, \tag{13.15}$$

with equality for some rows. All we can conclude from this is that

$$r \leq 1,$$

which is not sufficient to establish convergence. Fortunately, there are some rows of $\mathbf{B}$ for which strict inequality in (13.15) is attained. This gives us the hope that condition (13.14) could hold. A detailed analysis shows that this is indeed so, and that for Jacobi's method

$$r = 1 - |\eta|, \tag{13.16}$$

where $\eta = O(h^2)$. A similar result holds for the Gauss-Seidel method, with η about twice the magnitude of that for Jacobi's method. From this we conclude that both iterations converge, with better results to be expected from the Gauss-Seidel method.

However, (13.16) also shows that r is quite close to one, and suggests that convergence will be slow and deteriorate with decreasing h. While each iteration can be carried out quickly, usually many iterations have to be done before the desired accuracy is achieved. This has spurred on the study of alternatives that accelerate the convergence. One of a number of methods that will do this is the *over-relaxation* method, which is an instance of extrapolated iteration. If we can estimate the contraction factor r, the extrapolated value can be used as the new iterate, leading to the over-relaxation formula

$$\mathbf{U}^{[k+1]} = \frac{1}{1-r}\left\{-(\mathbf{D}+\mathbf{L})^{-1}\mathbf{R}\mathbf{U}^{[k]} + (\mathbf{D}+\mathbf{L})^{-1}\mathbf{y}\right\} - \frac{r}{1-r}\mathbf{U}^{[k]}. \quad (13.17)$$

If we have a good guess for r, this can improve the speed of convergence considerably.

The topic of iterative methods for these systems is an extensive and complicated subject that is explored in the relevant literature. Varga's book [25] is a classic in this area.

EXERCISES

1. To solve a finite difference system on a three by four grid, which of the two ordering schemes below would be preferable?

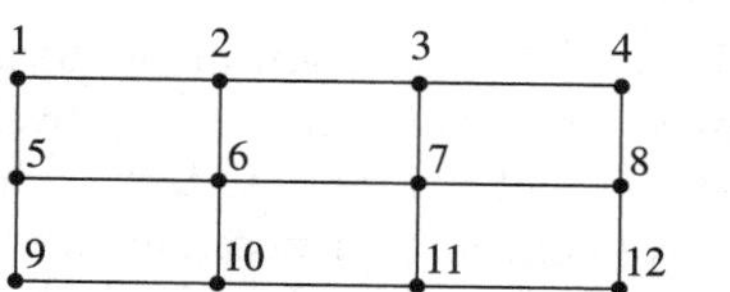

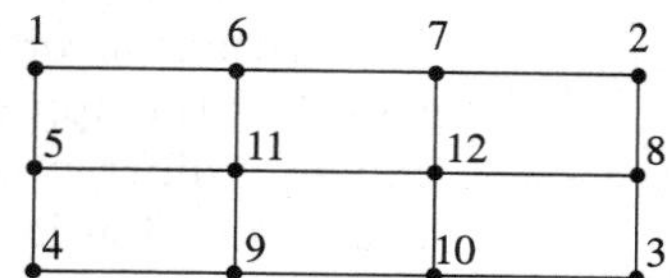

2. Suppose that Ω is a rectangular region twice as long in the x-direction as in the y-direction. If we use a finite difference discretization in which $\Delta x = \Delta y$ and a natural ordering for the grid points, is there any advantage of ordering by columns as opposed to by rows?

3. Verify that (13.13) is the correct matrix form of the Gauss-Seidel method.

4. Approximately how many more Jacobi iterations than Gauss-Seidel iterations are needed to reduce the error by two orders of magnitude?

5. Show that even in the presence of curved boundaries (13.15) holds and that strict inequality is attained for some rows.

13.4 The Finite Element Method

The finite element method is a popular approach to the numerical solution of elliptic partial differential equations. Much has been written about it in numerical analysis literature and it has also been widely used and discussed by structural engineers. We can characterize the finite element method as a Galerkin-type method, with local, often very simple bases, and some manipulations to reduce the smoothness needed for the basis of the approximations. While our discussion will focus on Poisson's equation, much of what we say is applicable to more general elliptic partial differential equations.

Galerkin's method involves writing the approximation as

$$u_n(x, y) = \sum_{i=1}^{n} c_i \phi_i(x, y), \tag{13.18}$$

and then selecting the expansion coefficients so that the residual of the approximation is orthogonal to all the expansion functions; that is, that

$$\int_\Omega \{\nabla^2 u_n(x, y) - g(x, y)\} \phi_i(x, y) dxdy = 0 \tag{13.19}$$

for all $i = 1, 2, \ldots, n$.

In addition to the orthogonality constraints, we also need to address the boundary conditions. For simplicity we will assume that the Dirichlet conditions are homogeneous, so the boundary values in (13.2) are

$$\alpha(x, y) = 0. \tag{13.20}$$

If we choose the basis functions ϕ_i so that each of them is zero on the boundary, the linear combination (13.18) satisfies (13.20). As we will see, this is generally not hard to do.

Writing the orthogonality condition in the form (13.19) apparently implies that the approximation, and hence every basis function, is twice continuously differentiable in both variables. But such high continuity requirements are often inconvenient and, if possible, we try to reduce them. The trick is integration by parts, using *Green's theorem*. One of the forms of Green's theorem is as follows: If Ω is a closed and bounded region in the plane, and if u and v are both twice continuously differentiable functions that vanish on the boundary Γ, then

$$\int_\Omega \left(\frac{\partial^2 u}{\partial s^2} + \frac{\partial^2 u}{\partial t^2} \right) v d\Omega = - \int_\Omega \left(\frac{\partial u}{\partial s}\frac{\partial v}{\partial s} + \frac{\partial u}{\partial t}\frac{\partial v}{\partial t} \right) d\Omega. \tag{13.21}$$

Applying this to the Laplacian and the expansion functions in (13.18), we have

$$\int_\Omega \left(\frac{\partial^2 u_n}{\partial x^2} + \frac{\partial^2 u_n}{\partial y^2} \right) \phi_i d\Omega = - \int_\Omega \left(\frac{\partial u_n}{\partial x}\frac{\partial \phi_i}{\partial x} + \frac{\partial u_n}{\partial y}\frac{\partial \phi_i}{\partial y} \right) d\Omega. \tag{13.22}$$

The right side not only replaces second derivatives by first order differentiation, but remains sensible as long as the first partial derivatives of all the functions are integrable. It can be shown that extending the approximation in this way does not change the solution, so we will use the right side of (13.22) instead of the left one. The implication is that we can use basis functions that are piecewise differentiable, with possibly bounded jump discontinuities in the first derivatives.

With this modification, Galerkin's method then reduces to a linear system

$$\mathbf{Ac} = \mathbf{y}, \tag{13.23}$$

where

$$[\mathbf{A}]_{ij} = -\int_{\Omega} \left(\frac{\partial \phi_i}{\partial x} \frac{\partial \phi_j}{\partial x} + \frac{\partial \phi_i}{\partial y} \frac{\partial \phi_j}{\partial y} \right) d\Omega, \tag{13.24}$$

and

$$[\mathbf{y}]_i = \int_{\Omega} g \phi_i d\Omega. \tag{13.25}$$

In the finite element literature the matrix $\mathbf{A}$ is called the *stiffness matrix.* Choosing the basis elements so that the homogeneous boundary conditions (13.20) are satisfied, and so that the stiffness matrix can be computed without too much difficulty, is the primary challenge in the practical use of the finite element method.

One of the simplest and most popular choices is a piecewise planar approximation on triangles. Suppose for the moment that Ω is a polygonal region. Then it can be *triangulated,* meaning that it can be partitioned into a set of triangles (Figure 13.6).

The approximation u_n should be a continuous, planar function on each triangle. This can be achieved by using basis elements which have the value one at one vertex P of a triangle and zero at all other vertices, that are planar in all triangles with P as a vertex (Figure 13.7), and that are zero everywhere else. A linear combination of such basis functions will then have the required continuity properties. We also note that if we use basis

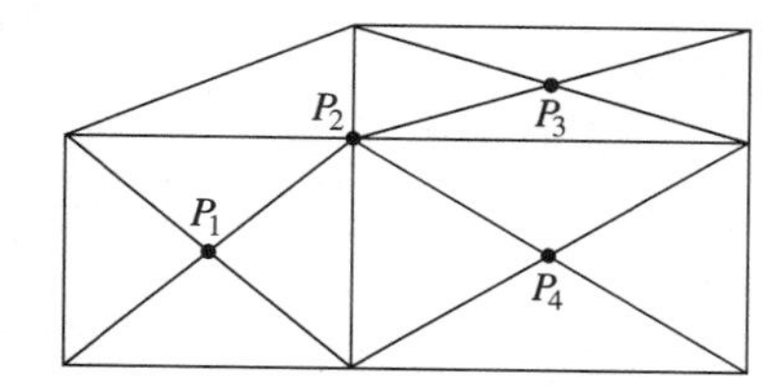

Figure 13.6
A triangulation of a polygonal region.

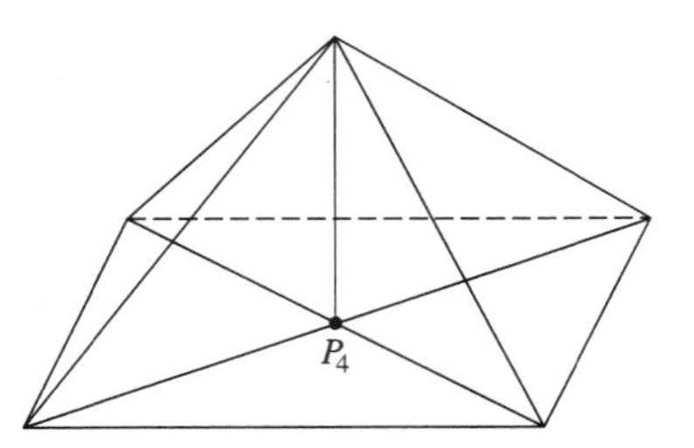

Figure 13.7
A basis element for piecewise planar approximations.

functions only for the interior points, the homogeneous boundary conditions (13.18) are automatically satisfied for any linear combination. For such piecewise planar elements, the matrix entries $[\mathbf{A}]_{ij}$ are easily computed analytically. The computations of the elements in (13.25) may have to be done numerically, but normally this does not cause any great difficulty.

We rarely know of an obvious and efficient triangulation that will solve a problem to a specified accuracy. Usually, we start with some rough triangulation, then subdivide those triangles that are likely to give the best improvement. Some care needs to be taken when subdividing a triangle. If the subdivision tends to make the triangles skinnier, the accuracy will suffer. A way of avoiding this is shown in Figure 13.8, where a triangle is divided into four smaller ones, each similar to the original one. In this way, we can adaptively place smaller triangles into regions where the solution changes rapidly and use larger triangles where the solution is very smooth. Re-computing the stiffness matrix after subdivision is relatively straightforward. Unlike the finite difference method, having triangles of widely varying size does not affect the method significantly. This is a strong advantage for the finite element method.

Flexibility is one of the main attractive features of the finite element method. Not only is subdividing a region easy, the triangular elements are very suitable for complex geometry. In addition, the method rests on strong physical and mathematical foundations, making it suitable for automatic implementation. There are also some disadvantages. Writing a computer program that finds a good triangulation can be challenging. The stiffness matrix is more complicated than the corresponding matrix for the finite

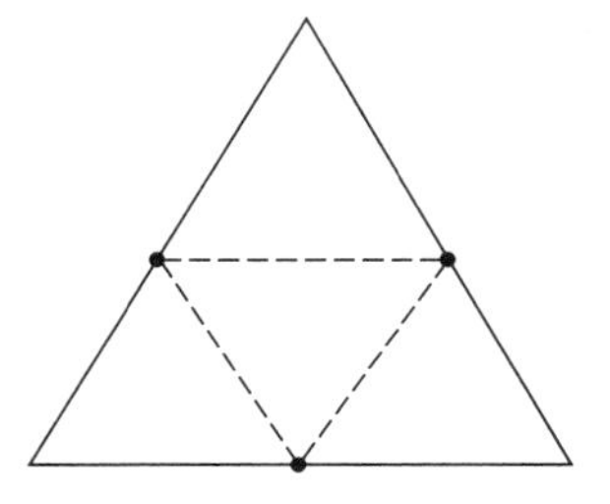

Figure 13.8
Subdividing a triangle, using the midpoints of each side.

difference methods. Although generally sparse, the sparsity pattern in the stiffness matrix is not as predictable as it is for the finite difference case. All of this makes it quite difficult to implement the finite element method. But most of these difficulties have been overcome. For many important elliptic partial differential equations there exist programs that produce good triangulations, construct the stiffness matrix, and solve the resulting sparse linear system efficiently. The widespread availability of such software is a major factor in the popularity of the finite element method.

Much has been written about the theory and the practice of finite element methods, their connection with variational principles, their convergence, and their stability. All of this is quite complicated, but there are a number of books that give a fairly accessible presentation. Good sources for further reading are Johnson [13] and Prenter [21].

EXERCISES

1. Suppose that in the Dirichlet problem the boundary conditions are not homogeneous. What can be done to make it suitable for a finite element treatment?

2. Find an explicit expression for

$$[\mathbf{A}]_{ij} = -\int_{\Omega} \left(\frac{\partial \phi_i}{\partial x} \frac{\partial \phi_j}{\partial x} + \frac{\partial \phi_i}{\partial y} \frac{\partial \phi_j}{\partial y} \right) d\Omega,$$

   where the ϕ_i are piecewise planar elements and Ω is a single triangle.

3. How does the triangulation affect the sparsity pattern of the stiffness matrix? What are typically the average number of elements in a row of the sparsity matrix?

4. Show that the subdivision suggested in Figure 13.8 produces four congruent triangles, each similar to the original triangle.

13.5 Multidimensional Equations

The partial differential equations that one encounters in practice are usually quite a bit more complicated than the simple prototypes we have studied here. Frequently, there are more than two dimensions involved. This introduces new complications coming from the geometry, increases the computer resources required, and often precludes any practically useful error analysis. Still, with the guidelines that we get from the simpler models, we can devise methods and successfully solve many very complicated equations.

As one step in the direction of more complicated models, we consider the *two-dimensional heat equation*

$$\frac{\partial u(x, y, t)}{\partial t} = \frac{\partial^2 u(x, y, t)}{\partial x^2} + \frac{\partial^2 u(x, y, t)}{\partial y^2}, \tag{13.26}$$

in some spatial region Ω with boundary Γ, in a time interval $0 \le t \le T$. There are initial conditions

$$u(x, y, 0) = g_1(x, y),$$

for all $(x, y) \in \Omega$, and boundary conditions

$$u(x, y, t) = g_2(x, y, t)$$

for all $(x, y) \in \Gamma$.

Equation (13.26) is a parabolic equation. It is an initial value problem that behaves very much like the one-dimensional heat equation (12.1). We can therefore model our numerical methods on the one-dimensional case. But in the steady-state case

$$\frac{\partial u(x, y, t)}{\partial t} = 0,$$

(13.26) reduces to Laplace's equation, so we should expect some aspects of boundary value problems to become significant.

Extending the framework of Section 12.1, we introduce a two-dimensional space grid, defined by grid points (x_i, y_j), $i = 0, 1, \ldots, n_1$, $j = 0, 1, \ldots, n_2$. The grid will be assumed to be uniform in each direction, with respective grid spacing Δx and Δy. A time grid is defined by time lines $t_0, t_1, \ldots,$ spaced at a distance Δt. The approximation to $u(x_i, y_j, t_k)$ will be denoted by $U_{i,j,k}$. To construct a finite difference method, we replace the Laplacian by its approximation on a five-point star (Figure 13.2) and the time derivative by a difference. As before, the question is what time difference to use.

The analog of the explicit method (12.10) is obvious; we simply use a forward time difference to get

$$\begin{aligned} U_{i,j,k+1} = U_{i,j,k} &+ \Delta t \frac{U_{i-1,j,k} + U_{i+1,j,k} - 2U_{i,j,k}}{(\Delta x)^2} \\ &+ \Delta t \frac{U_{i,j-1,k} + U_{i,j+1,k} - 2U_{i,j,k}}{(\Delta y)^2} \end{aligned} \tag{13.27}$$

for regular interior points, with suitable changes for the points near the boundary.

With the explicit method the solution can be computed easily with a forward marching process. We need to be concerned about stability and, not surprisingly, it turns out that the method is stable only for the condition

$$\frac{\Delta t}{(\Delta x)^2 + (\Delta y)^2} \le \frac{1}{2}. \tag{13.28}$$

The one-dimensional case suggests that we can alleviate the stability condition and increase the accuracy by going to an implicit scheme, such as an analog of the Crank-Nicholson method (12.14). Unfortunately, when we do this we find that we have to solve a complete Laplace equation at each time-step. This is quite expensive, so we need to look for cheaper alternatives.

One successful strategy is the *alternating direction implicit* (ADI) method. Each time-step is split into two equal parts. In the first part, integrating from $k\Delta t$ to $(k+\frac{1}{2})\Delta t$, x is taken to be implicit while y is treated explicitly. This gives

$$\begin{aligned} U_{i,j,k+1/2} = U_{i,j,k} &+ \frac{\Delta t}{2(\Delta x)^2}\left(U_{i-1,j,k+1/2} + U_{i+1,j,k+1/2} - 2U_{i,j,k+1/2}\right) \\ &+ \frac{\Delta t}{2(\Delta y)^2}\left(U_{i,j-1,k} + U_{i,j+1,k} - 2U_{i,j,k}\right). \end{aligned} \tag{13.29}$$

In the next step, from $(k+\frac{1}{2})\Delta t$ to $(k+1)\Delta t$, the method is made explicit in x and implicit in y, so that

$$\begin{aligned} U_{i,j,k+1} = U_{i,j,k+1/2} &+ \frac{\Delta t}{2(\Delta x)^2}\left(U_{i-1,j,k+1/2} + U_{i+1,j,k+1/2} - 2U_{i,j,k+1/2}\right) \\ &+ \frac{\Delta t}{2(\Delta y)^2}\left(U_{i,j-1,k+1} + U_{i,j+1,k+1} - 2U_{i,j,k+1}\right). \end{aligned} \tag{13.30}$$

Each time step now involves solving a number of implicit equations, but all of them are essentially tridiagonal. This is much cheaper than solving a complete Laplace equation at each step. The ADI method is known to be stable for all Δx, Δy, and Δt.

Chapter 14

Ill-Conditioned and Ill-Posed Problems*

In developing numerical methods, we normally make some implicit assumptions. It is generally taken for granted that the problem has a solution, that the solution is unique (at least locally), and that small perturbations in the problem do not lead to large changes in the solution. Such conditions are considered necessary for the construction of stable and convergent numerical algorithms. But we have already encountered some exceptions in problems that either had no solution or had many solutions. For example, an overdetermined linear system of equations may have no solution. In this case, we can get a solvable problem by looking for a least squares solution instead. A second case is data fitting, where many possible solutions exist. By choosing a particular type of approximation, we restrict the problem so that it has a unique solution. In both these cases, the reformulation makes it possible to solve the problem in a stable manner. There are, however, many practical situations where such a restating is not so obvious and where a great deal of care has to be taken to get meaningful results.

14.1 The Numerical Rank of a Matrix

The definition of the rank of a matrix is unambiguous. If $\mathbf{A}$ is an $m \times n$ matrix, then the rank of $\mathbf{A}$ is the dimension of the largest square nonsingular submatrix. For the linear system

$$\mathbf{A}\mathbf{x} = \mathbf{b} \tag{14.1}$$

the rank of $\mathbf{A}$ determines the nature of the solution set by the following well-known rules:

- If $m = n$ and the rank of $\mathbf{A}$ equals m, then (14.1) has a unique solution.
- If $m > n$ the system is *overdetermined,* and (14.1) may not have a solution. If the rank of $\mathbf{A}$ equals n and the system has a solution, then the solution must be unique.
- If $m < n$, then (14.1) cannot have a unique solution.
- The system always has a least squares solution. The least squares solution is unique if and only if the rank of $\mathbf{A}^T\mathbf{A}$ equals n.
- The system always has a unique generalized solution (see Chapter 8).

If we know the rank of the matrix $\mathbf{A}$, we can then stably compute whatever solutions of (14.1) might exist. Unfortunately, determining the rank of a matrix is not trivial. The reason for this lies in rounding which blurs the distinction between truly singular and nonsingular, but highly ill-conditioned matrices.

There are several ways in which we can approach this subject. Here we work primarily with singular value decomposition which exhibits the nature of the difficulties in a very explicit fashion.

Example 14.1 A 12×12 matrix was created as follows:

$$[\mathbf{A}]_{ij} = 10i\delta_{ij} + \frac{1}{\sqrt{ij}}$$

for $i = 1, 2, \ldots, 10$ and $j = 1, 2, \ldots, 12$, where

$$\delta_{ij} = \begin{cases} 1 \text{ if } i = j, \\ 0 \text{ if } i \neq j, \end{cases}$$

is the Kronecker delta. The eleventh row was taken as the sum of rows 1 and 2, and the twelfth row as the sum of rows 1 and 3. If all arithmetic were done exactly, the rank of the matrix would be 10.

Table 14.1
Computed singular values of a 12×12 matrix with rank 10.

i	λ_i
1	1.00×10^2
2	9.01×10^1
3	8.01×10^1
4	7.02×10^1
5	6.02×10^1
6	5.03×10^1
7	4.47×10^1
8	4.02×10^1
9	3.09×10^1
10	1.41×10^1
11	5.98×10^{-15}
12	4.88×10^{-16}

The computed singular values are shown in Table 14.1. Since two of the singular values are of the order of the rounding error, we can reasonably conjecture that the matrix has rank 10. When we extract the 10×10 leading principal minor, we find its condition number to be about 10, confirming our conclusion. MATLAB gets it right also: `rank(A)` gives the value 10. ■

Example 14.2 Although Hilbert matrices become ill-conditioned very quickly, strictly speaking, any Hilbert matrix is nonsingular. But this is hard to demonstrate numerically. For example, the condition numbers for the matrix $\mathbf{A}$ in Example 14.1 and the 12×12 Hilbert matrix are of the same order of magnitude, so this tells us little. A better idea can be gained by looking at the singular values. The singular values of a 12×12 Hilbert matrix are shown in Table 14.2.

The situation here is not so clear. If we use an estimate of rounding errors as a criterion for setting a singular value to zero, we might guess that the rank of the matrix is 10 or 11. This is in fact what MATLAB does; the expression `rank(A)` has the value 11. But the situation is clearly different from that in Example 14.1. If we proceed on the assumption that the rank is 11 and extract an 11×11 submatrix for further computation, the computations are very likely ill-conditioned and will yield poor results.

Table 14.2
Singular values of a 12×12 Hilbert matrix.

i	λ_i
1	1.8×10^{0}
2	3.8×10^{-1}
3	4.5×10^{-2}
4	3.7×10^{-3}
5	2.3×10^{-3}
6	1.1×10^{-5}
7	4.1×10^{-7}
8	1.1×10^{-8}
9	2.3×10^{-10}
10	3.1×10^{-12}
11	2.6×10^{-14}
12	1.1×10^{-16}

The gradual decay of the singular values is an indication that we are not dealing with a rank-deficient matrix, but with a situation that is inherently ill-conditioned.

■

We see from these two examples that there are serious difficulties in finding the rank of a matrix numerically. It is possible to work out some quite complicated schemes to try to decide between a very small number and a zero contaminated by rounding, but none of them always work. We have to conclude that it is not possible to compute the rank of a matrix numerically without the risk of an occasional failure. The implications of this inherent ambiguity are explored in the next few sections.

EXERCISES

1. Show that the rank of a matrix is equal to the number of nonzero singular values.
2. In terms of the singular values and singular vectors, justify the rules stated at the beginning of this section, relating the rank to the existence and uniqueness of a solution.

3. An $n \times n$ Wilkinson matrix for any odd n is defined by

$$[\mathbf{A}]_{i,j} = \begin{cases} \frac{(n+1)}{2} - i & \text{if } i = j, \\ 1 & \text{if } |i-j| = 1, \\ 0 & \text{otherwise.} \end{cases}$$

Show that the 1×1, 3×3, and 5×5 Wilkinson matrices are of ranks 0, 2 and 4, respectively.

4. The numerical approximations for the smallest two singular values of the 11×11 Wilkinson-matrix to five significant digits are 1.0002 and 1.2797×10^{-16}. What conjecture about the rank of an odd dimensional Wilkinson-matrix can you make from this observation and Exercise 3?

14.2 Computing Generalized Solutions

To avoid nonexistence and nonuniqueness issues in solving (14.1), we consider the computation of the generalized solution $\mathbf{x}_0^*$. As we know from Chapter 8, a unique generalized solution of (14.1) exists for all $\mathbf{A}$ and $\mathbf{b}$. From (8.30) we have the expression

$$\mathbf{x}_0^* = \sum_{i=1}^{p} \frac{\mathbf{v}_i^T \mathbf{b}}{\lambda_i} \mathbf{u}_i, \tag{14.2}$$

where $\lambda_1, \lambda_2, \ldots, \lambda_p$ are all the nonzero singular values, with $\mathbf{u}_i$ and $\mathbf{v}_i$ the corresponding right and left singular vectors of $\mathbf{A}$. Because the singular value decomposition of a matrix is readily computed, (14.2) is a viable method for computing the generalized solution.

The practical difficulty is that we cannot always distinguish between a zero singular value and one that is of the order of magnitude of the rounding error. This makes the choice of p, and hence (14.2), ambiguous. A simple way around this difficulty is to chose some small value α and treat all $\lambda_i < \alpha$ as zero, thus cutting off all singular values below a certain threshold. The generalized solution $\mathbf{x}_0^*$ is then replaced by

$$\mathbf{x}_\alpha = \sum_{i=1}^{p(\alpha)} \frac{\mathbf{v}_i^T \mathbf{b}}{\lambda_i} \mathbf{u}_i, \tag{14.3}$$

where $p(\alpha)$ is the largest integer such that

$$\lambda_{p(\alpha)} \geq \alpha. \tag{14.4}$$

The immediate question is how to choose the cut-off level.

Table 14.3 Dependence of the error on α in Example 14.3.

α	Max. error
10^{-15}	4.2×10^{-3}
10^{-12}	6.3×10^{-5}
10^{-11}	4.1×10^{-6}
10^{-7}	1.6×10^{-4}
10^{-3}	2.7×10^{-2}

Example 14.3 For the matrix in Example 14.1, choosing the cut-off level does not present much of a problem. Any value of α between 10^{-2} and 10^{-15} will do equally well and make the computation of the generalized solution stable. The situation in Example 14.2 is much more ambiguous, with different values of α giving very different results. In Table 14.3 we show the error in computing the generalized solution of $\mathbf{Ax} = \mathbf{b}$, with $\mathbf{A}$ the 12×12 Hilbert matrix, and $\mathbf{b}$ chosen so that $\mathbf{x} = (1, 1, \ldots, 1)^T$.

For very small α we cut off the worst singular values, but the ones remaining still cause considerable instability. If we increase α, we reduce the instability but modify the original problem so much that the solution is affected. Even with the best choice of α much accuracy is lost.

This example again shows that rounding contamination is much more easily dealt with in matrices that are truly rank-deficient than for those that in some way are inherently ill-conditioned.

Ill-conditioning does occur in practice. We have already seen how a Hilbert matrix can arise when polynomial approximation is approached naively. In this case the difficulty can be avoided by a reformulation, but unfortunately this is not always so. There are many practical situations where the ill-conditioning is inherent in the problem and cannot be avoided by a better numerical method. Before we study what can and cannot be done in such situations, it is instructive to investigate the origins of such inherently unpleasant problems.

EXERCISES

1. The MATLAB statement

```
A=rosser
```

creates a special 8×8 matrix that has properties that make it a good test case for eigenvalue algorithms. MATLAB tells us that the rank of this matrix is 7. Do you think that this is a reasonable answer?

2. Consider a small ill-conditioned linear system $\mathbf{Ax} = \mathbf{b}$ with

$$\mathbf{A} = \begin{bmatrix} 1 + 5 \times 10^{-15} & 1 & 0 \\ 1 & 0 & 1 \\ 0 & 1 & -1 \end{bmatrix},$$

and

$$\mathbf{b} = \begin{bmatrix} 3 + 5 \times 10^{-15} \\ 4 \\ -1 \end{bmatrix}.$$

The system has a theoretical solution $\mathbf{x} = (1,\ 2,\ 3)^T$.

(a) Find generalized solutions using cut-offs $\alpha = 10^{-15}$ and 10^{-14}.

(b) Suppose that the vector $\mathbf{b}$ is perturbed by adding 2×10^{-5} and -1×10^{-5} to the second and third components, respectively, so that

$$\mathbf{b} = \begin{bmatrix} 3 + 5 \times 10^{-15} \\ 4 + 2 \times 10^{-5} \\ -1 - 1 \times 10^{-5} \end{bmatrix}.$$

Examine how the generalized solution changes under such a small perturbation.

14.3 Inverse and Ill-Posed Problems

As we know, numerical differentiation is inherently ill-conditioned. A small perturbation in the value of a function can cause a large change in its derivative. Computing the first derivative of a function is unstable by any algorithm, although the instability is often manageable. But for higher derivatives, the instability is more pronounced and computing high derivatives numerically from physical data is generally a hopeless task. Similar problems arise in connection with a class of problems that we call *inverse*.

What is an inverse problem? While one can now find frequent references to inverse problems in the applied mathematics literature, there is no precise definition of this term. Although such problems are clearly different from the problems we are used to, and which we can call *direct* problems, inverse problems occur in a variety of circumstances and ostensibly have few characteristics in common. There is one thing, though, that all problems classified as inverse seem to share. In an inverse problem we almost always try to find the cause of an observed effect, while in the more traditional direct problems we try to predict the effect of a known cause. A direct problem involves a fully specified mathematical model whose solution can be used to predict the behavior of the physical system it represents. In an inverse problem, on the other hand, the model is incomplete, with some parameters or other essential information missing. The solution of an inverse problem involves using the observed behavior of the system to determine this missing information. Inverse problems are therefore generally associated with experimental inference, remote sensing, systems identification, image reconstruction, and similar concerns.

Example 14.4 Consider the heat conduction equation

$$\frac{\partial u(x,\, t)}{\partial t} = \frac{\partial}{\partial x}\left(\kappa(x)\frac{\partial u(x,\, t)}{\partial x}\right), \tag{14.5}$$

with boundary conditions

$$u(0,\, t) = u(1,\, t) = 0.$$

This form of the heat equation is a little more complicated than the one we studied in Chapter 12 because it provides for variable thermal conductivity. The function $\kappa(x)$ specifies how the thermal conductivity varies over the length of the wire. This complication is of little consequence to either the theory or the numerical solution of the equation, but is of obvious practical importance.

What has received the most attention is the direct problem. We know the thermal conductivity $\kappa(x)$ and the initial conditions $u(x,\, 0)$ for $0 \le x \le 1$, and we want to find the value of $u(x,\, T)$ for some subsequent time $T > 0$. In this form the problem has been studied extensively and much is known about it. In Chapter 12 we explored some of the numerical methods and showed that, with a certain amount of care, stable numerical algorithms can be found.

One inverse problem is the *backward heat equation*: Given $u(x,\, T)$, find the initial conditions $u(x,\, 0)$. We know much less about this inverse problem than we know about the direct problem, but we do know that it is difficult to solve numerically. A second inverse problem is the *heat equation identification problem*: Given $u(x,\, 0)$ and $u(x,\, T)$, find the thermal conductivity $\kappa(x)$. This is also a difficult problem. In both cases the construction of numerical methods leads to inherently ill-conditioned systems. ■

Example 14.5 A thin strip of material of density $\rho(x)$ is located on a line between $x = -1$ and $x = 1$, as shown in Figure 14.1. It exerts a gravitational attraction in the surrounding plane which is recorded by measuring the vertical component of the force along the line $-1 \le y \le 1$ at a distance d from the strip.

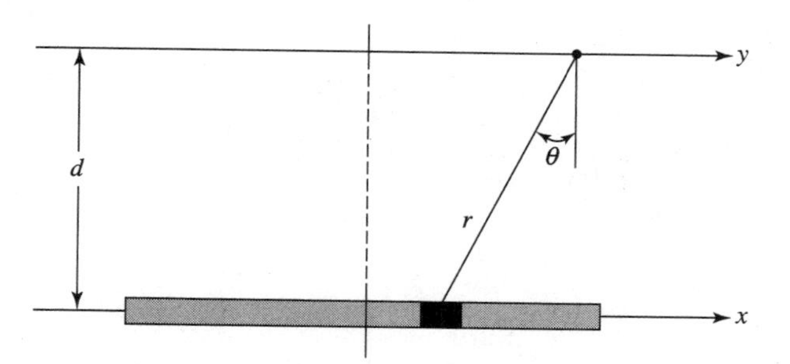

Figure 14.1
Gravitational force of a thin strip.

From the law of gravitational attraction, we know that the vertical force at y due to a small element of length dx located at x is

$$dF = \frac{g}{r^2}\rho(x)\cos\theta dx,$$

where g is the gravitational constant. Using $r = \sqrt{(y-x)^2 + d^2}$, $\cos\theta = d/r$, and integrating over the interval $-1 \le x \le 1$, we see that the total force at y is

$$F(y) = gd\int_{-1}^{1} \frac{\rho(x)dx}{[(y-x)^2 + d^2]^{3/2}}. \tag{14.6}$$

If we know the density of the strip, the computation of F is a simple direct problem, involving only an integration. On the other hand, if we measure the vertical force at various values of y and want to infer from these measurements the unknown density, then the problem is an inverse one.

This example is a simplified version of a situation that is often encountered in geophysical exploration where one tries to determine the location and size of underground mineral or oil deposits from gravitational, magnetic, or seismic surface measurements.

That both the above problems should show some sort of ill-conditioning becomes plausible if we examine the underlying physical principles. Consider the density of the strip in the last example. In Figure 14.2 we have two different scenarios. In the first, the density is nearly constant over the interval, while in the second, the density oscillates rapidly around the mean. When we measure the effect at a distance, the gravitational attraction is averaged and both situations give roughly the same contribution to the total force. In other words, when we measure the gravitational attraction from far away, we cannot tell whether the first or the second scenario holds. We cannot expect equation (14.6) to resolve the situation. Since there are some significant changes in ρ that cause insignificant changes in F, it is possible that small changes in F may result in large changes in the density. It is therefore reasonable to expect that Equation (14.6) is inherently unstable.

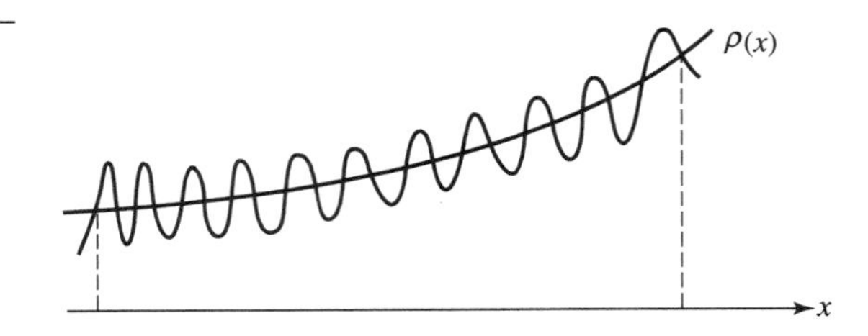

Figure 14.2 Density distributions essentially indistinguishable from a distance.

In his work on partial differential equations, the French mathematician J. Hadamard originated the concept of a *well-posed problem.* According to Hadamard, an equation is considered well-posed if it is uniquely solvable and is such that small perturbations in the parameters of the equation have a small effect on the solution. If an equation does not satisfy these conditions, it is said to be *ill-posed.* Hadamard and his contemporaries felt that equations that model physical situations must be well-posed, because the underlying physical systems are insensitive to small perturbations. Problems that were classified as ill-posed therefore seemed of theoretical interest at best and were neglected by mathematicians for some time. In the last few decades, though, it has become apparent that there are in fact many circumstances, primarily in connection with inverse problems, where the mathematical formulation leads to equations that are not well-posed in the Hadamard sense. As a result, the approximate solution of ill-posed problems has become an important topic in numerical analysis.

EXERCISES

1. Explain why you might suspect that the backward heat equation is not well-posed.

2. Do you think that the uncertainty with which we can find ρ from Equation (14.6) decreases or increases with d?

3. Inverse problems occur frequently in connection with image reconstruction. The situation that follows is an abstracted and simplified version of this.

   The light from an object O, whose light intensity is given by $o(x)$, passes through a lens to produce an image I, with light intensity $i(x)$. To find the relation between the object and its image, consider the small element of O in the figure on page 393.

   If the lens were perfect, the light from this small element would simply produce a corresponding concentrated spot in the image, but generally it is "smeared out" to produce an image intensity $k(x,\ y)o(x)\Delta x$. By adding up all the small pieces and taking the limit as $\Delta x \to 0$, we find that the relation between the object and its image is given by

$$i(y) = \int k(x,\ y)(x)o(x)dx.$$

   By observing the image intensity $i(y)$, we can reconstruct the object by solving this integral equation for $o(x)$.

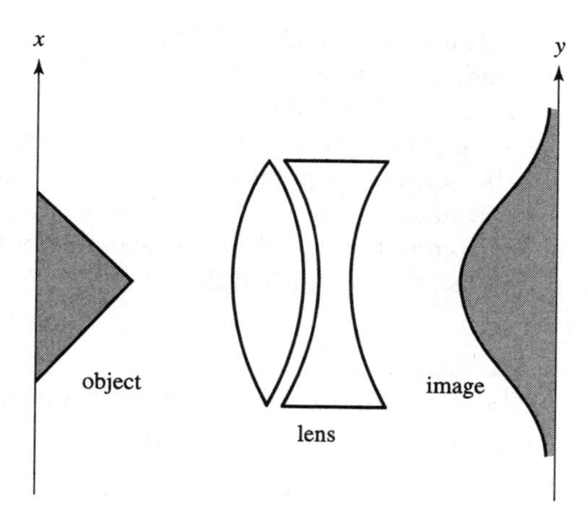

(a) Give reasons for why one might suspect this to be an ill-posed problem.

(b) Suppose that

$$k(x, y) = e^{-\alpha(x-y)^2}.$$

How would you expect α to affect the difficulty of reconstructing the object reliably?

4. Collect other examples that might be considered inverse problems. Can you see reasons why they might not be well-posed?

14.4 Regularization

The fact that most inherently ill-conditioned systems arise from applications that result in ill-posed equations has important consequences for their solution. Problems of this kind can be solved successfully only if we manage to inject into the process the concept of plausibility, based on what we know from the underlying physical situation and what we expect of the solution. Often the origins of the equation put restrictions on what solutions are reasonable.

Suppose we knew an exact solution $\mathbf{x}$ of (14.1). How could we demonstrate that it is in fact a true solution? We can substitute $\mathbf{x}$ into the original equation and compute the residual

$$\rho(\mathbf{x}) = \mathbf{A}\mathbf{x} - \mathbf{b}. \tag{14.7}$$

If all computations were done exactly, the residual would be zero. But because of rounding, even the exact solution will not produce a zero residual. Thus, we cannot distinguish an exact solution $\mathbf{x}$ from any approximation $\hat{\mathbf{x}}$ that gives a similarly small residual. From a practical point of view, $\mathbf{x}$ and $\hat{\mathbf{x}}$ must both be considered possible solutions. When $\mathbf{A}$ is well-conditioned then $||\mathbf{x} - \hat{\mathbf{x}}\,||$ is small, so the uncertainty is of little consequence. However, when $\mathbf{A}$ is badly conditioned this is not necessarily so and we can get many possible solutions that are quite different.

Example 14.6 In Example 14.3, an approximate solution $\hat{\mathbf{x}}$ was computed with the standard MATLAB system solver. The resulting solution gave

$$||\rho(\hat{\mathbf{x}})\,||_2 = 6.6 \times 10^{-16},$$

which is within machine accuracy and so effectively zero. Yet $\hat{\mathbf{x}}$ is quite inaccurate, with

$$||\mathbf{x} - \hat{\mathbf{x}}\,||_2 \cong 0.28.$$

Since even very accurate solutions cannot give a significantly smaller residual, the residual computation alone gives no indication that $\hat{\mathbf{x}}$ is quite inaccurate. We are left with a great deal of uncertainty about the solution of the system. ■

The only way we can get rid of the large uncertainties in the solution of ill-conditioned systems is to use our expectations of the true solution. By constraining the solutions in some way, we can eliminate those we consider undesirable and, we hope, get a solution that represents the desired solution well. This process is called *regularization*.

There are many different ways by which we can regularize an unstable process. Ideally, the regularization should take into account what we know about the solution and should also be justifiable from a formal standpoint. To begin, let us look at system (14.1) again, this time under small perturbations. For simplicity, assume that $\mathbf{A}$ is a square, full-rank matrix, so (14.1) has a unique solution $\mathbf{x}$. This assumption will be made for the rest of this chapter. Suppose now that all the computational errors can be accounted for by adding a term $\Delta\mathbf{b}$ to the right side of (14.1), so that the computed solution $\hat{\mathbf{x}}$ actually satisfies

$$\mathbf{A}\hat{\mathbf{x}} = \mathbf{b} + \Delta\mathbf{b}. \tag{14.8}$$

The solution of this system, in terms of the singular vectors of $\mathbf{A}$, is

$$\hat{\mathbf{x}} = \sum_{i=1}^{n} \frac{1}{\lambda_i}\mathbf{v}_i^T(\mathbf{b} + \Delta\mathbf{b})\mathbf{u}_i,$$

so that

$$\hat{\mathbf{x}} - \mathbf{x} = \sum_{i=1}^{n} \frac{1}{\lambda_i} \mathbf{v}_i^T \Delta \mathbf{b} \mathbf{u}_i. \tag{14.9}$$

This indicates that the small errors $\Delta\mathbf{b}$ can be magnified considerably for small singular values, resulting in large errors in the solution. Normally, since the singular vectors associated with small singular values are very oscillatory, the components of the computed $\hat{\mathbf{x}}$ tend to vary wildly. When highly oscillatory solutions are not physically plausible, we want to regularize by eliminating the irregular behavior. In Section 14.2 we suggested that computational error effects could be limited by choosing some cut-off level α and dropping terms for which $\lambda_i < \alpha$. This method, usually referred to as *singular value decomposition (SVD) with cut-off*, is perhaps the simplest way to regularize an ill-conditioned linear system. But as we see from Example 14.3, the solution is affected by the choice of the cut-off value. Sometimes, the choice of α can have a major effect on the solution.

Example 14.7 The computations of Example 14.3 were repeated using the right side vector $\mathbf{b}$ with an addition of the random vector $\Delta\mathbf{b}$ that has a relative size about 10^{-4} to $\mathbf{b}$, where[1]

$$\Delta\mathbf{b} = \begin{bmatrix} -0.0005621 \\ -0.0009059 \\ 0.0003577 \\ 0.0003586 \\ 0.0008694 \\ -0.0002330 \\ 0.0000388 \\ 0.0006619 \\ -0.0009309 \\ -0.0008931 \\ 0.0000594 \\ 0.0003423 \end{bmatrix}.$$

[1] The reason for giving $\Delta\mathbf{b}$ explicitly, instead of just the order of magnitude, is so that the results can be verified. For ill-conditioned matrices even small statistical fluctuations can change results quite a bit.

Table 14.4 Dependence of the error on α in Example 14.3 with a 0.1% random perturbation on the right side vector $\mathbf{b}$.

α	Max. error
10^{-15}	7.6×10^9
10^{-12}	1.9×10^8
10^{-11}	9.7×10^5
10^{-7}	7.5×10^2
10^{-3}	1.4×10^{-1}

The results in Table 14.4 show the error in the generalized solution with such a small perturbation on the right side vector $\mathbf{b}$. The solutions of all the five cut-off values α are essentially useless. ■

14.5 The Tikhonov Method

An exact or a least squares solution of (14.1) can be obtained by minimizing

$$\Phi(\mathbf{x}) = ||\mathbf{A}\mathbf{x} - \mathbf{b}||_2^2. \tag{14.10}$$

When $\mathbf{A}$ is rank-deficient or very nearly singular, standard algorithms often give solutions that vary rapidly, with large positive and negative values. To stabilize the computation, we add to (14.10) a term that penalizes the large components and thereby reduces them. Thus, instead of (14.10), we minimize the expression

$$\Phi_\alpha(\mathbf{x}) = ||\mathbf{A}\mathbf{x} - \mathbf{b}||_2^2 + \alpha\,||\mathbf{x}||_2^2 \tag{14.11}$$

for some positive α. A little algebra shows that the minimizing solution $\mathbf{x}_\alpha$ is given by the nonsingular linear system

$$(\mathbf{A}^T\mathbf{A} + \alpha\,\mathbf{I})\mathbf{x}_\alpha = \mathbf{A}^T\mathbf{b}. \tag{14.12}$$

It is then easy to show that the solution of (14.12) can be expressed as

$$\mathbf{x}_\alpha = \sum_{i=1}^{n} \frac{\lambda_i}{\alpha + \lambda_i^2}\,\mathbf{v}_i^T\mathbf{b}\mathbf{u}_i. \tag{14.13}$$

The effect of the addition in (14.11) is to dampen the contributions of the terms involving small singular values, so that instead of cutting them off altogether, we modify the method to reduce their impact. A small α has very little effect on the components associated with large λ_i, since for $\alpha << \lambda_i^2$

$$\frac{\lambda_i}{\alpha + \lambda_i^2} \cong \frac{1}{\lambda_i}.$$

On the other hand, if λ_i^2 is much smaller than α, then

$$\frac{\lambda_i}{\alpha + \lambda_i^2} \cong \frac{\lambda_i}{\alpha} << \frac{1}{\lambda_i},$$

so that the magnification of the components associated with the small singular values is reduced. With a good choice of α one can then hope to get a relatively smooth solution that is still a reasonably good approximation to the true solution. This is called *SVD with damping.*

Regularization by (14.12) dampens components that are large in magnitude, but it may not inhibit components that oscillate with moderate amplitudes. If such components are undesirable, we may need a stronger regularization. For this, we can add a penalty term that is large for rapid changes in the solution, such as

$$\Phi_\alpha(\mathbf{x}) = ||\mathbf{Ax} - \mathbf{b}||_2^2 + \alpha \sum_{i=2}^{n} ([\mathbf{x}]_i - [\mathbf{x}]_{i-1})^2. \tag{14.14}$$

This expression is minimized by the solution of

$$(\mathbf{A}^T\mathbf{A} + \alpha\, \mathbf{B}_1^T \mathbf{B}_1)\mathbf{x}_\alpha = \mathbf{A}^T\mathbf{b}, \tag{14.15}$$

where $\mathbf{B}_1$ is the $(n-1) \times n$ matrix

$$\mathbf{B}_1 = \begin{bmatrix} 1 & -1 & 0 & \cdots & 0 \\ 0 & 1 & -1 & & \vdots \\ \vdots & \ddots & 1 & \ddots & 0 \\ & & & \ddots & -1 \\ 0 & \cdots & & 0 & 1 \end{bmatrix}.$$

An even stronger regularization is based on minimizing

$$\Phi_\alpha(\mathbf{x}) = ||\mathbf{Ax} - \mathbf{b}||_2^2 + \alpha \sum_{i=2}^{n-1} ([\mathbf{x}]_{i+1} - 2[\mathbf{x}]_i + [\mathbf{x}]_{i-1})^2, \tag{14.16}$$

which leads to the system

$$(\mathbf{A}^T\mathbf{A} + \alpha\, \mathbf{B}_2^T \mathbf{B}_2)\mathbf{x}_\alpha = \mathbf{A}^T\mathbf{b}, \tag{14.17}$$

where $\mathbf{B}_2$ is an $(n-2) \times n$ matrix with elements

$$[\mathbf{B}_2]_{ij} = \begin{cases} -2, & \text{if } i = j, \\ 1, & \text{if } i = j+1 \text{ or } i = j-1, \\ 0, & \text{otherwise.} \end{cases}$$

Methods of type (14.12), (14.15), and (14.17) are called *Tikhonov regularization methods*, after the Russian mathematician A.N. Tikhonov, who first studied them. Because the term added in (14.14) reduces first differences, we refer to it as a Tikhonov regularization of *order one*. By a similar reasoning, (14.17) is said to be of *order two*. The SVD with damping methods is regularization with *order zero*.

Example 14.8 Here we use again the 12×12 Hilbert matrix as $\mathbf{A}$, with $\mathbf{b}$ chosen so that

$$[\mathbf{x}]_i = i^2.$$

If the system is solved without any regularization, the maximum error is about 9.1. Tikhonov regularization reduces this considerably. Figures 14.3 to 14.5 show the results for regularization of order zero, one, and two, respectively. In all three cases we used $\alpha = 10^{-9}$. The correct solution is marked with +, while the approximate solution is represented by a continuous graph through the computed values. While the regularized results are all much better than those without regularization, there are discernible differences for different regularization orders. Regularization with order zero still shows the kind of wiggly behavior one usually tries to eliminate. Regularization with order one improves the situation, but in this case, second order regularization gives by far the most satisfactory result. ■

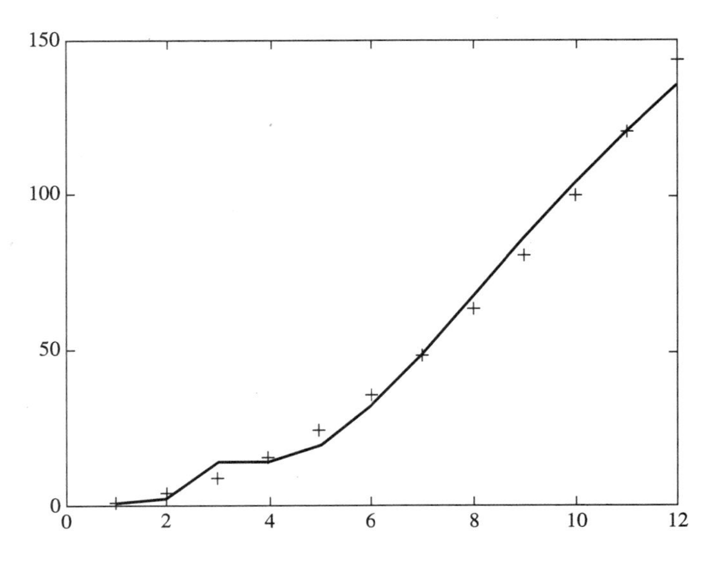

Figure 14.3 Results for Example 14.8 with regularization of order zero.

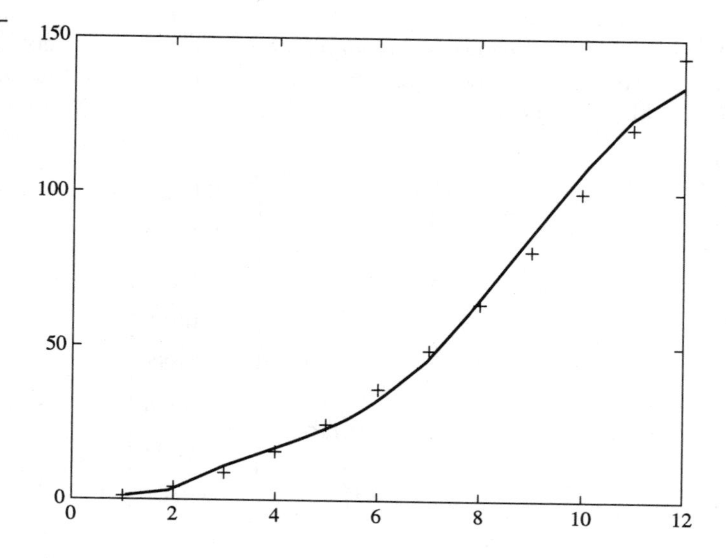

Figure 14.4 Results for Example 14.8 with regularization of order one.

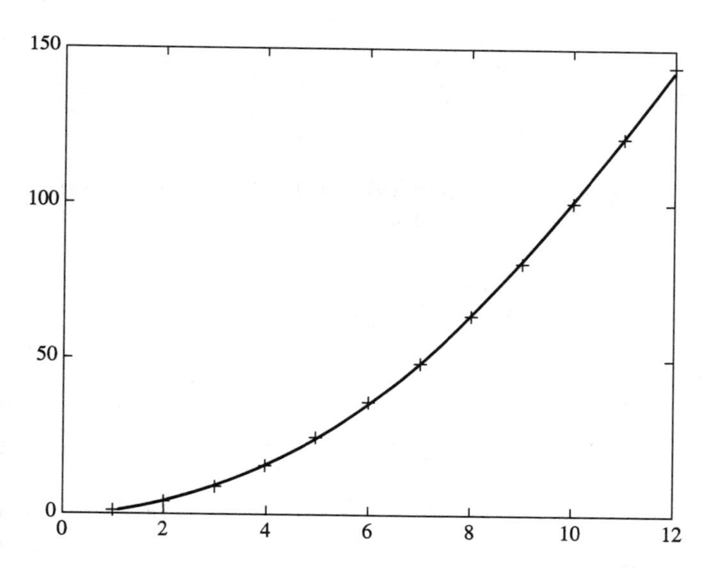

Figure 14.5 Results for Example 14.8 with regularization of order two.

Example 14.8 might lead us to suspect that with just the right kind of regularization the true solution can be recovered at least to plotting accuracy. But this is too optimistic, as the example unrealistically assumes that rounding is the only source of difficulty. In practice, ill-conditioned matrices often come from inverse problems in which experimental observations are made, and hence inaccurate data is used. This usually makes $||\Delta \mathbf{b}||$ in (14.8) much larger than rounding error, and recovery of the unknown becomes less certain. A good selection of the regularization constant is essential for good results.

Example 14.9 The computations of Example 14.8 were repeated, now adding a random error $\Delta\mathbf{b}$ of relative size roughly 10^{-3} to $\mathbf{b}$. The actual perturbation for this example was

$$\Delta\mathbf{b} = \begin{bmatrix} -0.0025 \\ -0.0009 \\ 0.0038 \\ 0.0054 \\ -0.0057 \\ -0.0026 \\ 0.0002 \\ -0.0101 \\ 0.0044 \\ -0.0037 \\ 0.0039 \\ -0.0026 \end{bmatrix}.$$

The results of second order regularization with several values of α are shown in Figures 14.6 to 14.8.

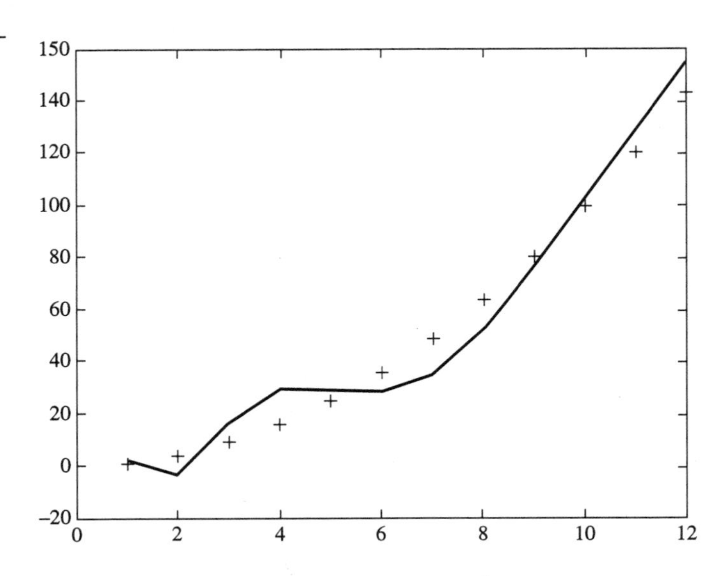

Figure 14.6 Regularization in Example 14.9 with $\alpha = 10^{-9}$.

Figure 14.7
Regularization in Example 14.9 with $\alpha = 10^{-8}$.

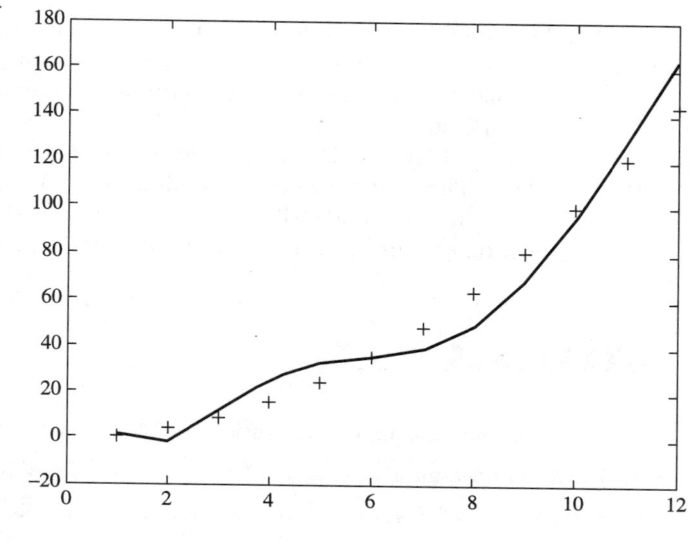

Figure 14.8
Regularization in Example 14.9 with $\alpha = 10^{-6}$.

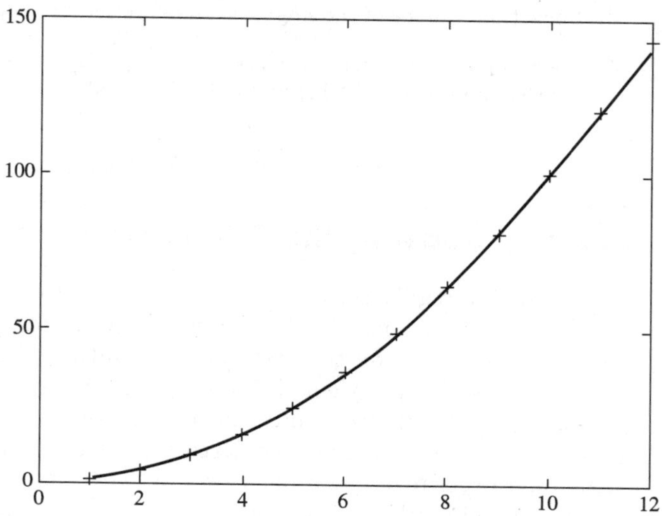

In practical situations obviously we do not know the exact answers, but only the approximate solution, so that judging the solution can be done only by inspecting the results. The somewhat erratic aspect of the curve in Figure 14.6 suggests that we are overfitting the data, making the solution dependent on the experimental errors. Reducing the size of the regularization constant confirms this: Increasing α to 10^{-8} improves the situation, and $\alpha = 10^{-6}$ gives quite good results.

■

Simple numerical case studies such as these give us a feeling for the performance as well as the limitations of regularization, but representative examples are hard to find. For well-posed problems, a few tests often give us good insight into the workings of a method, but for ill-posed problems one case may help little in predicting the performance of the algorithm for another problem. The results are often ambiguous, and even implementation errors may not be clearly identifiable. Solving ill-posed problems requires much more care and insight than it does for more traditional cases.

EXERCISES

1. Show that minimizing the expression in (14.11) leads to (14.12).
2. Prove that the matrix $\mathbf{A}^T\mathbf{A} + \alpha\mathbf{I}$ is nonsingular for all $\alpha > 0$.
3. Show that (14.13) gives the correct solution of (14.12).
4. Show that, if none of the singular values are zero, then in (14.13), $\lim_{\alpha \to 0} \mathbf{x}_\alpha = \mathbf{x}$.
5. Show that minimizing the expression in (14.14) leads to (14.15). Prove this for the expressions (14.16) and (14.17) as well.

14.6 Choosing the Regularization Parameters

For Tikhonov regularization we have to make two choices: the order of the regularization and the regularization constant α. Not much formal work has been done on the first question, but there are some simple rules of thumb. Normally, when the solution can be expected to have a great deal of smoothness, second order regularization is preferred. As our examples illustrate, second order regularization tends to perform a little better than lower order regularization. However, when the exact solution can have points where its values or its derivatives change rapidly, second order regularization has a tendency to smooth the solution too much. In such cases, a lower order method can do better, although in some cases none of the orders give very good results.

The selection of the regularization constant α has received much more attention and several different proposals have been made. In practice, the selection of the best α involves trying several values. The various suggestions that have been made differ primarily in how "best" is defined, although in all cases this has to do with what we know about the problem and what we see as a good solution. As we have already pointed out, it is undesirable to place too much emphasis on the true solution of (14.1) because the various errors involved make an exact answer inaccessible in any case. If the computational and data errors are represented by a term of size ε, then it seems reasonable to consider any answer $\mathbf{x}$ that satisfies

$$||\mathbf{Ax} - \mathbf{b}|| \leq \varepsilon \tag{14.18}$$

as indistinguishable from an exact solution, and therefore *acceptable.* It is normally not too difficult to find some acceptable solution, but in ill-conditioned problems we want to pick one that somehow embodies our idea of a reasonable solution. This idea is incorporated in the regularization matrix $\mathbf{B}$ which is chosen so that $||\mathbf{Bx}||$ is large for all the undesirable $\mathbf{x}$. We formalize this by saying that an $\mathbf{x}$ is a *plausible* solution only if

$$||\mathbf{Bx}|| \leq M, \tag{14.19}$$

where M is some chosen positive number. Our task then is to select α so that the regularized solution $\mathbf{x}_\alpha$ is both acceptable and plausible.

The requirement that a solution be both acceptable and plausible is generally not enough to yield a unique answer, so we still have some choices. One choice is to minimize the residual

$$\rho(\mathbf{x}) = ||\mathbf{Ax} - \mathbf{b}||, \tag{14.20}$$

subject to

$$||\mathbf{Bx}|| \leq M. \tag{14.21}$$

This is a constrained optimization problem whose solution is the *best plausible* solution. Alternatively, we can try to find $\mathbf{x}$ that minimizes the functional

$$\Phi_B = ||\mathbf{Bx}||, \tag{14.22}$$

subject to

$$||\mathbf{Ax} - \mathbf{b}|| \leq \varepsilon. \tag{14.23}$$

We call this the *most plausible acceptable* solution.

When we use the 2-norm, the solution of the optimization problems can be achieved with standard minimization techniques. Generally, the extremum of the functional to be minimized is unique and occurs on the boundary defined by the constraints. In this case we can use the method of Lagrange multipliers to find the solution. The minimization of the residual in (14.20) subject to (14.21) can be replaced by finding an α and an $\mathbf{x}_\alpha$ such that

$$(\mathbf{A}^T\mathbf{A} + \alpha\mathbf{B}^T\mathbf{B})\mathbf{x}_\alpha = \mathbf{A}^T\mathbf{b}, \tag{14.24}$$

with

$$||\mathbf{Bx}_\alpha|| = M. \tag{14.25}$$

The minimization problem (14.22) and (14.23) is equivalent to solving

$$(\mathbf{A}^T\mathbf{A} + \alpha\mathbf{B}^T\mathbf{B})\mathbf{x}_\alpha = \mathbf{A}^T\mathbf{b}, \tag{14.26}$$

subject to

$$||\mathbf{A}\mathbf{x}_\alpha - \mathbf{y}|| = \varepsilon. \tag{14.27}$$

In both cases we have a one-parameter root-finding problem. If we have a value for M, we solve equation (14.24) repeatedly with various values of the regularization constant until we find a value of α such that (14.25) is satisfied. In (14.26) and (14.27) we do the same, starting with a given value of ε. Since in practice, it seems a little easier to get a reasonable value for ε than for M, the second method is often preferred. Selecting α so that (14.27) is satisfied is referred to as the *Morozov discrepancy principle.*

Example 14.10 In this example, we solved the equation with $\mathbf{A}$ chosen as the 20×20 Hilbert matrix. For $\mathbf{b}$ we used $\mathbf{b} = \mathbf{b}_0 + \Delta\mathbf{b}$, with $\Delta\mathbf{b}$ a perturbation whose elements are random numbers normally distributed with mean zero and variance 10^{-5}. The $\mathbf{b}_0$ was taken so that the unperturbed equation has the solution

$$[\mathbf{x}]_i = \sin\left(\frac{i-1}{19}\pi\right), \qquad i = 1, 2, \ldots, 20.$$

The approximate solution $\hat{\mathbf{x}}_\alpha$ was computed by second regularization with the perturbed right side. Using the assumption that $\varepsilon = 5 \times 10^{-5}$ and $M = 0.1$, (14.20) and (14.21) gave a value of $\alpha \cong 4 \times 10^{-5}$. On the other hand, (14.22) and (14.23) suggested a regularization constant approximately 5×10^{-9}. This rather large discrepancy may seem disturbing, but in this case there is not very much difference in the result. The two computed solutions are shown in Figure 14.9. In this example, the regularization constant can range over several orders of magnitude without significantly changing the results. Unfortunately, this is not always so. ■

In the methods above we assumed that we can assign reasonable values to ε and M, but in practice this may be difficult, especially for M. A number of strategies have been devised to not require that we have reasonable values for these quantities. A simple way of choosing the regularization parameter is to compute the results for various α and inspect the results. We know what to expect: as α decreases, the residual of $\mathbf{x}_\alpha$ decreases while $||\mathbf{B}\mathbf{x}_\alpha||$ increases. If we find an α for which the residual is sufficiently small and which satisfies our intuitive idea of a plausible solution, we can accept that solution. In many cases the behavior of $||\mathbf{A}\mathbf{x}_\alpha - \mathbf{b}\,||$ gives a good indication of the proper α. The residual will often decrease steadily with α until some critical value α_0, after which further decreases in α have little effect on the

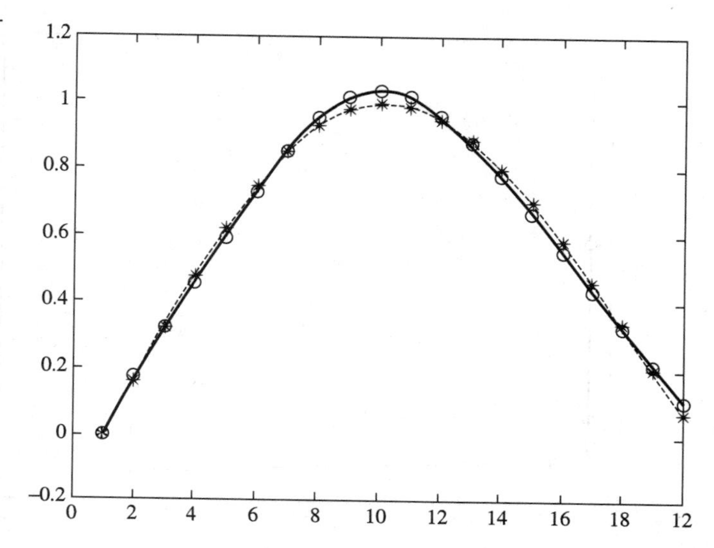

Figure 14.9 Solution of Example 14.10 with different regularization constants. The line marked with o represents $\alpha = 5 \times 10^{-9}$, the line marked * is for $\alpha = 4 \times 10^{-5}$.

residual. At the same time, decreasing α will simply make $||\mathbf{Bx}_\alpha||$ larger; that is, reduce the plausibility of the solution. It would seem then that the critical value α_0 is the proper choice for the regularization parameter.

Example 14.11 With $\mathbf{A}$ and $\mathbf{b}$, as in Example 14.10, the values of $||\mathbf{Ax}_\alpha - \mathbf{b}||$ and $||\mathbf{Bx}_\alpha||$ were computed for a range of α. The results are shown in Table 14.5, where the predicted behavior shows up very clearly. Around $\alpha = 10^{-6}$ the norm of the residual settles down and further decreases do not help much. At the same time, $||\mathbf{Bx}_\alpha||$ starts to increase significantly. Based on this, we might select $\alpha = 10^{-6}$.

Table 14.5 Effect of α on second order regularization in Example 14.11.

α	$\|\|\mathbf{Ax}_\alpha - \mathbf{b}\|\|_2$	$\|\|\mathbf{Bx}\|\|_2$
10^{-2}	1.3×10^{-3}	7.5×10^{-2}
10^{-3}	4.8×10^{-4}	7.8×10^{-2}
10^{-4}	9.2×10^{-5}	8.1×10^{-2}
10^{-5}	4.8×10^{-5}	8.2×10^{-2}
10^{-6}	3.9×10^{-5}	8.3×10^{-2}
10^{-7}	3.8×10^{-5}	8.4×10^{-2}
10^{-8}	3.8×10^{-5}	9.1×10^{-2}
10^{-9}	3.8×10^{-5}	1.1×10^{-1}
10^{-10}	3.7×10^{-5}	3.0×10^{-1}
10^{-11}	3.7×10^{-5}	7.8×10^{-1}

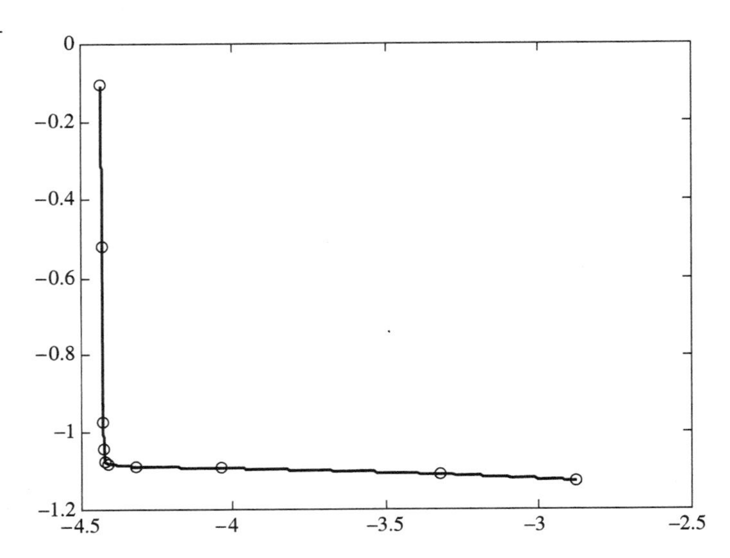

Figure 14.10
A plot of $log_{10}(||\mathbf{Ax}_\alpha - \mathbf{b}||_2)$ vs. $log_{10}(||\mathbf{Bx}||_2)$ for Example 14.11.

It turns out that this critical value is most easily identified if we plot $||\mathbf{Ax}_\alpha - \mathbf{b}||$ against $||\mathbf{Bx}_\alpha||$ on a logarithmic scale. Often the result is an L-shaped curve with a very sharp corner at the critical value of the regularization parameter. This is the so-called *L-method* for the selection of the regularization parameter. For this example we can see from Figure 14.10 that the plot is indeed L-shaped with a sharp corner between $\alpha = 10^{-6}$ and $\alpha = 10^{-7}$. A word of caution is in order: Such clear-cut results as in Figure 14.10 are not always achievable and small changes can affect the L-curve method considerably.

■

In addition to these methods for selecting the regularization constants, there are several effective strategies based on statistical reasoning. None of them are elementary, and they require some knowledge of probability and statistics. For this reason they are beyond the scope of this book.

EXERCISES

1. Show how the minimization problem (14.20) and (14.21) is solved by (14.24) and (14.25).
2. Show why the choice of α by (14.22) implies (14.26) and (14.27).

14.7 The Numerical Solution of Integral Equations of the First Kind

The most widely studied ill-posed problem is the integral equation

$$\int_a^b k(s,\, t)x(t)dt = y(s). \tag{14.28}$$

The function $k(s,\, t)$, the *kernel* of the equation, is given, as is the right side $y(s)$. The unknown function $x(t)$ is to be determined for $a \le t \le b$. The equation is called a *Fredholm equation of the first kind.* As illustrated by Example 14.5, many inverse problems lead to such equations.

It is not hard to see that Fredholm equations of the first kind could be ill-posed. Take for example the simple case $k(s,\, t) = 1$, $x(t) = \cos(nt)$, and $a = 0,\ b = 1$. Since

$$\int_0^1 \cos(nt)dt = \frac{1}{n}\sin(1),$$

an $x(t)$ of order unity produces a $y(s)$ of order $1/n$. Because this is true for arbitrarily large n, an appreciable change on the left of (14.28) can create an imperceptible effect on the right. The converse implies that a small change on the right side may produce a large change in the solution. This behavior holds for any continuous kernel, so that all equations of type (14.28) with continuous kernels are ill-posed.

Numerical methods for Fredholm integral equations seem quite easy to invent. For the simplest algorithm we replace the integral with a numerical quadrature on the points $t_1,\ t_2,\ \ldots,\ t_n$ with weights $w_1,\ w_2,\ \ldots,\ w_n$ to get

$$y(s) \cong \sum_{j=1}^{n} w_j k(s,\, t_j)x(t_j). \tag{14.29}$$

To make this into a solvable finite system, we satisfy the equation exactly at the quadrature points, giving the system

$$y(t_i) = \sum_{j=1}^{n} w_j k(t_i,\, t_j)x_j\,,\ \ i = 1,\, 2,\, \ldots,\, n, \tag{14.30}$$

for the unknowns $x_1,\ x_2,\ \ldots,\ x_n$. It seems reasonable to expect that x_j will be a good approximation to $x(t_j)$. But unfortunately, because (14.28) is ill-posed, there are some serious difficulties.

Table 14.6
Results for Equation (14.31) by the trapezoidal method.

t	$n = 5$	$n = 9$	$n = 17$
0.000	0.5602	0.5306	−47.3089
0.125		1.7584	−316.2421
0.250	1.5182	−0.5601	102.0745
0.375		3.4871	49.3657
0.500	0.4213	−1.8604	−8.5007
0.625		3.3715	−24.4726
0.750	1.4807	−0.4432	−33.7697
0.875		1.7075	30.1161
1.0000	0.5991	0.5480	3.0331

Example 14.12 As is easily verified, the integral equation

$$\int_0^1 e^{(s+1)t} x(t)dt = \frac{e^{s+1} - 1}{s+1} \tag{14.31}$$

has a solution $x(t) = 1$. In an attempt to solve this equation numerically we used (14.30) with an n-point composite trapezoidal rule with uniformly spaced quadrature points. Results for several n are shown in Table 14.6.

The results in Table 14.6 are not encouraging. While for small n the computed solution has some resemblance to the true one, the error increases rapidly with n, and we cannot get anything that approximates the solution with acceptable accuracy. The reason for this is quite clear when we look at the condition numbers for the matrices in (14.30). As n gets larger, we expect that the discretization error—that is, the accuracy with which (14.30) represents (14.28)—will get smaller. Unfortunately, the condition numbers grow much more rapidly, destroying any gain from the more accurate discretization. Table 14.7 shows the condition numbers for some n; it is quite obvious from these that any computations with n larger than eight are quite meaningless.

■

Table 14.7
Condition numbers for Example 14.12.

n	Condition number
5	1.2×10^6
8	4.3×10^{12}
10	4.5×10^{18}

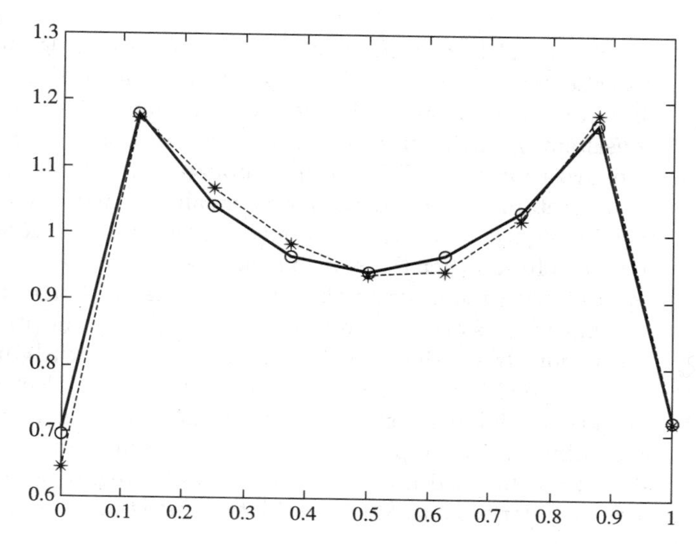

Figure 14.11 Regularization of order zero in Example 14.13. $\alpha = 10^{-12}$ gives the solid line, $\alpha = 10^{-8}$ gives the dashed line.

Generally, when we attempt to solve ill-posed problems in this way the matrices we get are inherently ill-conditioned, and the condition numbers grow very rapidly. We must therefore use some kind of regularization if we are to produce reasonable approximations. There are two avenues of approach. The first is to regularize (14.28) to produce a well-posed equation which is then discretized.[2] This has many theoretical advantages, but poses some problems. The more obvious approach is to discretize first, and then apply regularization to the resulting finite systems, so that we can apply all the regularization methods we have discussed. Some care has to be taken, though, since not all regularization methods produce satisfactory results.

Example 14.13 Equation (14.31) was discretized, using the trapezoidal approximation with $n = 9$. The resulting system was then solved by Tikhonov regularization with order zero and $\alpha = 10^{-8}$ and $\alpha = 10^{-12}$. The results are shown in Figure 14.11. While better than the unregularized solution, the answers are still not particularly satisfying. The graphs for $\alpha = 10^{-8}$ and $\alpha = 10^{-12}$are quite close (relative to what we can expect for ill-posed problems), so we might be tempted to accept them. But a comparison with the true value indicates a rather large error, particularly near the end points of the interval. ■

[2] How to regularize a continuous equation such as (14.28) is not entirely apparent from what we have said here. It can be done by formally adapting the matrix methods to continuous systems. We do not pursue this matter as it is unimportant for our discussion.

Figure 14.11 can be understood by remembering the interpretation of order zero regularization. Equation (14.11) tells us that solutions with large $||\mathbf{x}||$ are penalized, but this does not preclude rapid jumps. The results are consistent with this interpretation. We can also view order zero Tikhonov regularization as an SVD with damping that reduces the effect of the singular vectors associated with very small singular values. Since these are usually highly oscillatory, damping removes their undesired effect. Getting a good solution then depends on the smoothness of the singular vectors associated with the large singular values. It is sometimes assumed that these singular vectors do not have rapidly varying components, but unfortunately this is not always the case. It can be shown that with quadrature methods of the type (14.30), some of the singular vectors reflect the patterns in the weights, explaining why the solution has about half its true value at the end points. Such simple examples indicate that regularization of order zero should not be used for solving discretized integral equations. Since higher order regularization reduces oscillations, it is more promising.

Example 14.14 The equation discretized with $n = 9$ was solved, using order one regularization and $\alpha = 10^{-8}$. The results in Table 14.8 show much improvement. ■

The errors in the last example are still quite large and stem in part from a substantial discretization error. By increasing n, we can get better results.

Table 14.8 Results for Example 14.12 with $n = 9$, regularization of order one, and $\alpha = 10^{-8}$.

t_i	x_i
0.000	1.0067
0.125	0.9841
0.250	0.9672
0.375	0.9816
0.500	1.0237
0.625	1.0638
0.750	1.0580
0.875	0.9796
1.000	0.8775

Table 14.9 Solution for Example 14.12 with $n = 17$.

t_i	x_i
0.000	1.0074
0.125	0.9983
0.250	0.9882
0.375	0.9901
0.500	1.0046
0.625	1.0203
0.750	1.0199
0.875	0.9930
1.000	0.9632

Example 14.15 When the computations in Example 14.14 are repeated with $n = 17$, we get the improved results in Table 14.9.

From Table 14.9, it looks like we have a convergent process: as n increases the results improve. By increasing n or using more accurate quadratures, the approximation can be improved further. ■

It is possible to justify theoretically that, in the absence of computational errors and with proper regularization, a convergent discretization error leads to a convergent solution. This result, however, is of less significance than might appear. First, rounding errors can never be avoided and are always magnified, so there is a limit, several orders of magnitude above rounding, to the achievable accuracy even in the most idealized problems. More importantly, ill-posed problems occur in experimental situations where there is a significant observational error. Reducing the discretization much below the errors in the observations is of little use. The kinds of accuracy that were observed in Table 14.9 are typical. In most real situations an error of 1 percent is considered remarkable.

14.8 Conclusions

The few examples we have presented here give a good indication of the problems we face when we solve ill-posed and inverse problems. Because of the severe ill-conditioning, systems always have to be stabilized by some form of regularization. In regularizing we are often faced with choices that may significantly affect the answers. Reliable answers should be expected only if we have enough outside information about the expected solution.

Otherwise, even though we get a reasonable answer, we may not be sure that there are not other solutions that are equally satisfactory.

Writing and debugging software for ill-posed equations is also more challenging than writing software for more routine well-posed problems. Sometimes the answers are very sensitive to small changes in the program and the data, making it hard to judge the correctness of an implementation. While there are supporting theoretical results, such as convergence theorems, they are rarely useful in practice.

The overall conclusion is that one has to be extremely careful when dealing with ill-posed problems and treat all numerical results with some degree of skepticism.

Chapter 15

Numerical Computing in MATLAB

In this chapter we describe briefly some of the functions of MATLAB that are useful in numerical computations.[1] MATLAB is a very complicated language, designed primarily for scientific computation, so it contains many numerical algorithms. We make no attempt to provide an exhaustive list of all of these features, but instead limit ourselves to those methods that were described in the text and that were used in the examples. For a more thorough treatment, you need to refer to texts specifically aimed at MATLAB, such as [9].

15.1 MATLAB Arithmetic and Standard Functions

Standard implementations of MATLAB use the IEEE double format, so we normally work with a 15-digit accuracy. There are various ways we can look at the internal accuracy of MATLAB. The system variable `eps` defines the value of an ulp, so with IEEE double format the value is

```
eps= 2.2204e-016.
```

[1]The next three chapters were developed and tested with MATLAB, Version 4.1. There are changes in later versions that make some of the examples give slightly different answers. This is particularly so in the solution of ill-posed problems.

The range of representable numbers is as discussed in Chapter 2. A check on this is available via the variables `realmax` and `realmin`. The first of these is the largest possible number, and the second is the smallest positive number that can be represented with the convention decribed in Section 2.2.[2]

As part of its syntax and semantics, MATLAB provides for exceptional values. Positive infinity is represented by `Inf`, negative infinity by `-Inf`, and not-a-number by `NaN`. These exceptional values are carried through the computations in a logically consistent way. For example, the expression `(1.0e300)^2` gives `Inf`, while `Inf- Inf` yields `NaN`.

MATLAB actually lets us look at the exact bit pattern stored for a number. If we type

```
format hex
```

all subsequent output shows the internal bit patterns in hexadecimal form. With this we can examine how numbers, including exceptional values, are stored. We can also follow how arithmetic is done because we see exactly what happens on each step, without the additional errors that occur when a binary number is converted to decimal for printout.

Although internally MATLAB stores all numbers to full accuracy, it uses a default format of five significant digits in either fixed-point or floating-point format to display a decimal output. By typing

```
format long
```

we can switch the output display to fixed-point format with 15 significant digits. Similarly, MATLAB will switch the display to floating-point format with 15 significant digits if we type

```
format long e.
```

A complex number in MATLAB is represented by

```
a+ bi.
```

Complex arithmetic has the same syntax as real arithmetic. Most of the matrix, polynomial, root-finding, minimization, and eigensolution functions in MATLAB that will be discussed in the next few sections work on complex inputs. In addition, there are three simple but useful complex functions, `real(z)`, `imag(z)`, and `conj(z)`, corresponding to the real part, imaginary part, and conjugate of complex number z.

[2]For an elaboration on this point, see Exploration 3 in Section 17.1.

MATLAB provides for a large number of elementary and special functions. These include

- trigonometric functions and their inverses
- hyperbolic functions and their inverses
- logarithmic and exponential functions
- special functions, such as Bessel and Gamma functions.

These functions appear to be carefully implemented and are claimed to be accurate to within a few `eps`. We are, however, given little information on how these functions are constructed, so it is hard to verify the accuracy. (For an indirect way of doing this, see for example Explorations 5 through 9 in Section 17.1).

In testing numerical software it is often very convenient to have random number generators.[3] The command

```
a=rand(m,n)
```

generates an $m \times n$ matrix of random numbers, distributed uniformly in $[0,\ 1]$, while

```
a=randn(m,n)
```

produces a matrix of random numbers distributed normally with mean 0 and variance 1.

While the numbers produced by `rand` and `randn` have properties of random numbers, they are reproducible. Whenever you start a session, the sequence of numbers generated is the same. This allows you to reproduce results even when they have random numbers in them. This can be done even within a session by using the "seed" feature (a seed defines the starting value for a pseudo-random sequence). The statement

```
n=rand('seed')
```

saves the current seed, while

```
rand('seed',n)
```

sets the seed to n and restarts the sequence. Every seed value gives a different sequence.

[3] Computer-generated random numbers are rarely truly random, but are generated by deterministic algorithms. However, the generated sequences have statistical distributions that make them essentially indistinguishable from random sequences.

Example 15.1 The instructions

```
rand
rand
```

will produce two different random numbers, but

```
n=rand('seed');
rand
rand('seed',n);
rand
```

will give two identical results. ■

15.2 Solving Linear Systems

MATLAB started as a linear algebra extension of Fortran. Since its early days, MATLAB has been extended far beyond its initial purpose, but linear algebra methods are still one of its strongest features.

The linear system

$$\mathbf{Ax} = \mathbf{b}$$

can be solved by

```
x=A\b.
```

When $\mathbf{A}$ is a nonsingular matrix, this will produce the solution by GEM. If $\mathbf{A}$ is not square or is singular, $\mathbf{x}$ will be a least squares solution.[4]

The inverse of a nonsingular matrix $\mathbf{A}$ is obtained by

```
Ainv=inv(A)
```

and its determinant by

```
dtA=det(A).
```

[4]If the least squares solution is not unique, this will produce just one of them. For more details, see Section 8.3.

The rank of a matrix **A** can be obtained by

```
rk=rank(A).
```

MATLAB has provisions for computing various norms. The expression

```
norm(A,2)
```

gives the Euclidean norm of **A** while

```
norm(A, Inf)
```

gives the maximum norm. Here **A** can be a vector or a matrix. The 1-norm (see Exercise 8, Section 3.5) of a vector or a matrix can be obtained by

```
norm(A,1).
```

The condition number of a matrix **A** is computed by `cond(A)`. This is equivalent to `norm(A,2)*norm(inv(A),2)`.

15.3 Polynomials and Polynomial Approximations

MATLAB provides some tools for working with polynomials. Polynomials are given by their coefficients and stored as vectors, so that

$$p(x) = a_n x^n + \ldots + a_1 x + a_0$$

can be represented as

```
p=[an,...,a1,a0].
```

Addition and subtraction of polynomials p and q can be done by vector addition and subtraction, but if the polynomials are not of the same degree, one of them has to be "padded" with zeros to make the two vectors the same length. Polynomial multiplication and division is actually more straightforward. The statement

```
prod=conv(p,q)
```

produces the product of polynomials p and q, while

```
[qot,rem]=deconv(p,q)
```

gives the quotient and remainder when p is divided by q. The function `polyval` evaluates polynomials at a specified set of points:

```
v=poyval(p,x)
```

gives a vector `v` of the same length as `x`, so that `v(i)` is `p` evaluated at `x(i)`.

Example 15.2 Let $p = x^3 - 2x + 1$ and $q = 2x^2 + x - 6$. Plot the product of p and q in the interval $[0,\ 1]$.

Solution:

```
p=[1 0 -2 1]; q=[2 1 -6];
x=[0:0.01:1];
plot(x,polyval(conv(p,q), x)).
```

■

Approximation by polynomials can be done with the function `polyfit`. If `x` and `y` are two vectors of length `n`, then

```
p=polyfit(x,y,n-1)
```

is the polynomial of degree `n` $-$ 1 that passes through all the $x - y$ data points. If `m` < `n` $-$ 1, then

```
p=polyfit(x,y,m)
```

produces a polynomial of degree m that gives the best fit of the data in the least squares sense.

15.4 Interpolation and Data Fitting

The main MATLAB function for interpolation and data fitting in one dimension is `interp1`. This function can be called by a statement of the form

```
intfc=interp1(x,y,z,method).
```

Here `x` and `y` are two vectors both of length n, specifying the x and y coordinates of the data points. The variable `method` can have one of three string values and determines how the interpolating function is constructed.

(a) `method='linear'`. The interpolating function is piecewise linear with knots at the data points.

(b) `method='spline'`. The interpolating function is a cubic spline with a special end condition

$$y'''(x_1) = y'''(x_2) \text{ and } y'''(x_{n-1}) = y'''(x_n).$$

(c) `method='cubic'`. The values of the interpolating function at any point z will be constructed by cubic interpolation, using the four points closest to z.

The variable `intfc` is a vector whose elements are the values of the interpolating function at the points specified by `z`.

Note that all three approximations are strictly interpolating; that is the approximation passes through all given data points.

MATLAB also has Fourier interpolation functions. The statement

```
z=interpft(y,m)
```

produces a trigonometric interpolation on the data set `y`. For the somewhat complicated rules relating `z`, `y`, and `m`, refer to a MATLAB manual or the HELP section in the MATLAB program.

Two-dimensional interpolation can be done by

```
zz=interp2(x,y,z,xx,yy).
```

The variables `x`, `y`, `z` are the data set, with `x` and `y` defining a two-dimensional grid at which the corresponding `z` values are given. The variables `xx` and `yy` define another grid at which a two dimensional linear interpolation is used to evaluate the elements of `zz`.

15.5 Numerical Integration

There are several numerical quadrature functions in MATLAB. The simplest is trapz, which is an implementation of a composite trapezoidal method with nonuniform panels.

The command

```
S=trapz(x,y)
```

simply computes

$$S=\tfrac{1}{2}(x(2)-x(1))*(y(1)+y(2))+\ \tfrac{1}{2}(x(3)-x(2))*(y(2)+y(3))+\ \ldots\ +\ \tfrac{1}{2}(x(n)-x(n-1))*(y(n-1)+y(n)).$$

This gives the user control over panel selection that is sometimes useful.

Example 15.3 The poor behavior of the integrand near $x = 0$ in the integral

$$I = \int_0^1 \sqrt{x}e^{-x}dx$$

suggests that we split the integral

$$I = \int_0^{0.1} \sqrt{x}e^{-x}dx + \int_{0.1}^1 \sqrt{x}e^{-x}dx.$$

We can then use a small panel size, say 0.01, on the first part and a larger panel size 0.1 on the second part. In MATLAB this can be done with

```
x=[[0:0.01:0.1],[0.2:0.1:1]];
I=trapz(x,sqrt(x).*exp(-x)).
```

■

Two more sophisticated MATLAB functions for the numerical integration over a finite line segment are `quad` and `quad8`. Both are adaptive[5] algorithms based on composite Newton-Cotes formulas. The first uses Simpson's rule, the second is an implementation of a more complicated higher order method. The most general usage is through

```
quad(fname,a,b,tol,trace)
```

and

```
quad8(fname,a,b,tol,trace),
```

respectively. Here `fname` is the character string giving the name of the function to be integrated. This function must be stored as an M-file. The input values `a,b` specify the limits of integration, and `tol` sets the maximum permissible relative error. The parameter `trace` gives us an opportunity to look into the inner workings of the adaptive algorithm. When `trace` is zero, the function just returns the approximate value of the integral. When `trace` is not zero, the function also prints a graph showing the location of the quadrature points. On the whole, the performance of `quad8` is better than that of `quad`.

Example 15.4 The integral in Example 15.3 can be evaluated by

```
I=quad8('g',0,1,1e-3,0)
```

[5] In later versions of MATLAB, quad8 has been replaced by quadl, a Lobatto-type quadrature.

with M-file

```
function res=g(x)
  res=sqrt(x).*exp(-x);
```

It produces answer

```
I =
  0.3789
```

Since the fourth parameter is set to zero, it does not give a trace. ■

15.6 Root-Finding

The function `fzero` is MATLAB's main tool for finding real roots in one dimension. It can be invoked by

```
fzero(fname,x,tol,trace).
```

Here `fname` is a character string giving the name of function whose root is to be found. The function is to be defined as an M-file. `x` is an initial guess, `tol` is the error tolerance for accuracy, and `trace` allows the user to follow the iterations. If `trace` is set to a nonzero value, successive iterates will be displayed.

The result is only one root, even though there may be many others. MATLAB uses an algorithm somewhat like that described in Section 7.8. To bracket the root, it first searches the real line for a change of sign. MATLAB does not recognize roots of even multiplicity and may go into a loop, vainly trying to find a change of sign. Also, failure to define the function `fname` for all values of the independent variable can cause problems. In fact, it is quite easy to "confuse" `fzero` and have it give incorrect answers. This is not a criticism of `fzero`, but a reflection of the difficulty one encounters when one constructs software for nonlinear problems.

MATLAB also provides a special function `roots` to find all the roots of polynomials. This function employs the companion matrix method and an eigenvalue solver. Because the eigenvalues of a matrix can be complex, roots delivers complex results. To compute all the `roots` of a polynomial, we simply write

```
roots(coef)
```

where `coef` is the vector of the polynomial coefficients, using the normal MATLAB convention for polynomials.

There is no multidimensional root-finder in standard MATLAB, but nonlinear systems can be solved by minimization. The function call

```
fmins(fname,x)
```

results in a local minimization of the function `fname`, starting at `x`. The algorithm employs a simplex searching strategy.

Example 15.5 To find a solution to the nonlinear system

$$x_1^2 + x_2^3 = 9$$
$$x_1 + x_2^4 = 17$$

we define

$$\Phi(x_1, x_2) = (x_1^2 + x_2^3 - 9)^2 + (x_1 + x_2^4 - 17)^2$$

and implement the latter function as the M-file

```
function val=phi(x);
  val=(x(1).^2+x(2).^3-9).^2+(x(1)+x(2).^4-17).^2;
```

Then the statement

```
fmins('phi',[0 0])
```

gives the answer $x_1 = 1$, $x_2 = 2$. By substitution into the original system, we can verify that this is indeed a solution. Unfortunately, not all initial guesses work out as satisfactorily. ■

15.7 Matrix Functions in MATLAB

MATLAB provides functions for all the methods discussed in Chapter 8. The LU decomposition of a matrix **A** can be done with

```
[L,U,P]=lu(A)
```

where `L`, `U`, and `P` are the matrices in Theorem 8.1. The QR decomposition is carried out by

```
[Q,R]=qr(A).
```

For symmetric positive-definite matrices a Cholesky factorization is achieved by

```
R=chol(A)
```

where `R` is an upper triangular matrix equal to $\mathbf{L}^T$ in Definition 8.1. MATLAB recognizes cases not suitable for the Cholesky method and issues an error message.

For singular value decomposition we write

```
[Q,S,P]=svd(A)
```

which gives the matrices in (8.22).

Normally, the command

```
x=A\b
```

solves the linear system

$$\mathbf{Ax} = \mathbf{b}$$

by the GEM as mentioned in Section 15.2. But MATLAB recognizes special circumstances. When a matrix is symmetric, MATLAB will try the Cholesky method. If the matrix is not positive-definite, this will fail and MATLAB reverts to the GEM. If the matrix is tridiagonal, the fast method of (8.31) will be used.

As was also mentioned in Section 15.2, when $\mathbf{A}$ is not square the command

```
x=A\b
```

produces the least squares solution, computed by a QR decomposition. But this works only when the least squares solution is unique. When this is not the case, MATLAB tells us that the matrix is singular. In this case, we can get the minimum norm least squares solution by

```
x=pinv(A)*b.
```

Here `pinv(A)` stands for the pseudo-inverse of `A` introduced in Exercise 8, Section 8.3.

In computing the pseudo-inverse, a decision has to be made when a small singular value is to be considered zero. When used in the above form, MATLAB will use some default criterion, but it also gives the user control over this. In

```
pinv(A,tol)
```

all singular values whose magnitude is smaller than `tol` will be set to zero. A similar approach is used in `rank`, which determines the rank of a matrix, using singular value decomposition. In

```
rank(A,tol)
```

the rank is computed by setting all singular values below `tol` to zero. If the second parameter is omitted, MATLAB uses its default value.

MATLAB also provides support for sparse matrices. The function `sparse` converts a matrix in row-column form into sparse format, squeezing out the zeros and representing the result as a list. Its inverse is `full`, which translates a sparse matrix back into a two-dimensional array form. There are several reordering methods available. The Reverse Cuthill-McKee method, referred to in Section 8.4, is done by

```
p=symrcm(A)
```

which yields a permutation that then can be applied with

```
B=A(p,p)
```

to give a matrix `B` with a reduced bandwidth. To see the structure of a sparse matrix, we write

```
spy(A).
```

This gives a picture of the location of the nonzero elements of `A` via an xy plot.

Matrix arithmetic and most matrix functions are available for sparse matrices and are often done quite efficiently. Full and sparse matrices can be mixed in expressions and normally yield full results.

15.8 Software for Eigensolutions

The main MATLAB function for computing eigenvalues and eigenvectors is `eig`. The implementation of `eig` uses an algorithm based on the ideas described in Section 9.4, although there are quite a few fine points that we have not brought up. The command

```
eig(A)
```

gives a vector with the eigenvalues of `A`, while

```
[evl,evc]=eig(A)
```

delivers the eigenvalues as a diagonal matrix `evl` and the eigenvectors as the columns of `evc`. This will work for nonsymmetric as well as symmetric matrices. When the eigensolutions are complex, `eig` will give complex results.

It is also possible to compute the characteristic polynomial by

```
poly(A).
```

For any square matrix, `poly` returns that characteristic polynomial in MATLAB form. The eigenvalues of `A` can then be found by

```
roots(poly(A)),
```

but this is not very efficient because both `roots` and `poly` use eigenvalue programs to get their results.

MATLAB also has a number of eigenvalue functions for problems not described here. For example, `eig` can be used to solve the general eigenvalue problem (9.2).

MATLAB eigenvalue functions are well implemented and generally give reliable results. On the negative side they are implemented as "black boxes" that give the user little control or insight into their inner workings.

15.9 Software for ODE Initial Value Problems

MATLAB provides two functions for the numerical solution of initial value problems: `ode23` and `ode45`. Both use two different Runge-Kutta-Fehlberg methods for error estimation and control. The first uses a second and a third order method, while the second uses a fourth and fifth order one. Both are adaptive; that is, they estimate the error at each step and adjust the mesh in an attempt to satisfy the stated accuracy criterion. The commands

```
[x,y]=ode23(fname,x0,xf,y0,tol)
```

and

```
[x,y]=ode45(fname,x0,xf,y0,tol)
```

are used to call these two functions. `fname` is the function on the right side of (10.1), `x0` and `xf` are the starting and ending values of x, respectively, `y0` specifies the initial condition, `tol` specifies the absolute error that the function tries to meet, `x` is the set of points at which the approximation is evaluated, and `y` is the set of corresponding approximation values.

The dimensions of the vectors `y0` and the results delivered by `f` must match. For details, see a MATLAB manual.

Example 15.6 The solution of the system

$$y_1'(x) = \cos(y_1(x)) + y_2(x) - 1,$$
$$y_2'(x) = -y_1(x) + \sin(y(x)),$$

with $y_1(0) = 1$, $y_2(0) = 0$, in the interval $(0, 2)$, with an accuracy of 10^{-6} can be obtained and plotted by

```
[x,y]=ode45('f',0,2,[1 0],1e-6);
plot(x,y)
```

with M-file

```
function res=f(x,y)
  res(1)=cos(y(1))+y(2)-1;
  res(2)=-y(1)+sin(y(2));
```

Chapter 16

NASOFT User Manual and Example Scripts

NASOFT is a collection of functions that complements the standard MATLAB software for numerical computation. One purpose of NASOFT is to extend the capability of MATLAB for the solution of a wider variety of problems. A second purpose is to allow the student to experiment with algorithms and to investigate their efficiency. Finally, because all NASOFT code is written in MATLAB and is available as M-files, it provides for experience in modifying existing programs to suit different purposes.

16.1 Piecewise Polynomial Approximations

hat

Purpose

To compute the hat function basis $H_i(x)$ for a piecewise linear approximation, as defined by Equation (5.10).

Usage

```
y=hat(mesh,x)
```

where

> `mesh` = three element vector `[x1,x2,x3]`, representing the mesh points x_{i-1}, x_i, x_{i+1}, respectively in (5.10),
>
> `x` = vector of points at which `hat` is evaluated,
>
> `y` = result vector of the same size as `x`.

Conditions and Restrictions

The inequality `x1<x2<x3` must hold.

quad_q

Purpose

To compute the basis functions $Q_i(x)$ for a piecewise quadratic approximation, as defined by Equation (5.12).

Usage

```
y=quad_q(mesh,x)
```

where

> `mesh` = two element vector `[x1,x2]`, representing the mesh points x_i, x_{i+1}, respectively in (5.12),
>
> `x` = vector of points at which `quad_q` is evaluated,
>
> `y` = result vector of the same size as `x`.

Conditions and Restrictions

The strict inequality `x1<x2` must hold.

quad_r

Purpose

To compute the basis functions $R_i(x)$ for a piecewise quadratic approximation, as defined by Equation (5.13).

Usage

```
y=quad_r(mesh,x)
```

where

> `mesh` = three element vector `[x1,x2,x3]`, representing the mesh-points x_{i-1}, x_i, x_{i+1}, respectively in (5.13),
>
> `x` = vector of points at which `quad_r` is evaluated,
>
> `y` = result vector of the same size as `x`.

Conditions and Restrictions

The strict inequality `x1<x2<x3` must hold.

Example

A function f has given values $f(1) = f1$, $f(2) = f2$, and $f(3) = f3$. Find an approximate value for $f(1.5)$ by quadratic interpolation at the three given values.

Solution:

```
f15=f1*quad_r([0,1,3],1.5)+f2*quad_q([1,3],1.5)+f3*quad_r([1,3,4],1.5)
```

spl_b3

Purpose

To compute cubic B-splines defined by Equation (5.14).

Usage

```
y=spl_b3(mesh,x)
```

where

> `mesh` = five element vector `[x1,x2,x3,x4,x5]`, representing the mesh points x_{i-2}, x_{i-1}, x_i, x_{i+1}, x_{i+2}, respectively in (5.14),
>
> `x` = vector of points at which `spl_b3` is evaluated,
>
> `y` = result vector of the same size as `x`.

Conditions and Restrictions

The strict inequality `x1<x2<x3<x4<x5` must hold.

spl_b31

Purpose

To compute the first derivative of cubic B-splines defined by Equation (5.14).

Usage

```
y=spl_b31(mesh,x)
```

where

`mesh` = five element vector `[x1,x2,x3,x4,x5]`, representing the mesh points x_{i-2}, x_{i-1}, x_i, x_{i+1}, x_{i+2}, respectively in (5.14),

`x` = vector of points at which `spl_b31` is evaluated,

`y` = result vector of the same size as `x`.

Conditions and Restrictions

The strict inequality `x1<x2<x3<x4<x5` must hold.

spl_b32

Purpose

To compute the second derivative of cubic B-splines defined by Equation (5.14).

Usage

```
y=spl_b32(mesh,x)
```

where

`mesh` = five element vector `[x1,x2,x3,x4,x5]`, representing the mesh points x_{i-2}, x_{i-1}, x_i, x_{i+1}, x_{i+2}, respectively in (5.14),

`x` = vector of points at which `spl_b32` is evaluated,

`y` = result vector of the same size as `x`.

Conditions and Restrictions

The strict inequality `x1<x2<x3<x4<x5` must hold.

16.2 Data Fitting

This group of programs is used for fitting a data set consisting of n pairs $\{(x_1, y_1), (x_2, y_2), \ldots, (x_n, y_n)\}$, using various piecewise polynomials on a knot sequence $\{k_1, k_2, \ldots, k_m\}$. Although there is generally no connection between the knots and the data points, results can be expected to be reasonable only if there are a significant number of data points in each knot interval $[k_i, k_{i+1}]$. If there are some knot intervals that contain less than a minimum number of data points (say 2 or 3), the numerical process may become unstable and the results unsatisfactory. This problem arises in particular if the knot interval is smaller than the spacing of the data set.

lsq_lin

Purpose

To compute the least squares piecewise linear approximation to a given data set.

Usage

```
y=lsq_lin(knots,xdata,ydata,x)
```

where

`knots` = vector defining the knot sequence,

`xdata` = x-coordinates of the data points,

`ydata` = y-coordinates of the data points,

`x` = vector of points at which `lsq_lin` is evaluated,

`y` = resulting approximation, evaluated at the points of vector `x`.

Conditions and Restrictions

`xdata` and `ydata` must be vectors of the same size. `knots` must be a strictly increasing sequence.

lsq_quad

Purpose

To compute the least squares piecewise quadratic approximation to a given data set.

Usage

```
y=lsq_quad(knots,xdata,ydata,x)
```

where

`knots` = vector defining the knot sequence,

`xdata` = x-coordinates of the data points,

`ydata` = y-coordinates of the data points,

`x` = vector of points at which `lsq_quad` is evaluated,

`y` = resulting approximation, evaluated at the points of vector `x`.

Conditions and Restrictions

`xdata` and `ydata` must be vectors of the same size. `knots` must be a strictly increasing sequence.

lsq_cspl

Purpose

To compute the least squares cubic spline approximation to a given data set.

Usage

```
y=lsq_cspl (knots,xdata,ydata,x)
```

where

`knots` = vector defining the knot sequence,

`xdata` = x-coordinates of the data points,

`ydata` = y-coordinates of the data points,

`x` = vector of points at which `lsq_cspl` is evaluated,

`y` = resulting approximation, evaluated at the points of vector `x`.

Conditions and Restrictions

`xdata` and `ydata` must be vectors of the same size. `knots` must be a strictly increasing sequence.

16.3 Numerical Integration

gauss

Purpose

To compute the points and weights for the n-point Gauss-Legendre quadratures over finite interval $[a,\ b]$.

Usage

```
[points, weights]=gauss(n,a,b)
```

where

$\texttt{n}$ = number of quadrature points,

$\texttt{a}$ = left end point of interval,

$\texttt{b}$ = right end point of interval,

$\texttt{points}$ = computed quadrature points,

$\texttt{weights}$ = corresponding quadrature weights.

Conditions and Restrictions

$$2 \le \texttt{n} \le 10 \text{ and } \texttt{a}<\texttt{b}.$$

quad_cg

Purpose

To compute an approximation for the definite integral

$$I = \int_{a_1}^{a_m} f(x)dx$$

using a composite Gaussian quadrature. The interval $[a_1,\ a_m]$ is subdivided into panels $[a_i,\ a_{i+1}]$ for $i = 1,\ \dots,\ m-1$ with $a_1 < a_2 < \ \dots \ < a_m$, and an n-point Gauss-Legendre rule is applied over each panel $[a_i,\ a_{i+1}]$.

Usage

```
ival=quad_cg(fname,panels,n)
```

where

$\texttt{fname}$ = character string giving the name of integrand f,

$\texttt{panels}$ = vector specifying the panels,

$\texttt{n}$ = number of quadrature points used in each panel,

$\texttt{ival}$ = approximation to the integral.

Conditions and Restrictions

fname must be a MATLAB function that accepts vectors as input and delivers a vector-valued result. The points in panels must be strictly increasing. The number of quadrature points in each panel is limited by $2 \le \texttt{n} \le 10$.

Example

Approximate

$$I = \int_1^3 \frac{\sin(x)}{1+x^2} dx$$

using a composite Gauss-Legendre rule with panels [1, 2] and [2, 3], using six points in each panel.

Solution:

```
ival=quad_cg('ifcn',[1 2 3],6)
```

with

```
function val=ifcn(x)
  val=sin(x)./(1+x.^2);
```

quad_dbl

Purpose

To compute an approximate value for the double integral

$$I = \int_a^b \int_{h(x)}^{g(x)} f(x, y) dy dx \tag{16.1}$$

to a specified tolerance.

Usage

```
ival=quad_dbl(fname,gname,hname,a,b,tol)
```

where

- `fname` = character string giving the name of f,
- `gname` = character string giving the name of g,
- `hname` = character string giving the name of h,
- `a` = lower limit of outer integration,
- `b` = upper limit of outer integration,
- `tol` = desired absolute error,
- `ival` = approximation to the integral.

Conditions and Restrictions

The function $f(x, y)$ must be implemented in a special way. First, it is to be defined only as a function of the single variable y. Second, to account for the variable x, the statement

```
global xvar;
```

must be included in the M-file for f and `xvar` used in place of x.

Example

Compute

$$I = \int_0^1 \int_0^{x^2} x^3 y dy dx$$

to an accuracy of 10^{-6}.

Solution:

```
ival=quad_dbl('fxy_func','g_func','h_func',0,1,1e-6)
```

with

```
function val=fxy_func(y)
  global xvar;
  val=xvar.^3.*y;

function val=g_func(x)
  val=0;

function val=h_func(x)
  val=x.^2;
```

quad_sim

Purpose

To compute an approximation for the definite integral

$$I = \int_{a_1}^{a_m} f(x)dx$$

using Simpson's rule. The interval $[a_1,\ a_m]$ is subdivided into intervals $[a_i,\ a_{i+1}]$ for $i = 1,\ \ldots,\ m-1$ with $a_1 < a_2 < \ \ldots \ < a_m$, and an adaptive or a nonadaptive method will be applied over each panel $[a_i,\ a_{i+1}]$.

Usage

```
[ival, ncount]=quad_sim(fname,panels,tol,nLevel)
```

where

`fname` = character string giving the name of integrand f,

`panels` = vector specifying the panels,

`tol` = if `tol` > 0, the adaptive Simpson's method will be applied over each panel and `tol` is the desired absolute error for the approximate integral. If `tol` ≤ 0, the nonadaptive Simpson's method will be applied over each panel.

`nLevel` = number of levels for interval splitting allowed in each panel when adaptive method is used,

`ival` = approximation to the integral,

`ncount` = number of function calls used to compute the approximate integral.

Conditions and Restrictions

The points in `panels` must be strictly increasing. The number of levels for interval splitting for adaptive method, `nLevel`, is an optional input parameter with a default and minimum value 50.

Example

Approximate the same integral as the example given in `quad_cg` using adaptive and nonadaptive Simpson's method over panels [1, 2] and [2, 3].

Solution:

```
Ival_non_adaptive=quad_sim('ifcn',[1 2 3],-1)
Ival_adaptive=quad_sim('ifcn',[1 2 3],1e-8,55)
```

16.4 Solving Systems of Nonlinear Equations

muller

Purpose

To carry out a single iteration of Muller's method for approximating solution of a single real or complex variable function or polynomial

$$f(z) = 0.$$

Usage

```
xout=muller(fname,xin)
```

where

fname = character string giving the name of f or vector giving the coefficients of polynomial f,

xin = vector giving the three initial guesses for the solution,

xout = the approximate solution created by a single iteration of the Muller's method.

Conditions and Restrictions

When fname is a character string giving the name of function f, it must be a complex function that accepts a single value as input and delivers a single value as its output. When fname is a vector, it represents the coefficients of a polynomial. For example, fname=[1 -6 2+3i -4] defines the third degree polynomial $f(z) = z^3 - 6z^2 + (2+3i)z - 4$. The result xout will be a real or complex value.

Example

Carry out one iteration of the Muller's method to approximate solutions of

$$z^3 - 6z^2 + (2+3i)z - 4 = 0$$

with starting guesses xin = [1 2 3]. The two equivalent statements illustrate the use of character string and vector formats for fname.

Solution:

```
z1=muller('poly_func',[1 2 3])
z2=muller([1 -6 2+3i -4],[1 2 3])
```

with

```
function fval=poly_func(z)
  fval=((z-6)*z+2+3i)*z-4;
```

newton

Purpose

To approximate a real vector solution of the system

$$\begin{aligned} f_1(x_1, x_2, \dots, x_n) &= 0, \\ f_2(x_1, x_2, \dots, x_n) &= 0, \\ &\vdots \\ f_n(x_1, x_2, \dots, x_n) &= 0, \end{aligned} \tag{16.2}$$

by carrying out several iterations of a modified Newton's method. The algorithm used is the standard Newton's method in which derivatives are approximated by finite differences. Specifically, the method uses

$$\frac{\partial f_i(x_1, x_2, \dots, x_j, \dots, x_n)}{\partial x_j} \cong \frac{f_i(x_1, x_2, \dots, x_j + h, \dots, x_n) - f_i(x_1, x_2, \dots, x_j - h, \dots, x_n)}{2h},$$

where the value of h is chosen to take into account the machine accuracy limitation ε as well as the scaling of the problem. The rule we use here is

$$h = \sqrt[3]{\varepsilon} \max(1, ||\mathbf{x}||_\infty).$$

Usage

```
x=newton(fname,xin, tol,itmax,trace)
```

where

`fname` = character string giving the name of f that defines the system,

`xin` = initial guess for the solution,

`tol` = stopping criterion. The iterations are stopped whenever the 2-norm of `x` is less than `tol`,

`itmax` = stopping criterion. The computations are stopped whenever the number of iterations that have been carried out exceeds `itmax`,

`trace` = if `trace` is zero, the only output from `newton` will be the final approximation. If `trace` is any nonzero value, all iterates will be display,

`x` = computed approximate solution.

Conditions and Restrictions

The system in (16.2) must have the same number of unknowns as equations. The function `fname` must therefore accept a vector of size n as input and deliver an output vector of the same size. `xin` must be a vector of size n and the result `x` will have the same size as `xin`.

The method fails if, after the specified number of iterations, the desired accuracy has not been achieved.

Example

Explore the solutions of

$$\begin{aligned} x_1^2 + x_2^2 + x_3^2 - 1 &= 0, \\ x_1 - x_2 + x_3 &= 0, \\ \cos(x_1) + x_2 - x_3 &= 0, \end{aligned}$$

in the region $-5 \leq x_i \leq 5,\ i = 1,\ 2,\ 3$. Lacking any useful way for finding a good starting value, we carry out repeated applications of `newton` with `random` starting points.

Solution:

```
x=newton('fn_func',10*(rand(3,1)-0.5),1e-8,20,0).
```

with

```
function fval=fn_func(x)
fval(1)=x(1)^2+x(2)^2+x(3)^2-1;
fval(2)=x(1)-x(2)+x(3);
fval(3)=cos(x(1))+x(2)-x(3);
```

16.5 Eigenvalue Problems

eigpower

Purpose

To carry out an iterative matrix computation

$$\mathbf{v}_{k+1} = \mathbf{A}\mathbf{v}_k$$

and from the computation to estimate the dominant eigenvalue and the corresponding eigenvector of $\mathbf{A}$.

Usage

```
[lambda,v]=eigpower(A,vin,k)
```

where

`A` = input square matrix,

`vin` = original guess for `v`,

`k` = number of iterations to be carried out,

`lambda` = estimated dominant eigenvalue,

`v` = estimated corresponding eigenvector.

Conditions and Restrictions

The matrix **A** must be square but need not be symmetric. However, for unsymmetric matrices the power method does not necessarily converge to a dominant eigenvalue.

jacobi

Purpose

To carry out one Jacobi rotation on a square matrix **A**.

Usage

```
A1=jacobi(A,row,col)
```

where

`A` = initial square matrix,

`row` = row number of pivot,

`col` = column number of pivot,

`A1` = resulting matrix after rotation.

Conditions and Restrictions

The matrix **A** must be square but need not be symmetric. The program will carry out a Jacobi rotation on unsymmetric matrices, but the results are generally not useful.

hholder

Purpose

To carry out a single Householder rotation on a square matrix **A**.

Usage

```
A1=hholder(A,col)
```

where

`A` = initial square matrix,

`col` = column number all of whose elements below the codiagonal will be set to zero,

`A1` = resulting matrix after the transformation.

Conditions and Restrictions

The matrix **A** must be square but need not be symmetric.

16.6 Solution of Ordinary Differential Equations—Initial Value Problems

oderk4

Purpose

To compute an approximate solution for the system of initial value problems

$$\begin{aligned} y_1'(x) &= f_1(x, y_1(x), y_2(x), \ldots, y_m(x)), \\ y_2'(x) &= f_2(x, y_1(x), y_2(x), \ldots, y_m(x)), \\ &\vdots \\ y_m'(x) &= f_m(x, y_1(x), y_2(x), \ldots, y_m(x)), \end{aligned} \tag{16.3}$$

with conditions

$$y_i(x_0) = y_{0,i}, \quad i = 1, 2, \ldots, m, \tag{16.4}$$

using the classical Runge-Kutta method (10.32) with a fixed step size h.

Usage

```
[x,y]=oderk4(fname,x0,y0,h,n)
```

where

`fname` = character string giving the name of f,

`x0` = initial x-value,

`y0` = initial y-values,

`h` = step size,

n = number of steps to be taken,

x = vector specifying the mesh points such that `x(i)=x0+ih`,

y = approximate solution at the mesh points.

Conditions and Restrictions

The input values `y0` define the size of the system. The function `fname` must give a vector of the same size as `y0`. The values of `h` and `n` must both be positive.

Example

To solve

$$
\begin{aligned}
y_1'(x) &= y_2(x) - x, \\
y_2'(x) &= y_1(x) + 1,
\end{aligned}
$$

with $y_1(0) = 0$ and $y_2(0) = 1$, in $0 \le x \le 5$ with step size $h = 0.05$, use

```
[x,y]=oderk4('odef',0,[0,1],0.05,100)
```

and defined function

```
function res=odef(x,y)
  res(1)=y(2)-x;
  res(2)=y(1)+1;
```

odepc

Purpose

To compute an approximate solution to the system (16.3), (16.4), using the predictor corrector scheme

$$
\begin{aligned}
Y_{i+4} = Y_{i+3} + \frac{h}{24}[&55f(x_{i+3}, Y_{i+3}) - 59f(x_{i+2}, Y_{i+2}) \\
&+ 37f(x_{i+1}, Y_{i+1}) - 9f(x_i, Y_i)],
\end{aligned} \tag{16.5}
$$

$$
\begin{aligned}
Y_{i+4} = Y_{i+3} + \frac{h}{24}[&9f(x_{i+4}, Y_{i+4}) + 19f(x_{i+3}, Y_{i+3}) \\
&- 5f(x_{i+2}, Y_{i+2}) + f(x_{i+1}, Y_{i+1})]
\end{aligned} \tag{16.6}
$$

with `oderk4` to get the necessary starting values.

Usage

```
[x,y]=odepc(fname,x0,y0,h,n)
```

where

fname = character string giving the name of f,

x0 = initial x-value,

y0 = initial y-values,

h = step size,

n = number of steps to be taken,

x = vector specifying the mesh points such that x(i)=x0+ih,

y = approximate solution at the mesh points.

Conditions and Restrictions

Usage conditions are the same as oderk4.

16.7 Linear Two-Point Boundary Value Problems

bvp_fd

Purpose

To compute an approximate solution for the linear two-point boundary value problem

$$y_2'' - p(x)y_2' - q(x)y_2(x) = r(x) \tag{16.7}$$

subject to the general boundary conditions

$$\begin{aligned} s_1 y(a) + s_2 y'(a) &= s_3 \\ s_4 y(b) + s_5 y'(b) &= s_6 \end{aligned} \tag{16.8}$$

using the finite difference method (11.16).

Usage

```
y=bvd_fd(pname,qname,rname,s,mesh)
```

where

pname = character string giving the name of the function p,

qname = character string giving the name of the function q,

rname = character string giving the name of the function r,

s = six-component vector specifying s_1 to s_6,

mesh = vector specifying the mesh points,

y = approximate solution at the mesh points.

Example

Solve

$$y''(x) - xy'(x) - y(x) = xe^x$$

with boundary conditions $y(0) = 1$ and $y(1) = e$, on a mesh of width $h = 0.1$.

Solution:

```
y=bvp_fd('p_func','q_func','r_func',[1 0 1 1 0 exp(1)],[0:0.1:1]),
```

where

```
function res=p_func(x)
  res=x;
function res=q_func(x)
  res=1;
function res=r_func(x)
  res=x.*exp(x);
```

Conditions and Restrictions

The mesh points must be strictly increasing. Note that the boundary points are implicitly defined by the mesh; that is,

$$a = \min(\texttt{mesh}),$$
$$b = \max(\texttt{mesh}).$$

bvp_coll

Purpose

To compute an approximate solution for the linear two-point boundary value problem (16.7), (16.8), using least squares collocation with a cubic spline basis.

Usage

```
y=bvp_coll('pname','qname','rname',s,collpts,x)
```

with the same convention as in `bvp_fd`, except that

`collpts` = vector giving the collocation points. These points are also the knots of the spline,

`x` = set of points at which the approximation is evaluated,

`y` = vector of approximate values.

Conditions and Restrictions

The collocation points must form a strictly increasing sequence. The ends of the interval must be knots, so that the interval $[a,\ b]$ is implicitly defined by

$$a = \min\,(\texttt{collpts}),$$
$$b = \max\,(\texttt{collpts}).$$

16.8 Partial Differential Equations

heat

Purpose

To solve the standard heat equation with homogeneous boundary conditions using the two-level scheme (12.14).

Usage

```
[t,x,u]=heat(gname,delx,delt,r,nsteps)
```

where

`gname` = character string giving the name of function g in (12.2),

`delx` = Δx,

`delt` = Δt,

`r` = parameter in (12.14),

`nsteps` = number of time steps taken,

`t` = column vector specifying the time intervals, such that `t(i)` $= t_{i-1}$,

`x` = row-vector of the spatial grid values, such that `x(i)` $= x_{i-1}$,

`u` = two-dimensional array of approximate values, such that `u(i,j)` approximates $u(x_{i-1},\ t_{j-1})$.

Conditions and Restrictions

`delx` must divide the interval $[0, 1]$ into equal parts. `delt` must be larger than zero.

Example

The output `t` and `x` are useful for plotting of the answers. The commands

```
[t,x,u]=heat('g',0.01,0.01,0.5,10);
mesh(x,t,u)
```

applies ten steps of the Crank-Nicholson method and produce a three-dimensional surface plot of the approximation.

wave

Purpose

To solve the standard wave equation with homogeneous boundary conditions using the finite difference method (12.32).

Usage

```
[t,x,u]=wave(gname,hname,delx,delt, nsteps)
```

where

`gname` = character string giving the name of function g in (12.5),

`hname` = character string giving the name of function h in (12.6),

`delx` = Δx,

`delt` = Δt,

`nsteps` = number of time steps taken,

`t` = column vector specifying the time intervals, such that `t(i)` $= t_{i-1}$,

`x` =row-vector of the spatial grid values, such that `x(i)` $= x_{i-1}$,

`u` = two-dimensional array of approximate values, such that `u(i,j)` approximates $u(x_{i-1}, t_{j-1})$.

Conditions and Restrictions

`delx` must divide the interval $[0,\ 1]$ into equal parts. `delt` must be larger than zero.

16.9 Ill-Conditioned Linear Systems

tik_reg

Purpose

To compute an approximate solution to the linear system

$$\mathbf{Ax} = \mathbf{b} \tag{16.9}$$

using Tikhonov regularization of order zero, one, or two.

Usage

```
[xalpha,res, sm]=tik_reg(A,b,alpha,order)
```

where

`A` = matrix of the system in (16.9),

`b` = right hand side of (16.9),

`alpha` = vector of regularization parameters,

`order` = order of the Tikhonov regularization,

`xalpha` = regularized solution corresponding to the regularization parameter alpha,

`res` = 2-norm of the residual corresponding to `xalpha`,

`sm` = 2-norm of the smoothness measure.

Conditions and Restrictions

`alpha` can be a vector with n components. If `alpha` has more than one components, `xalpha` will be a matrix with n columns, each column representing solution with the corresponding `alpha`. `res` and `sm` will also be vectors with n components. `order` must be 0, 1, or 2.

Example

The command

```
[xalpha,res,sm]=tik_reg(A,b,10.^[-5:-1:-10],1)
```

computes the solution by first order regularization with $\alpha = 10^{-5}$, 10^{-6}, 10^{-7}, 10^{-8}, 10^{-9}, and 10^{-10}. Then

```
plot(log(res),log(sm))
```

can the be used to apply the L-curve method.

16.10 Example Scripts

In addition to the NASOFT software we also supply the source code for those examples that produce numeric or graphical output. The source code is given as MATLAB script files. This allows the reader to reproduce results, to check our implementation, or to examine similar problems or other data.

To run any example, you just need to type the example number in a special way. For example, to reproduce the numbers in Example 7.5, Table 7.4, just type

exm7_5

This will give what is in Table 7.4, with some identifying title.

To examine or to modify the script file for this example, use the "edit M-File" option. Just select "open M-file" and select exm7_5.

16.11 NASOFT

The source code for the NASOFT functions and example scripts are available on-line at: `http://math.jbpub.com/numericalmethods`. Avoid possible name conflicts when you download these files, by first creating a new MATLAB directory, and then transferring the entire folder that you download to this directory. After you change to this workspace, all functions and examples are accessible for running and editing.

Chapter 17

Explorations

In this last chapter we have collected problems on the themes of software construction, evaluation, and modification. The exercises given at the end of each section of the main text are fairly traditional, involving simple problems that illustrate important concepts, that ask for straightforward extensions of given ideas, or that fill in some technical gaps. The explorations in this chapter are a little different and generally more difficult. Some of them are quite extensive, others are purposely vague so that pinning the problem down becomes part of the exercise. For many of the explorations it is necessary to design methodologies for experiments, carry out tests, summarize the results, and draw conclusions. The explorations give a preview of the real-life difficulties one faces in the use and development of software for scientific computing.

17.1 Floating-Point Representation, Rounding, and Accuracy of Elementary Functions

1. Use `format hex` to determine if your version of MATLAB uses IEEE double format.

   (a) Find what representations are used for the following system-defined variables: `Inf`, `NaN`, `-Inf`, `eps`, `realmax`, `realmin`. Can you think of reasons why these patterns were chosen?

(b) How is zero represented? Is there a difference between 0 and –0?

2. Investigate the rules that are used for MATLAB arithmetic involving exceptional values. For example, what is produced by `Inf+Inf`, `Inf-Inf`, `0/Inf`, and `0/NaN`? Are the rules logically consistent and intuitively reasonable?

3. The description of IEEE floating-point representation in Section 2.2 is not entirely complete and breaks down at some point. To see this, consider the hexadecimal pattern

$$0000000000000000.$$

According to our description, this stands for 1×10^{-1023}. But the pattern is also used to represent a true zero.

Examine how MATLAB handles this discrepancy by looking at the hexadecimal representations of `realmin`, `realmin/2`, `realmin/4`, and so on.

4. Conduct an experiment to determine if the recurrence

$$\begin{aligned} f(n+1) &= f(n) + \frac{1}{n} f(n-1), \\ f(0) &= 0, \\ f(1) &= 1, \end{aligned}$$

is stable. Compute $f(50)$ and estimate its accuracy.

5. Carry out an investigation of the relative accuracy of the MATLAB function `exp` in the intervals $[0,\ 1]$ $[10,\ 20]$, and $[-10,\ -20]$. One way of approaching this problem is to use the relation between e^x, e^y, and e^{x+y}.

6. While MATLAB implements elementary functions, such as `sin` and `cos`, carefully so that the absolute error is smaller than 10^{-15}, this does not preclude a large relative error in `sin(x)` when `x` is very small. Design a test to see how the relative error in `sin` behaves for small x. Hint: use the Maclaurin series

$$\sin(x) = x - \frac{x^3}{3!} + \frac{x^5}{5!} - \cdots.$$

7. For $\alpha = O(10^{-6})$, write programs that evaluate the following functions with high relative accuracy.

(a) $1 - \cos(\alpha)$

(b) $\sin(x + \alpha) - \sin(x)$, $\frac{1}{2} \le x \le 1$

8. The MATLAB function `gamma` approximates the gamma function

$$\Gamma(a) = \int_0^\infty e^{-x} x^{1-a} dx$$

for all values $1 \leq a < \infty$. MATLAB also provides the natural logarithm of the gamma function via `gammaln`. Can one use the relation between these two functions to explore their accuracy?

9. The error function

$$erf(x) = \frac{2}{\sqrt{\pi}} \int_0^x e^{-t^2} dt$$

can be computed approximately with the MATLAB expression `erf(x)`. Devise some tests that allow one to conclude with reasonable confidence whether or not `erf(x)` evaluates the error function with an absolute accuracy of 10^{-14} or better for all values of x.

17.2 Solving Linear Systems

1. In testing matrix routines it is often convenient to have a set of singular matrices or matrices with a range of known condition numbers. Singular matrices can be generated easily, for example, by making two rows equal or by taking one row as the sum of two others. Generating nonsingular matrices with varying degrees of conditioning is not so obvious. One way is to start with a very ill-conditioned matrix, then change it to make it better conditioned. For example, we can start with a Hilbert matrix, then add something on the diagonal.

   If $\mathbf{H}_{10}$ denotes the 10×10 Hilbert matrix, examine the conditioning of the matrices

$$\mathbf{A}_\alpha = \mathbf{H}_{10} + \alpha \mathbf{I}$$

   as a function of α.

2. Conduct timing tests to

   (a) verify the $O(n^3)$ behavior of the GEM,

   (b) show that inverting a matrix is more time-consuming than solving a system of equations.

3. Conduct tests to examine the claim that the GEM often produces solutions with small residuals even when the matrix is very ill-conditioned.

4. Investigate the accuracy of the solution of least squares problems by `A\y` with that of the solution by the normal equations.

5. Implement the GEM *without pivoting* and examine its stability.

6. Use `inv` to compute approximate inverses for the $n \times n$ Hilbert matrices with $n = 5,\ 6,\ 7,\ 8$, then use Theorem 3.2 to bound the true inverses. Check this against the results obtained from `invhilb`.

7. Implement the Gauss-Jordan method (see Exercise 5, Section 3.1) and conduct timing tests to compare its efficiency with a similar implementation of the GEM.

8. Implement the GEM with complete pivoting (see Exercise 4, Section 3.1). Does complete pivoting give better results than row pivoting? How is the efficiency affected?

9. Investigate how scaling affects the MATLAB method for solving $\mathbf{Ax} = \mathbf{b}$. For a set of matrices (both well- and ill-conditioned), solve the equation. Then change the scaling by multiplying some rows and columns of $\mathbf{A}$ by large and small factors and making the proper changes to $\mathbf{b}$ to keep $\mathbf{x}$ (in theory) the same. How are the numerical results affected? Can you get any idea from these results on how MATLAB might do its internal scaling? In particular, do you think the program uses equilibration?

17.3 Approximation by Polynomials

1. Investigate the convergence of simple interpolation at uniformly spaced points for the function

$$f(x) = \frac{1}{1 + \alpha^2 x^2}$$

on $[-1,\ 1]$ for a range of α. How is the magnitude of α related to the difficulty with interpolation. In particular, what is the largest α for which you observe convergence? Are the results different if you interpolate at the roots of the Chebyshev polynomials?

2. Examine the performance of polynomial interpolation for functions that are not smooth, such as $f(x) = |x|$ and $f(x) = \sqrt{|x|}$ on $[-1,\ 1]$.

3. Compare the growth of the condition numbers of the interpolation matrices in (4.9) for

   (a) uniformly spaced interpolation points in $[-1,\ 1]$ and

   (b) interpolation at the Chebyshev roots.

4. Choose a set of test functions and examine the error produced in interpolation at the Chebyshev roots to see how closely the equi-oscillation conditions are satisfied.

5. In Section 15.3 it was stated that for $n < m - 1$, the function `polyfit` produces a discrete least squares solution. Implement the discrete least squares method as described in Section 4.3, using monomials. Do the results differ significantly from those of `polyfit`?

6. As m increases, one expects that the discrete least squares solution approaches the true least squares solution. Test this conjecture.

7. Use the polynomial discrete least squares method for the function in Exploration 1 of this section. Try $n = 5,\ 10,\ \ldots$ and see what values of m are needed to give good results.

8. Implement the discrete Fourier approximation described in Example 4.11 and test its effectiveness on various functions, including

   (a) $f(x) = x^2 \sin(x)$

   (b) $f(x) = e^{(x/\pi)^2}$

   (c) $f(x) = e^{(x/\pi)}$.

   What explains the differences in the performance in these examples?

9. Implement the Remez algorithm and use it to find the best uniform approximation to $\sin(x)$ in $[0,\ \pi/4]$ by a polynomial of degree six. Compare this with the approximation obtained by interpolation at the Chebyshev roots.

17.4 Interpolation and Data Fitting with Piecewise Polynomials

1. Examine the accuracy and order of convergence for the three MATLAB one-dimensional interpolation alternatives for $f(x) = \sin(x)$ in $[0,\ 2\pi]$.

2. Examine the accuracy and order of convergence for the three MATLAB one-dimensional interpolation alternatives for $f(x) = \sqrt{|x|}$ in $[-1,\ 1]$.

3. Implement the spline interpolation in (5.20) and study its performance.

4. Implement the spline interpolation in (5.23) and compare its effectiveness with the implementation in Exploration 3.

5. Implement a spline interpolation function in which the two end conditions are given by approximating the first derivatives of y by a three-point difference formula. Compare the effectiveness of this with the alternatives in Explorations 3 and 4.

6. Study what happens in `lsq_quad` when two data points are coincident or nearly so.

7. Examine what happens in `lsq_quad` when there are no data points in some knot interval. What is the minimum number of points that every knot interval should have?

8. Find a least squares cubic spline fit to the data

x	y	x	y
0.0	0.10	0.6	0.83
0.1	0.21	0.7	0.84
0.2	0.33	0.8	0.85
0.3	0.44	0.9	0.85
0.4	0.57	1.0	0.86
0.5	0.70		

Choose both the number of knots and their placement, then discuss why you consider your result reasonable.

9. How sensitive are the answers in the above exploration to knot placement?

10. In many practical applications, one encounters curves in the xy plane for which there is no suitable representation of y as a function of x (e.g., a spiral). Such curves can sometime be represented parametrically, by $x(k) = x_k$ and $y(k) = y_k$ for $k = 1, 2, \ldots, N$, as in the table below.

k	x	y	k	x	y
1	3.0	0.5	11	8.0	6.5
2	4.5	2.0	12	5.0	6.8
3	5.0	1.5	13	4.0	7.0
4	5.5	4.0	14	2.5	5.5
5	6.0	5.5	15	1.5	6.0
6	6.0	7.0	16	1.0	7.5
7	6.5	7.0	17	1.5	3.0
8	7.0	7.5	18	2.0	2.0
9	8.5	7.5	19	2.5	2.5
10	9.0	7.0	20	3.0	0.5

Using the data given in this table, fit x by a fifth degree polynomial and y by a sixth degree polynomial, respectively, then plot the result in xy plane.

17.5 Approximating Derivatives and Integrals

1. Find an approximation to $f'''(x)$, using the values $f(x-2h)$, $f(x-h)$, $f(x)$, $f(x+h)$ and $f(x+2h)$. Predict the accuracy limitation of this approximation, then carry out numerical experiments to verify this prediction.

2. Use the centered difference formula to compute

$$\frac{d}{dx}\left(\frac{\sin(x)}{x}\right)$$

at $x = 0.2$. Use various values of h and see if the observed and true optimal values are close.

3. Find the weights of the quadrature rule based on the zeros of T_5. Write a program for this quadrature and compare its efficiency with that of a quadrature rule based on a five-point Newton-Cotes rule. Compare also with `quad8` with $n = 5$.

4. Compare the performance of the functions `quad` and `quad8`. Can you find any examples where `quad` performs better?

5. Experimentally establish convergence rates for a composite quadrature based on the open Newton-Cotes formulas in Example 6.2.

6. Find cases where the relative accuracy requirements of `quad8` cause inefficiencies.

7. All numerical software can fail to varying degrees. For adaptive quadrature we can define several different outcomes:

   - *Success.* The result is delivered within the specified accuracy.
   - *Soft Failure.* The error is larger than specified by the tolerance, but is within a factor of ten.
   - *Hard Failure.* The error is more than ten times the specified tolerance.
   - *Crash.* The software causes some sort of systems failure, for example, the hardware tries to execute an illegal operation or the program goes into an infinite loop.

   Try to find examples in which these different types of situations occur for `quad8`.

8. An adaptive quadrature sometimes delivers results that are much more accurate than specified by the tolerance. This can be considered inefficient since the desired accuracy could have been obtained with less work. Examine results produced by `quad8` to see if it significantly overestimates the error.

9. Use `quad8` as a basis to write a new adaptive quadrature method that allows the user to specify a desired level of absolute accuracy.

10. Find the following integral to a relative accuracy of 10^{-6}.

    (a) $I = \int_1^{\infty} \frac{\log_e(x)}{1 - x + 2x^4} dx$

    (b) $I = \int_0^{\infty} \frac{e^{-x}}{1 + x^5} dx$

    Give convincing arguments to defend that the desired accuracy has been achieved.

11. Use the function `quad_cg` to build an approximate integrator for three-dimensional integrals of the form

$$\int_0^1 \int_0^1 \int_0^1 f(x, y, z) dx dy dz.$$

12. Write a MATLAB function to approximate

$$\int_0^1 \int_0^1 \int_0^1 \int_0^1 f(x_1, x_2, x_3, x_4) dx_1 dx_2 dx_3 dx_4$$

    by the Monte Carlo method. Study examples to characterize the convergence rates of such an approximation.

13. Implement a composite Filon method that evaluates

$$I = \int_0^{nk\pi/\alpha} \cos(\alpha\, x) f(x) dx.$$

    Compare the effectiveness of your program with that of `quad8`, using different functions f and a range of α.

14. Investigate the effectiveness of Gaussian quadrature for integrands that are not infinitely differentiable. Establish some guidelines for convergence and accuracy in terms of the differentiability of the integrand.

15. Extend the approach in `quad_dbl` for the approximate computation of three-dimensional integrals

$$I = \int_a^b \int_{g(z)}^{h(z)} \int_{r(y)}^{s(y)} f(x, y, z) dx dy dz.$$

16. Design and implement an effective adaptive quadrature for the evaluation of integrals of the form

$$I = \int_0^1 \ln x \, f(x) dx.$$

17.6 Solving Nonlinear Equations

1. Conduct experiments to verify the predicted order of 1.62 for the secant method.
2. Conduct experiments to establish the rate of convergence of Muller's method.
3. Do some analysis and carry out experimentation to explore the solution set of the system in Example 7.5.
4. Implement the suggestions made in Section 7.8 for a one-dimensional root-finder. Write your program so that it finds as many roots as possible in a given finite interval $[a, b]$.
5. Use `fmins` to explore the possible solutions of the system in Example 7.5.
6. For Newton's method the set of starting points that give a sequence of iterates converging to a solution is called the *domain of convergence* for that solution. Domains of convergence often have a rather complicated structure. Explore the domains of convergence for the system in Example 7.6.
7. Experimentally examine the convergence of the one-dimensional Newton's method near multiple roots. Do the convergence rates show the predicted behavior? Compare the convergence rate of the secant method with that of Newton's method.
8. Combine Muller's method with a strategy for bracketing a root. Examine the order of convergence of the resulting algorithm. Use the NASOFT function `muller` to make implementation easier.

9. Combine `newton` with a search strategy that systematically explores and finds roots in some given finite region of n-space.

10. Including the term involving $(x^* - x)^3$ in a Taylor expansion, derive the suggested root-finding method

$$x^{[i+1]} = x^i - \frac{f'(x^{[i]}) - \sqrt{\{f'(x^{[i]})\}^2 - 2f(x^{[i]})f''(x^{[i]})}}{f''(x^{[i]})}.$$

Implement this method for functions of one variable and explore the convergence rate of the resulting algorithm.

11. Investigate how one might combine `newton` with the MATLAB function `fmins` to create an effective multi-dimensional root-finder.

17.7 Matrix Methods

1. Examine the relative efficiencies of the LU and QR methods.
2. Explore the stability of the singular value decomposition. How sensitive are very small singular values to small changes in the matrix elements?
3. Use the singular value decomposition to write a program for the minimum norm least squares solution. Compare this with results using `pinv`. Can you guess from the results how MATLAB handles small singular values?
4. Find out if MATLAB solves systems with positive definite symmetric matrices more efficiently than it solves general systems.
5. Find out if MATLAB solves tridiagonal systems more efficiently than it solves general systems.
6. Investigate if MATLAB solves sparse linear systems more quickly than full ones.
7. Examine how MATLAB computes the rank of a matrix when `tol` is not specified. Is the way this is done reasonable? Can you find examples where MATLAB makes an apparent wrong decision?

17.8 Computing Eigensystems of Matrices

1. Investigate what happens when the power method is applied to nonsymmetric matrices. Can you draw any conclusions from this about the usefulness of the power method for nonsymmetric matrices?

2. Use `jacobi` for a full implementation of the Jacobi method for eigenvalues. Try various pivoting strategies and examine their effectiveness.

3. In the Jacobi method, how does the time required to achieve a given level of accuracy depend on the size of the matrix?

4. Modify `jacobi` so that it can be used to compute eigenvectors.

5. Use `hholder` for a full implementation of the Householder method for eigenvalues of symmetric matrices.

6. Extend the program in Exploration 5 to implement a function that computes both eigenvalues and eigenvectors.

7. Does the MATLAB function `eig` work more efficiently for sparse matrices than for full matrices?

8. Experiment with `eig` to see if it solves symmetric matrices significantly faster than nonsymmetric ones.

9. Use `eigpower` and deflation to construct a program that will compute the first p eigensolutions of a symmetric $n \times n$ matrix, with $n > p$. Compare the efficiency of this approach with `eig`.

10. Find the largest positive value of α for which the spectrum of

$$\begin{bmatrix} \alpha & 0 & 0.5 \\ 0 & \alpha^2 & 0.2 \\ 0.5 & 0.2 & \alpha^2 \end{bmatrix}$$

is contained in $(-1,\ 1)$.

11. Since eigenvalue algorithms often involve finding the roots of polynomials, and since finding roots of high multiplicity is unstable, one might suspect that finding eigenvalues of high multiplicity accurately is not easy. Examine how `eig` performs when there are eigenvalues of high multiplicity.

17.9 ODE Initial Value Problems

1. Carry out experiments to determine the order of convergence of `odepc`.

2. Determine the sensitivity of `oderk4` and `odepc` to the starting errors.

3. Determine the sensitivity of `oderk4` and `odepc` to rounding.

4. Compute the stability regions for the classical fourth-order Runge-Kutta method (10.32).

5. Apply `oderk4` to the problem

$$y'(x) = \lambda\, y(x),$$
$$y(0) = 1,$$

with various λ to see if you encounter any weak instability.

6. Compare the efficiency of `ode45` and `odepc` for problems whose solution does not require step size changes.

7. What happens when you apply `ode45` to situations like Example 10.2 where the solution goes to infinity?

8. Compare the performance of `ode45`, `oderk4`, and `odepc` in the solution of the system

$$y_1'(x) = -5.5y_1(x) - 2.25y_2(x) + 2.25y_3(x),$$
$$y_2'(x) = 90y_1(x) - 55y_2(x) - 45y_3(x),$$
$$y_3'(x) = 99y_1(x) - 49.5y_2(x) - 50y_3(x),$$

in the interval (0, 1), with $y_1(0) = 1$, $y_2(0) = y_3(0) = 0$. Explain the results you get.

9. Use `oderk4` to develop an adaptive method. Your method should increase and decrease the step size as required by the problem at hand. Test and evaluate the effectiveness of your adaptive strategy. Compare and evaluate the performance of your method with that of `ode45`.

10. Two objects of equal mass M are in a fixed position and another mass m moves between them under the force of gravitational attraction, as shown in the following diagram.

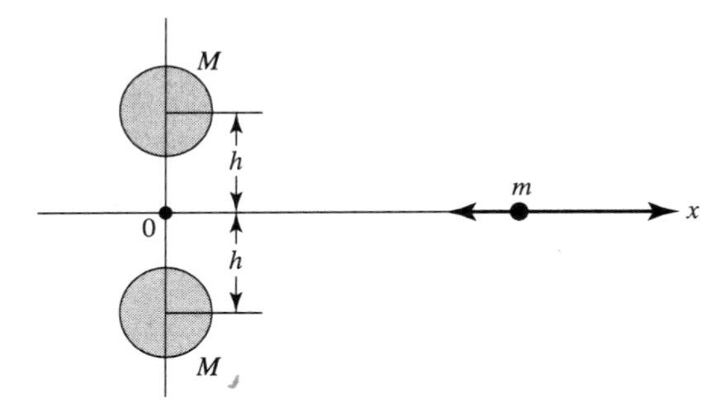

If m is at rest at time $t = 0$, show that its subsequent motion is determined by the equation

$$x''(t) = -cx(t)(x^2(t) + h)^{-3/2},$$

with

$$x(0) = x_0,$$

and

$$x'(0) = 0.$$

(a) For $c = 1$, $h = 1$, and $x_0 = 1$, solve the equation in the interval $0 \le t \le 30$. Does the solution agree with what we should expect from the physics of the situation?

(b) Find the time at which m first arrives at O, to an accuracy of four significant digits.

11. Design a method, analogous to Euler's method, for solving the integro-differential equation

$$\frac{dy(x)}{dx} = f(x, y(x)) + \int_0^x y(x)dx$$

in $0 \le x \le a$, with $y(0) = y_0$.

(a) Construct some test cases to investigate if the method works.

(b) What do you think the order of convergence of the method is? Verify your prediction experimentally.

12. Design and implement a method for solving the equation in Exploration 11 by converting it into a system of ordinary differential eqautions.

17.10 ODE Boundary and Eigenvalue Problems

1. Modify `bvp_fd` so that it can handle the general boundary conditions (11.2). Carry out experimentation to form conjectures on the stability and order of convergence of the modification. What happens when you use Richardson's extrapolation in this case?

2. One way to overcome the difficulty with the boundary conditions in the least squares method is to require that they only be satisfied approximately. With this, `bvp_coll` can easily be converted into a discrete least squares method: we simply use more collocation points, add the boundary conditions, and solve the resulting system in the least squares sense. Implement this suggestion and discuss your observations.

3. Implement a shooting method algorithm for solving the fourth-order equation in Exercise 8, Section 11.1.

4. Find the two lowest eigenvalues of

$$((1 + x)y'(x))' + \frac{\lambda}{1 + x^2}y(x) = 0$$

by the shooting method to four digit accuracy. Defend your claim that the results are good to the stated accuracy.

5. Implement a shooting method to find eigenvalues of (11.41) under the more general homogeneous boundary conditions

$$c_1 y(a) + c_2 y'(a) = 0,$$
$$c_3 y(b) + c_4 y'(b) = 0.$$

6. Investigate to what extent Richardson's extrapolation is effective in the solution of the eigenvalue problem in Example 11.11.

17.11 The Heat and Wave Equations

1. Modify `heat` for the solution of the diffusion–convection equation (12.30). Investigate the dependence of the stability on the size of the constant b.

2. Implement an implicit method for the wave equation (see Exercise 3 in Section 12.2) and investigate its stability.

3. Implement the method of lines for the heat equation and compare its performance with `heat`.

4. Modify `wave` for solving the somewhat more complicated wave equation

$$\frac{\partial^2 u(x,t)}{\partial t^2} = c(x)\frac{\partial^2 u(x,t)}{\partial x^2}.$$

5. Devise a method for solving the equation

$$\frac{\partial u(x,t)}{\partial t} = \frac{\partial^2 u(x,t)}{\partial x^2} - u(x,t)\frac{\partial u(x,t)}{\partial x},$$

with conditions

$$u(x,0) = x(x-1)$$

and

$$u(0,t) = u(1,t) = 0.$$

Use the method to get an approximate solution to three-digit accuracy.

17.12 PDE Boundary Value Problems

1. For $\Gamma = \{(x,\ y)\,|\,0 \le x \le 1,\ 0 \le y \le 1\}$, construct the matrix $\mathbf{A}$ in (13.6) and experimentally examine its condition number. What conclusions can you draw from these results abut the convergence and stability of the finite difference methodin this case?

2. Implement the explicit method (13.27) for the two-dimensional heat equation and examine its stability.
3. Implement the ADI method for the two-dimensional heat equation and verify that it is unconditionally stable.
4. For $\Gamma = \{(x,\ y) \,|\, 0 \le x \le 1,\ 0 \le y \le 1\}$, design a method for solving the eigenvalue problem

$$\begin{aligned} \nabla^2 u(x,y) &= \lambda\, u(x,y), \\ u(x,y) &= 0 \text{ for all } (x,y) \in \Gamma. \end{aligned}$$

17.13 Ill-Posed Problems and Regularization

1. Explore the answers produced by `A\b` with those produced by `tik_reg(A,b,0,0)`. Theoretically they should always be the same. Explain any observed differences.
2. What happens in Example 14.12 if we increase n, using zero order regularization?
3. Try to use higher order quadratures to solve Equation (14.31). Do higher order methods perform better?
4. For Equation (14.6), perform numerical experiments to see how the value of d affects the conditioning of the resulting discretized system. Is there a physical interpretation that makes your observations not unexpected?
5. See how well you can recover the solution of (14.31) if the right side is perturbed with a random error normally distributed with mean zero and variance 10^{-4}.
6. Explore Example 14.10 with different $\Delta\mathbf{b}$ that have the same statistical properties, but are not identical. How do the small differences affect the L-curve and the selection of α? What effect do small changes have on the actual solution?
7. Explore the potential solution of the integral equation

$$\int_0^1 \frac{x(t)\,dt}{1+s^2+t^2} = 1$$

for $0 \le s \le 1$. Do the numerical results give any indication of the qualitative nature of a solution?

References for Further Reading

[1] M. Abramowitz and I. Stegun: Handbook of Mathematical Functions, US Govt. Printing Office, 1964.

[2] J.H.E. Ahlberg, E.N. Nilson, and J.L. Walsh: The Theory of Splines and their Application, Academic Press, 1967.

[3] ANS/IEEE Std. 754/1985: IEEE Standard for Binary Floating-Point Arithmetic, Institute of Electrical and Electronics Engineering, 1985.

[4] K.E. Atkinson: An Introduction to Numerical Analysis, John Wiley, 1978.

[5] E.K. Blum: Numerical Analysis and Computation: Theory and Practice, Addison–Wesley, 1972.

[6] P.J. Davis: Interpolation and Approximation, Blaisdell, 1963.

[7] J.E. Dennis and R.B. Schnabel: Numerical Methods for Unconstrained Optimization and Nonlinear Equations, Prentice–Hall, 1983.

[8] I.S. Duff, A.M. Erisman, and J.K. Reid: Direct Methods for Sparse Matrices, Clarendon Press, 1986.

[9] E. Part-Enander, Anders Sjoberg, Bo Mellin, and Pernilla Isaakson: The MATLAB Handbook, Addison–Wesley, 1996.

[10] G.H. Golub and C.F. Van Loan: Matrix Computations, John Hopkins University Press, 1983.

[11] G. Hammerlin and K.H. Hoffmann: Numerical Mathematics, Springer–Verlag, 1991.

[12] E. Isaacson and H.B Keller: Analysis of Numerical Methods, John Wiley, 1966.

[13] C. Johnson: Numerical Solutions of Partial Differential Equations by the Finite Element Method, Cambridge University Press, 1987.

[14] D. Kahaner, C. Moler, S. Nash: Numerical Methods and Software, Prentice–Hall, 1989.

[15] H.B. Keller: Numerical Methods for Two-Point Boundary-Value Problems, Blaisdell, 1968.

[16] P. Lancaster and K. Salkauskas: Curve and Surface Fitting, Academic Press, 1986.

[17] L. Lapidus and J.H. Seinfeld: Numerical Solution of Ordinary Differential Equations, Academic Press, 1971.

[18] C. Lawson and R. Hanson: Solving Least Squares Problems, Prentice–Hall, 1974.

[19] W. Murray: Numerical Methods for Unconstrained Optimization, Academic Press, 1972.

[20] R. Piessens, et al: QUADPACK: A Subroutine Package for Automatic Integration, Springer 1983.

[21] P.M. Prenter: Splines and Variational Methods, John Wiley, 1975.

[22] J. Rice: Numerical Methods, Software, and Analysis: McGraw–Hill, 1983.

[23] R. Richtmyer and K. Morton: Difference Methods for Initial Value Problems, Wiley, 1967.

[24] A.H. Stroud and D. Secrest: Gaussian Quadrature Formulas, Prentice–Hall, 1966.

[25] R.S. Varga: Matrix Iterative Analysis, Prentice–Hall, 1962.

[26] V. Vemuri and W.J. Karplus: Digital Computer Treatment of Partial Differential Equations, Prentice–Hall, 1981.

[27] H.J. Weaver: Theory of Discrete and Continuous Fourier Analysis, John Wiley, 1989.

[28] J.H. Wilkinson: The Algebraic Eigenvalue Problem. Oxford University Press, 1965.

Subject Index

Index

M

N

O

P